AF388960

Electrical Power Engineering and Renewable Energy Technologies

Electrical Power Engineering and Renewable Energy Technologies

Editors

Najib El Ouanjli
Saad Motahhir
Mustapha Errouha

Basel • Beijing • Wuhan • Barcelona • Belgrade • Novi Sad • Cluj • Manchester

Editors
Najib El Ouanjli
Applied Physics Department,
Université Hassan 1er
Settat, Morocco

Saad Motahhir
National School of Applied
Sciences, Sidi Mohamed Ben
Abdellah University
Fez, Morocco

Mustapha Errouha
Laboratory of Plasma and
Energy Conversion
(LAPLACE), University of
Toulouse
Toulouse, France

Editorial Office
MDPI
St. Alban-Anlage 66
4052 Basel, Switzerland

This is a reprint of articles from the Special Issue published online in the open access journal *Energies* (ISSN 1996-1073) (available at: https://www.mdpi.com/journal/energies/special_issues/electrical_power_renewable_energy).

For citation purposes, cite each article independently as indicated on the article page online and as indicated below:

Lastname, A.A.; Lastname, B.B. Article Title. *Journal Name* **Year**, *Volume Number*, Page Range.

ISBN 978-3-0365-8890-2 (Hbk)
ISBN 978-3-0365-8891-9 (PDF)
doi.org/10.3390/books978-3-0365-8891-9

Contents

About the Editors

Najib El Ouanjli

Najib El Ouanjli is currently a professor at the Department of Applied Physics at the "Faculty of Sciences and Techniques, Hassan First University, Settat, Morocco". He received a Ph.D. degree in 2021 from "Sidi Mohammed Ben Abdellah University, Higher School of Technology, Fez, Morocco" and a master's degree in 2015 from the "Faculty of Sciences Dhar El Mahraz, Fez, Morocco". He was a professor of physics Science Ministry of Education from 2013 to 2021. His research interests include renewable energy systems, electrical and electronics engineering, rotating electric machines, system modeling, control techniques, optimization techniques, fault diagnosis, wind turbines, and solar energy. He has published over 72 papers (international journals, book chapters, and conferences/workshops). He serves as a reviewer of international journals and conferences, including as IEEE Transactions on Industrial Electronics, SN Applied Sciences, Energies, and Complex and Intelligent Systems. He is also a Guest Editor of Special Issues of the "Energies MDPI" and "Machines MDPI" journals.

Saad Motahhir

Saad Motahhir (Eng., PhD, IEEE Senior Member) has previous expertise working in industry as an embedded system engineer at Zodiac Aerospace Morocco, holding this job from 2014 to 2019. He has recently become a professor at ENSA, SMBA university, Fez, Morocco. He received an engineering degree in embedded systems from (ENSA) Fez in 2014. In addition, he has earned a PhD. He earned a degree in electrical engineering from SMBA University in 2018. He has published and contributed to numerous publications in different journals and conferences in the last few years, most of which were related to photovoltaic (PV) solar energy and embedded systems. In addition, he has had several patents published by the Moroccan Office of Industrial and Commercial Property. He edited many books and acted as a Guest Editor of different Special Issues and topical collections. He is a reviewer, and he is a member of the editorial boards of multiple journals. He is associated with more than 30 international conferences as a Program Committee/Advisory Board/Review Board member, and he is a member of the Arab Youth Center council. He is the (ICDTA) Conference Co-Chair.

Mustapha Errouha

Mustapha Errouha received a bachelor's (BTech.) and master's degree (MTech.) in electrical engineering from Sidi Mohamed Ben Abdellah University. He received his Ph.D. Degree in electrical engineering and renewable energy from the Faculty of Sciences and Technologies, Sidi Mohamed Ben Abdellah University, in 2020. He was an assistant researcher at GREEN laboratory of the University of Lorraine, France. He is currently an assistant professor at the Engineering School of Toulouse, "École Nationale Supérieure d'Electrotechnique, d'Electronique, d'Informatique, d'Hydraulique et des Télécommunications" (ENSEEIHT). In 2021, he joined the "Laboratoire Plasma et Conversion d'Energie" (LAPLACE), Toulouse, as a researcher. His research interests include motor drives, renewable energy, power electronics, smart grid, artificial intelligence, and embedded systems. He has published several papers in peer-reviewed journals and presented his work at various international conferences. He has served on the technical program committees of several international conferences, and he is a regular reviewer of top-tier scientific journals related to his field.

Article

Innovative Optimal Nonstandard Tripping Protection Scheme for Radial and Meshed Microgrid Systems

Salima Abeid [1],*, Yanting Hu [1], Feras Alasali [2] and Naser El-Naily [3]

[1] Electrical and Electronics Engineering Department, The University of Chester, Chester CH2 4NU, UK; y.hu@chester.ac.uk

[2] Department of Electrical Engineering, Faculty of Engineering, The Hashemite University, P.O. Box 330127, Zarqa 13133, Jordan; ferasasali@hu.edu.jo

[3] Department of Electrical Engineering, College of Electrical and Electronics Technology-Benghazi, Benghazi 23P7F49, Libya; naseralnaile222@gmail.com

* Correspondence: 1625439@chester.ac.uk; Tel.: +44-(0)1244511000

Abstract: The coordination of optimal overcurrent relays (OCRs) for modern power networks is nowadays one of the critical concerns due to the increase in the use of renewable energy sources. Modern grids connected to inverter-based distributed generations (IDGs) and synchronous distributed generations (SDGs) have a direct impact on fault currents and locations and then on the protection system. In this paper, a new optimal OCR coordination scheme has been developed based on the nonstandard time–current characteristics (NSTCC) approach. The proposed scheme can effectively minimize the impact of distributed generations (DGs) on OCR coordination by using two optimization techniques: genetic algorithm (GA) and hybrid gravitational search algorithm–sequential quadratic programming (GSA–SQP) algorithm. In addition, the proposed optimal OCR coordination scheme has successfully employed a new constraint reduction method for eliminating the considerable number of constraints in the coordination and tripping time formula by using only one variable dynamic coefficient. The proposed protection scheme has been applied in IEEE 9-bus and IEC MG systems as benchmark radial networks as well as IEEE 30-bus systems as meshed structures. The results of the proposed optimal OCR coordination scheme have been compared to standard and nonstandard characteristics reported in the literature. The results showed a significant improvement in terms of the protection system sensitivity and reliability by minimizing the operating time (OT) of OCRs and demonstrating the effectiveness of the proposed method throughout minimum and maximum fault modes.

Keywords: overcurrent relays; optimum coordination; microgrid; distributed generation; nonstandard time–current characteristics; tripping time

Citation: Abeid, S.; Hu, Y.; Alasali, F.; El-Naily, N. Innovative Optimal Nonstandard Tripping Protection Scheme for Radial and Meshed Microgrid Systems. *Energies* **2022**, *15*, 4980. https://doi.org/10.3390/en15144980

Academic Editors: Najib El Ouanjli, Saad Motahhir and Mustapha Errouha

Received: 19 May 2022
Accepted: 5 July 2022
Published: 7 July 2022

Publisher's Note: MDPI stays neutral with regard to jurisdictional claims in published maps and institutional affiliations.

1. Introduction

1.1. Background

One of the major technical challenges facing the microgrid (MG) is the relay coordination for the different operations of the network. The different operation modes of MG result in frequent variations, thereby leading to changes in the fault current magnitude and its direction. The protection scheme needs to guarantee the optimal operation in all network scenarios, which include meshed and radial configurations as well as the connected grid and islanded modes of MGs. In case of a fault in the grid connection mode, the fault current will be fed from distributed generators (DGs) and the utility grid. On the other hand, when the grid is in islanded mode, the supply of the fault current will result from DGs as the only connection present in the MG [1,2]. These changes in the fault current have a significant impact on the basic requirements of the protection system such as speed, reliability, and sensitivity, which then result in some protection issues such as delay in tripping, false tripping, loss of mains, and blinding of protection [3,4]. In recent years, the coordination problem of overcurrent relays (OCRs) has obtained wide attention in the protection system research area. These studies are focused on addressing and preserving the protection schemes' reliability in interconnected MGs to acquire the optimal configuration for protection approaches in MGs. Optimal relay coordination methods regarding whole previous studies can be divided into five methods, namely, optimization techniques, new constraints, dual-setting protection schemes, new or modifying objective functions, and nonstandard characteristics (NSCs) [3]. Figure 1 illustrates the classification of OCR coordination methods. The mutual goal among these methods is to obtain the optimal OCR coordination. Initially, the optimization techniques played an essential role in the optimal OCR coordination, starting with simple ways and passing through nature-inspired algorithms and advanced artificial intelligence [3].

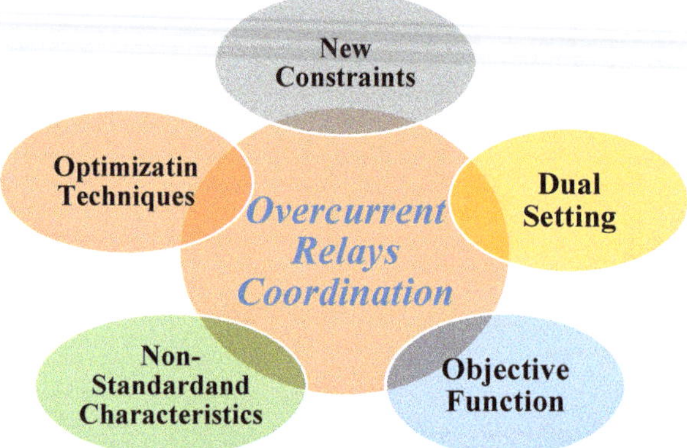

Figure 1. Overcurrent Relay Coordination Methods.

1.2. Literature Review

Metaheuristic techniques such as genetic algorithm (GA) have been applied successfully for reducing the tripping time of relays and avoiding miscoordination problems [5]. For miscoordination problems that are associated with both discrete- and continuous-time multiplier settings, in ref. [6], the particle swarm optimization (PSO) algorithm proved its effectiveness to deal with complex coordination problems. In addition, hybrid techniques have been established to improve metaheuristic techniques. For instance, a hybrid particle swarm optimization and moth–flame optimization (PSOMFO), which is a combination between the PSO and MFO to improve their achievement, and the outcomes proved its effectiveness compared to PSO and MFO algorithms individually [7]. In ref. [8], the authors proposed a hybrid gravitational search algorithm and sequential quadratic programming (GSA–SQP) algorithm that introduces a robust high-quality solution to solving the OCR coordination problem. It is effective due to it taking the pros of both GSA and SQP techniques; it has been tested and evaluated on various test systems. Secondly, some papers proposed and considered the effect of the new or reduced constraints on solving the OCR coordination problem. For the distribution system with a combination of the DGs in ref. [9], the study relieved about 43% of burdensome constraints from the process of coordination optimization compared to the two standards. A considerable number of constraints is an optimization problem; the inactive inequality constraints in the OCR coordination problem have been detected and removed by using a new proposed index. It is defined as a ratio for each OCR pair. The given results illustrate that the proposed index detected and removed more than 90% of unused constraints [10].

Dual-setting directional overcurrent relays (DOCRs) can operate in both directions which are forward and reverse; it is another solution that has been introduced for overcoming challenges relating to the MGs' protection systems. In ref. [11], a new coordination scheme using dual-setting DOCRs has been proposed to determine two optimum pairs of settings for each relay in connected and disconnected DGs. The obtained results appear to demonstrate the effectiveness of the proposed approach and reduce the overall operation time (OT) by roughly 50% compared to the traditional coordination scheme that utilizes the DOCRs with a pair of settings. A novel protection scheme for dual-setting DOCRs in MGs with DGs has been introduced in ref. [12]. This scheme uses a dual-protection setting in the main and backup operation topologies in one relay. Two strategy cases of one-point and three-point have been executed by the optimization. During utilization of the three-point coordination strategy, the total OT has been reduced by 57% compared with the conventional dual-setting method. The proposed method proves its applicability for both grid-connecting and islanding modes and improves the system's reliability by eliminating the demand for communication between relays. One of the suggestions made by many authors to contribute to preventing the miscoordinations is modifying the objective function. Multiple modifications in the objective function for DOCRs for meshed networks have been suggested by Alam et al. in ref. [13]; it minimized the OT for main and backup relays simultaneously. In ref. [14], a novel objective function is proposed to direct the settings of the OCRs towards optimal solutions suitably. The proposed approach showed a significant reduction in the OT of relays and avoided miscoordinations among them.

Investigation of alternative methods to standard protection schemes has required a considerable amount of effort by researchers by employing NSCs, which will be taken into consideration in this literature. There is limited literature on designing OCR coordination schemes based on NSCs for a power network equipped with DGs. For example, a hybrid application uses an inverse-time characteristic and a definite-time characteristic in parallel with the absence of the DG units has been proposed in ref. [15]. This novel hybrid NSC entails no complexity or additional cost, whereas the problem of dynamic coordination has appeared in these hybrid NSCs which require careful application. Thus, they can lead to miscoordination and less flexibility. In addition to the fault current characteristic, there are some NSCs based on the fault voltage that have been proposed and applied [16,17]. The authors in ref. [16] added per-unit voltage to measure the IEC standard characteristics (SCs) to reduce the overall OTs of the relays. In ref. [4], a significant reduction in the total OT of the OCRs has been achieved by employing the fault voltage natural logarithm function in the denominator of the characteristic equation. However, this logarithm function had a limited effect on the obtained characteristic. In another study [18], a new scheme excluded the time multiple setting (TMS) and utilized fault voltage in the numerator as a logarithm function. The relay OTs are not raised when the location of the relay moves to the source due to TMS elimination. To achieve lower OTs in comparison with SCs, an NSC-based algorithm function has been introduced in ref. [19]. However, the major disadvantage of NSCs in refs. [4,16,18] is the inability of applying the approaches within existing industrial OCRs. In addition, the new requirements of measuring the fault voltage beside the fault current will increase the problem's complexity.

To mitigate and avoid the limitations in the SCs such as the inverse time–current characteristics of MG systems, a new NSC based on adding the auxiliary variable to the conventional OCR OT equation is presented by ref. [20], which aims to achieve a fast protection scheme based on well-defined time–current characteristics (TCCs). All of these auxiliary variables are measured as coordination constraints; however, it is inapplicable in existing industrial OCRs. Therefore, a piece-wise linear (PWL) characteristic has been proposed in ref. [21] to preserve the coordination time interval (CTI) between the primary and the backup relay pairs for the entire range of the fault current. The successive straight lines were joined together to formulate the new characteristic curve and a tabular format is used to adjust the curve for the existing industrial relays. For obtaining the optimal coordination of OCRs in meshed networks, the piece-wise linear characteristic (PWLC) has been evolved as a novel method [22]. It used variables coefficients, namely A and B, of the PWLCs of the OCRs for adjusting purposes and obtaining a more flexible TCC with the normally inverse (NI) standard; however, the proposed method was not tested by changing the location and the size of the DGs to evaluate their effect on the performance of the proposed approach. There are some researchers who have considered the SCs' constant parameters as variable set points, which is another NSC format [23,24]. This NSC format aims to develop flexible TCCs by creating necessary CTI for the range of the entire fault current between the primary/backup pairs of the relay. To improve the selectivity and sensitivity in the OCR coordination and avoid the NI standard curve disadvantages, an optimal coordination scheme based on nonstandard time–current characteristics is presented in refs. [25,26]. However, the nonstandard time–current characteristics are created by using the logarithmic function and constant coefficients, which lead to significantly minimizing the overall operational time on maximum fault currents, but it showed limited behavior on minimum fault currents. Table 1 shows an overview of the literature review that has been introduced above.

Table 1. A Literature Review Comparison Analysis for Protection of Radial and Meshed Networks.

Ref.	Year	Coefficient Types of Coordination Protection Scheme		Number of Variables			Modes of the Operation			DNs Types		Curve Used	
		Constants	Variables	One Variable	Two Variables	Three Variables	Without DGs	Grid-Connecting	Islanding	Radial	Meshed	IEC NI Standard	Logarithmic Function
[15]	1993	✓	✗	✗	✗	✗	✓	✗	✗	✗	✓	✓	✗
[16]	2014	✓	✗	✗	✗	✗	✓	✓	✗	✗	✓	✓	✗
[17]	2017	✓	✗	✗	✗	✗	✓	✓	✗	✓	✗	✓	✗
[18]	2018	✗	✓	✗	✗	✓	✗	✓	✗	✓	✗	✓	✗
[19]	2007	✓	✗	✗	✓	✗	✓	✗	✗	✓	✗	✗	✓
[20]	2017	✗	✓	✗	✗	✗	✗	✓	✗	✓	✓	✓	✗
[21]	2017	✓	✗	✗	✗	✗	-	-	-	✓	✓	✓	✗
[22]	2022	✗	✓	✗	✓	✗	✓	✗	✗	✗	✓	✓	✗
[23]	2015	✗	✓	✗	✓	✗	✓	✓	✗	✗	✓	✓	✗
[24]	2017	✓	✗	✗	✗	✗	✗	✓	✓	✓	✓	✓	✗
[25]	2021	✓	✗	✗	✗	✗	✓	✓	✓	✓	✗	✗	✓
[26]	2022	✓	✗	✗	✗	✗	✓	✓	✗	✓	✗	✗	✓
Proposed approach		✗	✓	✓	✗	✗	✓	✓	✓	✓	✓	✗	✓

1.3. Contribution of The Paper

It can be seen that there are some drawbacks or imperfect points that create a research gap. There is a significant need to find an alternative to the NI standard curve for more flexible and dynamic protection coordination schemes for radial and ring distribution networks (DNs) and MGs, especially at far-end faults (minimum faults). In addition, ensuring the CTI selectivity between OCRs and minimizing the tripping time in different operating modes are required. For filling this discussed research gap, optimal OCR coordination based on a new nonstandard time–current characteristic (NSTCC) with dynamic coefficients to provide a fast and reliable response in different network scenarios is proposed. An optimization task to present the OCR coordination problem has been solved by applying two optimization techniques, namely GA and GSA–SQP algorithms. The main and key contributions of this article are ordered as follows:

- For improving the performance of the protection system by integrating the DG units and during the occurrence of different fault currents in the MGs, the novel NSTCC is created with consideration to the constraints of the existing model.
- In the OCR scheme, an optimum coordination approach is utilizing NSTCC to reduce the total OT compared to traditional SCs and other NSCs presented in the literature. This paper achieves a significant reduction in total OT without a miscoordination record.

- The proposed new optimal scheme based on NSTCC is developed with a lower number of constraints compared to the optimal coordination in the literature because it uses just one flexible coordination scheme. Therefore, the new proposed approach in this work achieves the optimal solution with minimum computational costs.
- There is no communication required between the OCRs in the proposed optimal coordination strategy in this work, where the measurement of the current is acquired locally. Therefore, the proposed approach provides adequate robustness to the OCR coordination approach, decreasing the demand for communication infrastructure. Moreover, it reduces the computational cost as well as the requirement to gain way in large quantities of the MGs and grid data.
- The sensitivity improvement for the proposed optimal scheme is shown by comparing the results of the NSTCC and the standard curves under different testing and MG operation modes.
- Finally, a comparison analysis has been provided for the proposed optimum coordination approach under various faulty conditions for two DNs types: radial networks (IEEE 9-bus test system and IEC MG benchmark) as well as meshed networks (IEEE 9- and 30-bus test system). This comparison proves the proposed approach's superiority over others, especially for minimum fault in islanding mode.

1.4. Outline of Paper

This paper is categorized as follows: Section 2 presents the problem statement and suggested solution. Section 3 illustrates the novel optimal OCR protection scheme methodology and the proposed NSTCC. The MG topology and the distribution grid with the simulation analysis and results are represented in Section 4. Finally, Section 5 presents the summary, conclusions, and the suggested future work of this paper.

2. Problem Statement: OCR Coordination

OCR is considered the most common apparatus for protection as applied in distribution systems. An OCR is used in measuring the current passing through it and also determines if a signal for opening a circuit breaker is to be sent or not [27]. Relays are of different types, some of which include definite-time OCRs and directional relays. However, one of the most preferred types is the inverse-time OCR since it is a protection relay with a time characteristic used for grading and, thus, can allow some loads to specifically draw more currents in a very short period of time [28]. OT is part of inverse-time OCRs, and is found to be in inverse proportionality with the fault current as indicated by the relay. OCRs are in two forms: electromechanical and digital. Electromechanical OCRs have dominated the market for the past two decades. This is because they were not expensive and had well-known performance, resulting from many years of application. The second form, which is the digital OCR, has several advantages over the electromechanical type and is more likely to be preferred in the future for the following reasons [29,30]: first, they are economically competitive, since they are cheap to acquire, similar to the electrochemical types. Second, they have increased reliability since they have properties that can detect and report any internal problems in the relay, thus avoiding any possible malfunction operation. Third, they have smart grid natives since they are compatible with the concept of the smart grid due to their digitalized nature. Further, they have a multifunctionality ability; thus, they can perform other added tasks such as measuring the current and voltage values as well as performing protection work. Finally, they have the flexibility ability, which arises from their capability to define TCCs, which are arbitrarily set by the user [28].

To achieve the maximum operation of OCRs, two parameters should be set. These consist of the TSM and plug setting multiplier (PSM). The former is determined based on the minimum load current as well as the maximum fault current. The required interrupting capacities of overcurrent protective devices can be determined by helping the maximum fault currents, while the minimum fault currents are utilized in overcurrent device coordination operations [31]. The maximum fault currents have ratings from 50 to 200%, at intervals of 25%, whereas minimum fault currents have ratings from 0.05 to 1, with 0.05 intervals. TSM is calculated in such a way that the system for protection can disconnect easily from the power system's faulty part [28,32]. The digital relays, which are new in the market, however, are able to make these parameters be set at intervals of 0.01 [33]. In general, the structure of OCR coordination problems in MGs is complex and intensive. This is especially seen in linked distribution systems, in which the burden of computation increases as the size and the network intricacy also increases. Figure 2 illustrates the coordination constraint between primary and backup relays, in which is horizontal axis represents the fault location and the vertical axis represents the tripping time. As ordinarily in a coordination approach, the fault is isolated firstly by using the primary OCR (Rp). If the Rp does not operate, the fault will be isolated by using the backup OCR (Rb) after a particular time, called CTI, which is represented in Figure 2 between the green curve and black curve [3].

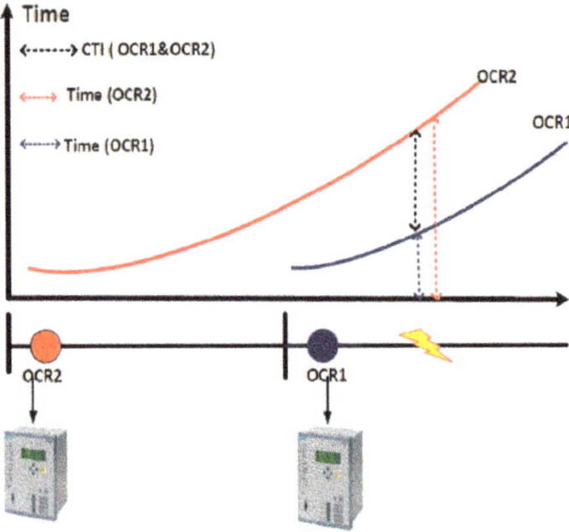

Figure 2. Coordination Constraint between Primary and Backup relay.

Most renewable energy sources such as wind turbines and photovoltaic systems have been used power electronic inverters for connecting to the MG system. In MGs, the inverter-based distributed generations (IDGs) have been used in protection; however, the inverters have a limitation of the generated fault current: 150% of the current rating [34]. This makes the conventional overcurrent devices either stop responding or respond at a much larger operating time [4]. New challenges and opportunities appeared due to the growing wind turbine energy share, which led to a preference in the use of doubly fed induction generators (DFIG) over fixed-speed wind turbine systems. The connection of the wind farm to the network contains a low-voltage ride-through (LVRT) ability that is the most important requirement. Furthermore, owing to the DFIG essentially working the same to synchronous distribution generations (SDGs), power factor control might be applied at a reduced cost [35]. The contribution of fault current from synchronous distribution generations (SDGs) can rise to about ten times the current rating [27]. In general, the fault currents in MGs are dependent on the ratio of ratings between SDGs and IDGs. Likewise, the fault current contribution ability (FCCA) of IDG is very low (110%). This means that the mode used is the islanding mode; the OCRs may be unsuccessful in the case the MG only has IDGs. This is because the ratio of IDGs to SDGs in the mixture causes difficulty in the protection coordination as well as low fault current. In general, the DOCR aims to deal with bidirectional flow of power failures evenly. In addition, if the MG is able to operate in the loop and radial topology, the relay coordination and the detection scheme under primary fault becomes complex using different types of DGs. In order to handle the protection challenges in MGs, a fast and robust optimal protection scheme is required. The main objective of this article is to present an optimal and fast coordination scheme that minimizes OCR operational times for all operation and fault scenarios in MGs.

2.1. Problem Description: Illustration-Based Analysis

MGs with sources of renewable energy must be protected to ensure that the operating conditions are reliable and safe [28–36]. The coordination of OCRs in distribution systems can be done easily, especially those in radial structures and weakly meshed MGs [37,38]. However, in interconnected and meshed systems, each of the given relays acts as backup relays, where several delays are set as a backup for one relay. Figure 3a shows the single-line diagram of DN with three distribution lines (DL_1, DL_2, DL_3) protected by overcurrent relays (R_1, R_2, R_3). Figure 3a illustrates the OCR coordination from the side of the load to the source. For instance, a fault at the F1 point results in R_1 primary relay and R_2 backup relay. If the R_1 is unable to detect the fault or tripping delays, the R_2 will have time delays. Figure 3b shows the effect of the fault location on the fault current. It can be noticed that the fault current is increased whenever closer to the main source. Figure 3c is an illustration of the OCR coordination curves in three modes, which are a conventional network (that has no sources of renewable energy), a power network with DG, and an islanding mode. Ordinarily, the OCR operating time, in the no-DG case, is high due to a CTI ranging from 0.2–0.5 s, stressing the network equipment, possibly causing the relay to fall into the precise time region. This is more so for maximal fault modes (when faults occur near the source) [29,34,36].

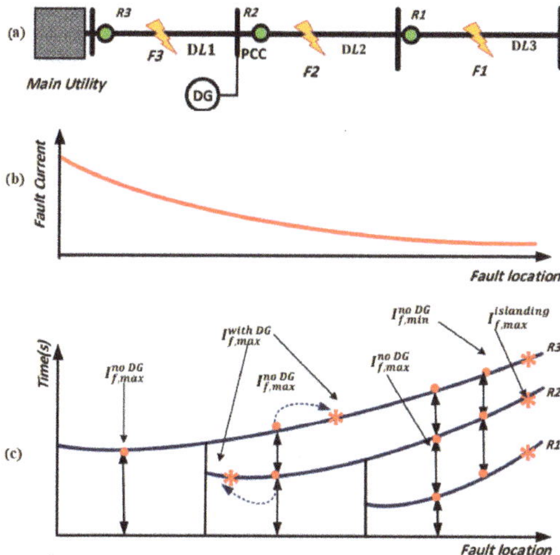

Figure 3. (**a**) The single-line diagram of the distribution grid with DG under three fault modes, (**b**) the relationship between fault current and fault location, and (**c**) the miscoordination between the primary and backup relays and fault characteristics with the absence and presence of DG.

The conventional protection scheme experiences more challenges because of the different fault characteristics between the distribution systems in the presence and absence of sources of renewable energy. The rising number of sources of renewable energy in the network has widened the variation range between the minimum and maximum levels of fault current. Consequently, the calculation of the traditional protection setting will not meet the system requirements of the main protection: sensitivity, speed, and selectivity [4,28]. For instance, when the fault occurs at the F2 point for a distribution grid that has DG as illustrated in Figure 3c, the fault current value will go up in the R_2 primary relay and go down in the R_3 backup relay, resulting in a time delay causing disconnection or OCR coordination failure [30,36]. Figure 3c illustrates the effect of the connection between the DG and the fault current, I_f. For R_2, maximum fault current for DG power network, $I_{f,max}^{with\,DG}$, goes up while it goes down for R_3 when comparing with the maximum fault current during the absence of DG, $I_{f,max}^{no\,DG}$. Generally, the fault characteristics of a DN with DG have been altered because of loading/generation level changes, variations in the network typology, islanding, fault point resistance, and the location of the fault point in relation to the main relay. Alterations of the fault current will result in OCR miscoordination; for instance, the R_3 will possibly not operate once there is a failure on R_2 for the minimum fault current, $I_{f,max}^{islanding}$, in which the current value will be reduced compared with the distribution network in the absence of DG, $I_{f,min}^{no\ DG}$. For the islanding case, the fault current is too low, making its detection based on the conventional scheme difficult [28,30]. As a result, the conventional coordination and protection scheme will not have the capacity to handle the issue, which makes the development of a new time–current characteristic to address the challenges of DG protection very important. Furthermore, a DG-equipped system of protection for DN is needed to respond to the faults in all modes of DG operation, e.g., grid and islanded. This study proposes and develops various quick and intelligent schemes of protection and coordination. The coordination schemes' main objective is calculating the PSM and coordination curves which reduce the OT.

2.2. The Coordination of Overcurrent Relay

The traditional coordination between OCRs is generally obtained with the assumption that the conditions and network parameters such as resistance, current, and voltage during a fault will remain constant [34,36]. Equation (1) describes the CTI between the OT for the primary relay and backup relay for short circuits that occurs, for instance, at the F2 point [34,36,39]. The CTI presentation is such that the time of coordination between the backup relay, t_{R3}, and the main relay, t_{R2}, is equal to or more than the designated CTI.

$$t_{R3} - t_{R2} \geq \text{CTI} \tag{1}$$

Figure 3b illustrates how DG addition to the distribution grid affects the scheme of coordination protection [40,41]. The fault current variations will result in a CTI between the primary relay and backup relay that is lesser than the chosen CTI, causing a miscoordination. Calculations of the OCR operating time, t, in traditional methods are based on constant fault currents as well as known fault currents, *Isc*, as shown in Equations (2) and (3).

$$t = \left[\frac{A}{\left(\frac{Isc}{Ip}\right)^B - 1} \right] \times \text{TMS} \tag{2}$$

$$t = \left[\frac{A}{\left(\frac{Isc}{Ip}\right)^B - 1} + C \right] \times \text{TMS} \tag{3}$$

where TMS represents the time multiplier setting, *Isc* represents the short-circuit current, and *Ip* represents the pickup current. Parameters *A*, *B*, and *C* in Equations (2) and (3) are related to a variety of relay characteristics that are defined on the basis of the standard of relay [42,43]. In general, numerical OCRs have the capability to update and modify the time operating characteristics based on real-time measurements. In this paper, the numerical OCRs provide the ability to use different time operating characteristics, such as the standard characteristics (IEC, ANSI), or generate new nonstandard operating characteristics. The proposed nonstandard time characteristic in this paper, NSTCC, aims to minimize the total tripping time and improve the performance of power protection in terms of selectivity and sensitivity.

3. The Proposed Methodology: Nonstandard Time Current Characteristics

This section aims to introduce optimal OCR coordination based on a new nonstandard time–current characteristic (NSTCC) with dynamic coefficients to reduce the tripping time associated with the value of fault currents. The proposed NSTCC scheme will be compared to the traditional OCR scheme (inverse definite minimum time (IDMT)) [29,36]. Further, there are different applications, such as the thermal stress issues occurring in the equipment (such as transformers and cables), that can use the proposed NSTCC. The next equation represents the proposed NSTCC, and the logarithmic function therein [19] is the basis of this equation.

$$t = \left(A - 1.35 \times \log_e \left(\frac{Isc}{Ip} \right) \right) \times \text{TMS} \qquad (4)$$

To ensure OCR coordination selectivity, the grading time should be kept constant and free from the network's location of the fault or the current level of the fault. Equation (4) represents the NSTCC for all relays through the use of logarithmic [19,36] and variable coefficients (A) with a range between 2 and 6.5; the time of grading will not be affected by the degree and point of fault. This will make the selectivity of the protection system better and independent of the fault location or current. Moreover, it was difficult for the normal inverse curves to detect the minimum fault. The NSTCC offers ample area for the detection and coordination of the OCRs in the minimum fault, as illustrated in Figure 4, to ensure selectivity without missing the tripping time. The following section describes an optimization task for determining the TMS based on Equation (4) that reduces the OT to a minimum. Therefore, coordination on the basis of nonstandard tripping characteristics will result in an optimal time of grading in relation to the time of tripping.

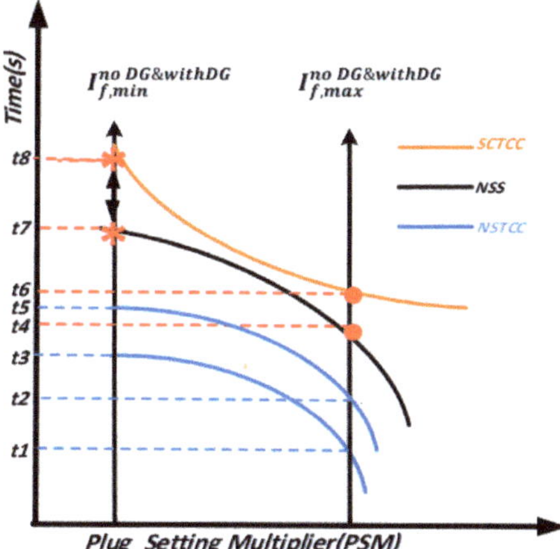

Figure 4. A Standard Tripping Current Characteristic (STCC), Nonstandard Scheme (NSS) and Proposed Nonstandard Tripping Current Characteristic (NSTCC).

The effect of adding renewable energy sources connected to the DN on the PSM and miscoordination problems that appear between OCRs during the maximum and minimum faults is shown in Figure 3b. Generally, the ratio between short-circuit current and the pickup current (Isc/Ip) is presented as a PSM. This section presents the importance of using NSTCC in coordinating the OCRs as illustrated in Figure 4. The fault's location near or at the end of the protected zone is responsible for obtaining the OCR coordination task. The fault's location near the protected line (maximum fault current) is covered by the F1 point, while the end of the protected zone (minimum fault current) is related to the F2 point. The two scenes were selected to attain the required CTI, cover on time attributes, and raise the OCR OT because of the addition of sources of renewable energy as shown in Figures 3 and 4.

It can be noticed that in Figure 4, the curve of the standard time–current characteristic (STCC) has high values of fault currents at both maximum and minimum fault currents. This curve represents the inverse definite minimum time (IDMT) overcurrent relay. As seen in Figure 4, the OTs of the STCC at the minimum and maximum fault current are t_6 and t_8, which are unchangeable values. Then, two variables' coefficients A and B for the maximum and minimum faults are required to reduce the tripping time effectively to control the two sides of the curve of the STCC. The researchers in the available literature such as [20,44] used this approach to reduce the operating time; however, this leads to increasing the number of constraints which is another disadvantage. In addition, the authors in ref. [25] have proposed the curve of the nonstandard scheme (NSS), which is represented by the black curve in Figure 4; they have used Equation (4) with constant coefficient A equal to 5.8. Yet, due to the curve being constant as seen in Figure 5, the fault currents at minimum faults, especially in islanding mode, are still slightly high-valued and there is a delay time that will lead to miscoordination problems between OCRs at t_7 as seen in Figure 4. In this work, this gap can be filled by using the NSTCC which reduces the tripping time compared with the STCC at maximum and minimum faults and nonstandard curve in ref. [25] by making the coefficient A a variable needed to have optimal value to achieve the best reduction in the total OT for relays. The coefficient A in the proposed NSTCC is controllable at both the maximum and minimum fault currents as shown in Figure 4; the blue curves represent the NSTCC and they illustrate how just the one variable coefficient A can control at both ends of the curve and reduce the OCR tripping time. It can be seen that the NSTCC can decrease the OT from t_5 to t_3 and from t_2 to t_1 at minimum and maximum faults.

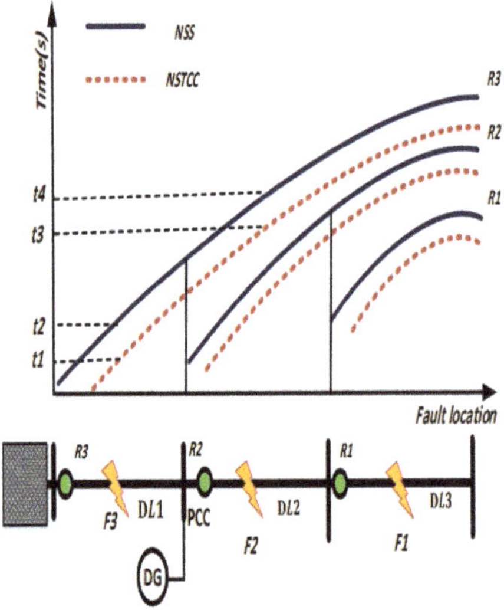

Figure 5. Proposed Nonstandard Time–Current Characteristics with variable coefficient A (NSTCC) and Nonstandard Scheme with constant coefficient A (NSS).

For the first point, the minimum tripping time (maximum fault current) must be guaranteed by each OCR in the DN. A lower OT is provided by using the NSTCC curve which is represented by the red dotted line compared to NSS which is demonstrated by the black line, as shown in Figure 5. With the existence of the DG units at the grid at the minimum fault current (islanding current) for the second point, the OCR OT will be increased more than the distribution grid in absence of the DGs. The avoidance of any miscoordination problem or nonoperational cases can be achieved by applying the proposed NSTCC curve, which minimizes the OCR tripping time as illustrated in Figure 5.

3.1. Formulating the OCR Coordination Problem

This study formulates the problem of OCR coordination in a DN that has DGs as a problem of optimization to discover the TMS which minimizes the OT for OCRs, insuring the selectivity between the primary relay and backup relay. This part introduces the suggested approach, mathematical formulae of the optimization for solving the issue of coordination, and the performance of the OCR optimization techniques in the MG protection in comparison to the traditional protection approaches.

Objective Function

The main variable for the OCR coordination issue in Equation (4) is the TMS which manages the OT of the relay. In this section, the operational time for OCRs and the coordination problem, as described in Section 2, is formulated as an objective function. This objective function (OF) minimizes the overall OT for the main relay and backup relay. The operational time, t, for the total number of relays, x, and total number fault locations, y, is formulated as follows [29,36]:

$$OF = \sum_{j=1}^{x} \sum_{k=1}^{y} t_{j,k} \tag{5}$$

where j is the relay number ($j = 1, 2, \ldots, x$), k is the fault location number ($k = 1, 2, \ldots, y$), and $t_{j,k}$ is the operational time for j relay at k fault. In this work, the NSTCC is used to calculate the total tripping time, OT, where the coefficient A is controllable at both the maximum and minimum fault currents. Equation (5) can be rewritten as follows:

$$\arg_{A} \min \sum_{j=1}^{x} \sum_{k=1}^{y} t_{j,k} \tag{6}$$

There are various constraints taken into consideration during application of the OF in Equations (5) and (6) as shown below:

- Coordination Criteria and Selectivity

The selectivity constraint for OCR coordination aims to add operational time delays between the primary and backup OCRs, to minimize power outages on the network based on the location of the fault. The backup OCR will not work except when the main OCR is nonoperational. The formulation of the criteria for selectivity can be done based on CTI as constraints of inequality:

$$t_b - t_p \geq \text{CTI} \tag{7}$$

where t_p represents the OT for primary relays and t_b represents the OT for backup relays. Generally, the CTI (in seconds) is between 0.2–0.5 to guarantee selectivity [29–37]. The value of CTI is dependent on various parameters like relay type and circuit breaker speed. This study works with a CTI from 0.2 to 0.5.

- Relay Setting, Operating Time Bounds:

To keep the limitations of operational time, the constraints should be presented for the minimal and maximal OCR operational time. Nevertheless, the protective relays should have quick operation taking the minimum possible time; if the OCR operation takes more time, there will be damage to the equipment and an unstable power system. The minimal and maximal operation time bounds are shown below:

$$TMS_{min} \leq TMS_j \leq TMS_{max} \tag{8}$$

$$OT_{min} \leq OT_j \leq OT_{max} \tag{9}$$

where TMS_{min} and TMS_{max} are the minimal and maximal TMS values of relay j and OT_{min} and OT_{max} represent the minimal and maximal time of operation needed for the relay j [28,30]. The OCRs' operation must be within the protection scheme's normal operation time. As a result, the PSM needs to be set in the domain of the minimal and maximal values in the minimum and maximum fault currents in the relay, even with light overloads.

- Proposed Setting of Coefficient Bounds:

In this study, the characteristic coefficient in Equation (4) is considered as a decision variable, as presented in Equation (6). In previous studies, the inverse curve requires more than one variable coefficient to shift it upwards and downwards [20,45], which leads to greater OCR tripping times with a slight shift in steepness and increase in the number of constraints. This causes miscoordination between primary and backup relays. For development purposes, the NSTCC is formulated in Equation (4) with just a single variable coefficient, which is A. The NSTCC tends to shift the curve downwards by changing A_i values; this leads to a reduction in OCR tripping times and the reduction of constraints. As a result, the OCR coordination performance is guaranteed. The following equation shows the maximum and minimum bounds of the variable coefficient A:

$$A_{min} \leq A_j \leq A_{max} \tag{10}$$

where A_{min} and A_{max} represent the minimal and maximal variable coefficient needed for relay j. It has been chosen to be between 2 and 6.5 in this study.

3.2. Optimization Methods for Solving the OCR Coordination Problem

In Section 3.2, the OCR coordination problem in a network connected to DGs is presented as an optimization task. This section presents two optimization algorithms, namely GA and GSA–SQP, as common and new powerful optimization algorithms for solving OCR coordination problems [8,37].

3.2.1. Genetic Algorithm Optimization (GA)

For solving complicated optimization problems, the genetic algorithm (GA) has been vastly used as an iterative optimization technique [46,47]. The GA technique considers various applicable solutions to obtain the best solution; it proves its worth in the power system protection coordination problem. Using GA in [37,47] was solely for power grids, without considering nonstandard OCR curves and renewable energy sources. In this study, the GA methodology is utilized as an innovative iterative optimization model to transact with the overcurrent coordination problem for a distribution system with DGs. Generally, the simulation of the GA is used with a specific population size, where the possible solutions for the proposed optimization problem are described by the population. Chromosome populations or individuals are the possible solutions in the population [37,39]. In the next step, an OF evaluates all solutions for the current generation; this step is called the fitness function (Equation (6)). The result of the fitness value is mainly associated with the proposed optimization problem for each solution. Creating a new population uses fitness evaluation by utilizing selecting, crossing, and mutating techniques.

In this paper, the general GA flowchart for the proposed ORC coordination problem is presented in Table 2. The process of the GA model launches for a profile group of first-generation OCR OTs. Therefore, for each OCR OT profile, Equation (6) of the objective function is used to evaluate fitness. Thus, an appropriate selection technique picks the parent OCR operating time profiles. For the next generation, the selection of the best performance (better fitness value or fittest solution) will be selected. These profiles are chosen for crossover as well as for creating a new generation (population); this step known as reproduction. From the profiles of the parent, the common genes of parents are retained to create the profile of a new generation of the OT whereas the residual genes are chosen at random from the parents. Nevertheless, the power network or relay constraint might be violated by the child OT profile; consequently, for examination purposes, whether the profile of the child is under the constraints or not, a feasibility test has been applied. Sometimes the child profiles are assigned to the impracticable zone. Then, the solution of the child, in this case, will be refused and an alternative one will be generated by randomizing the uncommon genes until the feasibility of the child profile is realized. At the initial time and for the initial iterations, the expectation about child profile will be varied and far away from parent OCR operation solutions. Nonetheless, in each iteration, the profiles of both children and parents are nearer to each other, and the search directs closely to the optimum OCRs operation time profile. Achieving the greatest number of iterations or reaching the proposed threshold is the goal of this process, which will be repeated many times until meeting this goal.

Table 2. The procedures of the GA technique.

Stages	Explanation
Stage 1: Initiate the inhabitants	Applying IEEE9 and IEC MG models and calculating the short circuits. Early possible profiles of the OCR operation time (1000 profiles).
Stage 2: Population assessment	For each solution, the prime assessment can be obtained via solving the optimum OF, Equation (5) under bounds from Equations (6) to (9).
Stage 3: Selecting	The two most excellent OCR OT profiles will be selected as a parent.
Stage 4: Crossing	For the following generation, a child profile will be launched, which will keep the popular genes while the rest of them will be chosen from the parent's profile at random. The child profile must be a solution within the bounds of possibility.
Stage 5: Mutating	In each iteration, a single gene is mutated to update the child profile randomly. In this study, the setting of mutation is 0.1
Step 6: The model output and optimal solution	Reaching the maximum number of iterations (200 iterations) by repeating the stages that have been mentioned above, starting with stage 2.

3.2.2. A Hybrid Algorithm Gravitational Search Algorithm–Sequential Quadratic Programming (GSA–SQP)

The GSA–SQP algorithm is presented by refs. [8,48] as a powerful optimization solver for OCR coordination. Firstly, the GSA–SQP algorithm is presented as a multipoint method of search that is based on probability through using the gravitational search algorithm (GSA). Secondly, non-linear programing (NLP) techniques such as sequential quadratic programming (SQP), which are the single-point method of search, have the disadvantage of getting trapped in a local optimum point when the first option is closer to the local optimum. The NLP techniques offer a globally optimal solution if the correct first choice is made [48]. The study by ref. [48] suggested a hybrid of GSA with SQP to take advantage of the methods while overcoming their drawbacks. The SQP routine is introduced in GSA as a local technique of search to boost the convergence. Initially, the GSA method is performed and the best fitness for each generation is chosen in each interaction. From this, the corresponding agent is set as the initial value of the variables in the SQPP technique. The SQP routine is then executed according to the local search's adopted probability of local search (α_{LS}), improving the best fitness obtained from GSA in the current interaction. This is how the algorithm of GSA–SQP offers the global optimal solution. For calculating the optimal setting of the OCRs, several agents that represent a complete solution set are presented as the control variable of the OCR coordination problem, X.

$$X = \left[TMS_j^1, \ldots, TMS_j^m, Ip_j^1, \ldots, Ip_j^m, A_j^1, \ldots, A_j^m \right] \tag{11}$$

where N is the population size (agents in the system), $j = 1, 2, \ldots$; N and m are the numbers of the relay in the grid. The process of the GSA–SQP started with a set of the first-generation OCR operation time profiles based on the power network and fault calculation data. Then, for each OCR operation time profile, the objective function (Equation (6)) is used as an evaluation and fitness approach in this work. Thus, the best and worst solutions will be selected. For the next generation, a random number will be generated and compared to the constant α_{LS}. In the case that the random number is less than the α_{LS}, the model will calculate the gravitational constant at time t, G(t), masses of agents, M(t), and the total force that acts on the i-th agent at time t, F(t), to update the velocity and position of the searching agent. In case of a random number larger than the α_{LS}, the SQP method will be used to update the next generation by selecting the new agent as the best agent. Finally, the GSA–SQP model will select the optimal solution among all solutions which is helped to achieve the global solution as shown in Figure 6. The parameters of the algorithm applied for the optimal coordination problem in this paper are: constant G0, the initial gravitational which is set to 100, and constant α, a user-specified value which is adjusted to 20. Both constants G0 and α control the GSA performance. t is the current iteration while t_{max} is the maximum iteration number which is set to 200; N is adjusted to 50; α_{LS} is 95. For miscoordination problems, β has been used. Miscoordination decreases with increasing β; however, the relay OTs rise. Thus, for omitting the miscoordination, a fit β value should be selected. In addition, there are other parameters related to local search that should be calculated throughout the process which are: G(t), the gravitational constant at time t, M(t), the masses of agents, F(t), the total force that acts on the i-th agent at time t, and a(t), the acceleration of the i-th agent. For meshed network case studies, the weighting factors are chosen as α_1 and α_2 which are set to 2 and 15, respectively. IEEE 9-bus system β is set to 500 and the IEEE 30-bus system is set to 1000.

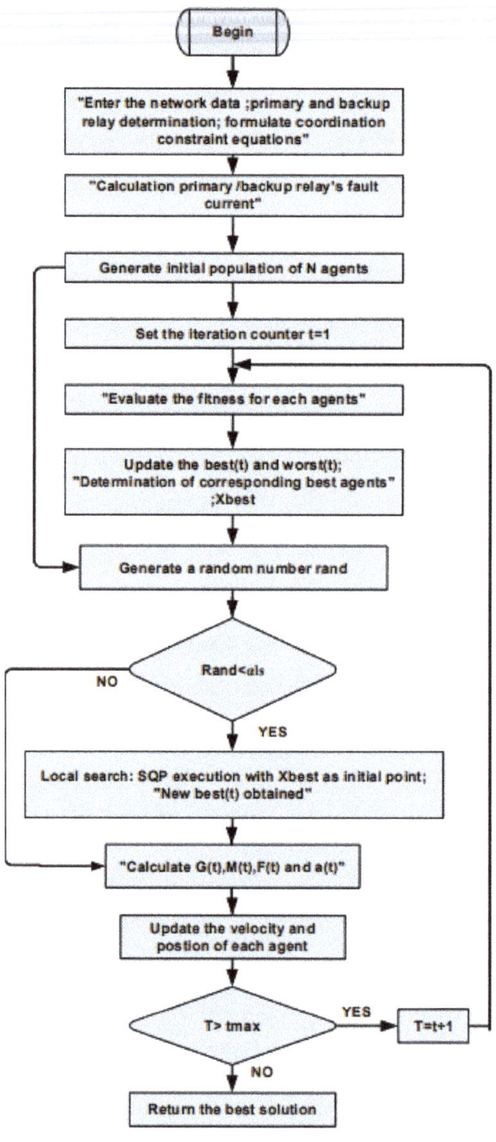

Figure 6. The flowchart of the GSA–SQP algorithm.

4. Simulation Results and Discussion

This section aims to present the results of the proposed nonstandard OCR coordination approach, NSTCC, using radial and meshed distribution systems and under different operating scenarios. Throughout this section, the NSTCC will be compared to the conventional OCR coordination scheme and nonstandard OCR scheme developed by ref. [25]. For solving the OCR protection coordination problem, GA and hybrid GSA–SQP optimization techniques are used in this study based on the following network scenarios:

- Radial Networks: IEEE 9-bus test system and IEC MG benchmark.
- Meshed Networks: IEEE 9-bus and IEEE 30-bus meshed networks.

4.1. Radial Networks

In this part, an IEEE 9-bus test system and IEC MG benchmark have been carried out as a radial network. Figure 7 shows the flowchart for the implementation of NSTCC in radial networks. Afterwards, the short-circuit calculations in various modes and locations were achieved. In this section, the GA method has been used to obtain the optimal setting (TMS and the coefficient A) for OCRs based on STCC [25], NSS [25], and the proposed NSTCC in this paper. Furthermore, the NSTCC has been compared with the STCC and NSS in terms of reducing the relays' operation time, OT, and ensuring the CTI selectivity. For solving the OCR protection coordination problem, the GA technique is used in this section based on the presented network configurations in Figure 7.

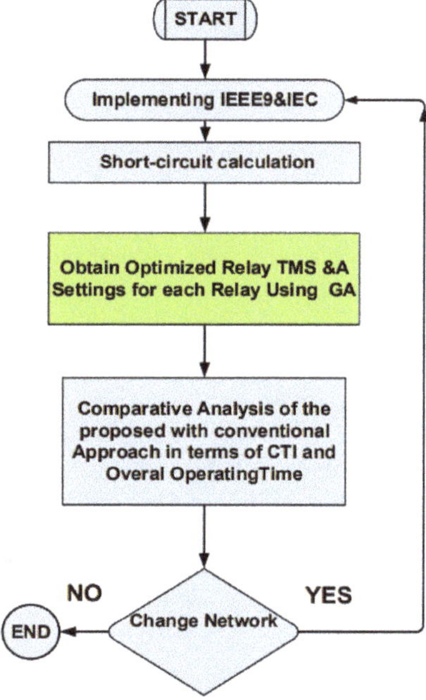

Figure 7. Flowchart of the Implementation of the NSTCC in Radial Networks.

4.1.1. The Radial 9-Bus Test Systems

The proposed network system is developed based on the Canadian Urban Benchmark 4-bus feeder distribution system [49]. As shown in Figure 8, the IEEE 9-bus consists of one DG and 10 OCRs as well as 2 directional OCRs, DOCRs, which are R8 and R10. A utility main source feeds this radial distribution capacity of short circuit = 500 MVA as well as the ratio of X/R = 6 and all lines with length = 500 m. The system is associated with the utility throughout a transformer of 20 MVA, 115 kV/12.47 kV. The simplified network, as shown in Figure 8, presents the OCRs and DOCRs. The plug setting (PS) and the current transformer ratio (CTR) for each OCR are stated as follows: for R1 and R5, the PS and CTR are 1.128 and 100/1, respectively, while 1.130 and 200/1 are the values of the PS and CTR for the R2 and R6. For all R3, R7, R8, and R9, the values of PS and CTR are 1.132 and 300/1, respectively. 1.135 and 400/1 are the PS and CTR values for all R4, R10, and R11. Lastly, the PS and CTR of the R12 are 1.140 and 600/1, respectively [25]. In the following subsection, the results of the proposed NSTCC, STCC, and NSS schemes are presented over different fault and power network model scenarios, as discussed in the previous section.

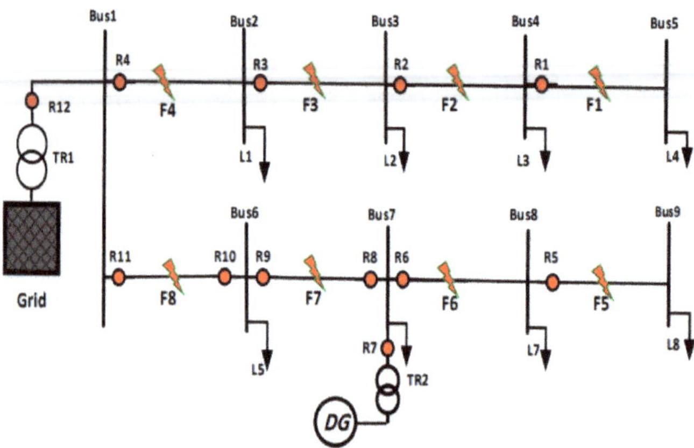

Figure 8. The IEEE 9-Bus MG system.

- Radial IEEE 9-Bus System without DGs: Mode 1 test results

This mode represents a conventional power network, which is fed only via the main utility feeder without DGs, as shown in Figure 8. The GA optimization technique is used to evaluate the performance of the NTSCC method and compare it with the conventional STCC and the NSS [25]. In general, the optimized values of TMS, the overall OT, and coefficient A in mode 1 for all OCRs are presented in Table 3. The acquired settings and the total tripping time were computed by utilizing MATLAB software and GA methodology. As shown in Table 3, the NSTCC approach achieved the minimum overall OT of all OCRs which equals 8.224 s, compared to the STCC and NSS methods which are equal to 9.352 and 8.848 s, respectively. In addition, the optimized value of coefficient A for the NSTCC approach has been chosen as approximately 5 for all OCRs as shown in Table 3, which is a suitable value for maximum current faults in this conventional power network case.

Table 3. The Overall OT for STCC, NSS, and NSTCC Curves in IEEE 9-Bus (DGs- Mode 1).

Relays	STCC [25] TMS	NSS [25] TMS	NSTCC	
			TMS	A
R1	0.010	0.010	0.010	5.003
R2	0.139	0.171	0.300	5.000
R3	0.239	0.264	0.395	5.002
R4	0.322	0.345	0.486	5.000
R5	0.010	0.010	0.010	5.001
R6	0.139	0.171	0.300	5.000
R7	0.235	0.264	0.396	5.000
R8	0.319	0.345	0.479	5.017
R9	0.367	0.384	0.501	5.000
R10, R11, R12				
Overall OT(s)	9.352	8.848	8.224	

- Radial IEEE 9-Bus System with DGs: Mode 2 test results

Integration of the DGs in the network leads to raising the complexity of obtaining the optimal OCR coordination. This mode tests and evaluates the NSTCC on the network that is fed by all types of DGs, as illustrated in Figure 8. Table 4 shows the optimized values of TMS, the overall OT, and coefficient A in mode 2 for all OCRs. The total OT of the NSTCC in mode 2 of all OCRs equals 9.327 s, while the overall OTs for the STCC and NSS are 11.282 and 10.353 s, respectively. As with mode 1, an appropriate optimized value of coefficient A has been selected in mode 2 for the NSTCC approach, which is approximately 5 for all OCRs as illustrated in Table 4. As a result, the NSTCC scheme has recorded the lowest overall OT in Table 4 and optimal optimized value of the coefficient A.

Table 4. The Overall OT for STCC, NSS, and NSTCC Curves in IEEE 9-Bus (DGs- Mode 2).

Relays	STCC [25] TMS	NSS [25] TMS	NSTCC	
			TMS	A
R1	0.010	0.010	0.010	5.001
R2	0.139	0.171	0.276	5.084
R3	0.237	0.264	0.397	5.001
R4	0.321	0.345	0.437	5.001
R5	0.010	0.0100	0.010	5.001
R6	0.139	0.166	0.277	5.022
R7	0.062	0.161	0.177	5.004
R8	0.237	0.258	0.373	5.017
R9	0.010	0.010	0.010	5.003
R10	0.010	0.010	0.010	5.002
R11	0.118	0.132	0.191	5.001
R12	0.367	0.384	0.466	5.001
Overall OT(s)	11.282	10.353	9.327	

- Radial IEEE 9-Bus System under the islanding condition: Mode 3 test results

The grid's operational way in this mode is called islanding mode. Applying the NSCC on the network with this mode shows the reliability and effectiveness of the proposed approach with a low fault current. Over the islanding mode, a comparison has been made between the proposed approach and other approaches in the Table 5, in terms of the optimized values of TMS and the overall OT. Similarly, the NSTCC approach achieves the minimum overall operational time of all OCRs compared to STCC and NSS approaches. The overall OT of all OCRs in mode 3 is 1.3728, 1.34, and 1.187 s for STCC, NSS, and NSTCC, respectively. It can be noticed that the optimized value of coefficient A in the Table 5 (islanding mode) for the NSTCC approach has been chosen to be approximately 2, which is suitable for detecting the minimum fault currents, and it is considered the key contribution of this NSTCC approach without delaying time during the optimization task.

Table 5. The Overall OT for STCC, NSS, and NSTCC Curves in Radial IEEE 9-Bus—Mode 3.

Relays	STCC [25] TMS	NSS [25] TMS	NSTCC	
			TMS	A
R1				
R2				
R3				
R4				
R5	0.010	0.010	0.010	2.000
R6	0.039	0.077	0.291	2.050
R7	0.137	0.129	0.438	2.049
R8				
R9				
R10				
R11				
R12				
Overall, OT(s)	1.373	1.34	1.187	

- Discussion of the Radial IEEE 9-Bus System results

In this section, the performance of proposed NSTCC, NSS, and STCC approaches on the radial IEEE 9-bus system over the different operation modes is presented. The overall OT for operation modes shown in Figure 9 was obtained by using the GA algorithm. The previous subsections show that the NSTCC approach reduced the overall OT of OCRs for all modes compared to NSS and STCC approaches. For example, the NSTCC reduced the overall OT in Mode 2 by 17.32% and 9.91% compared to NSS and STCC approaches, respectively.

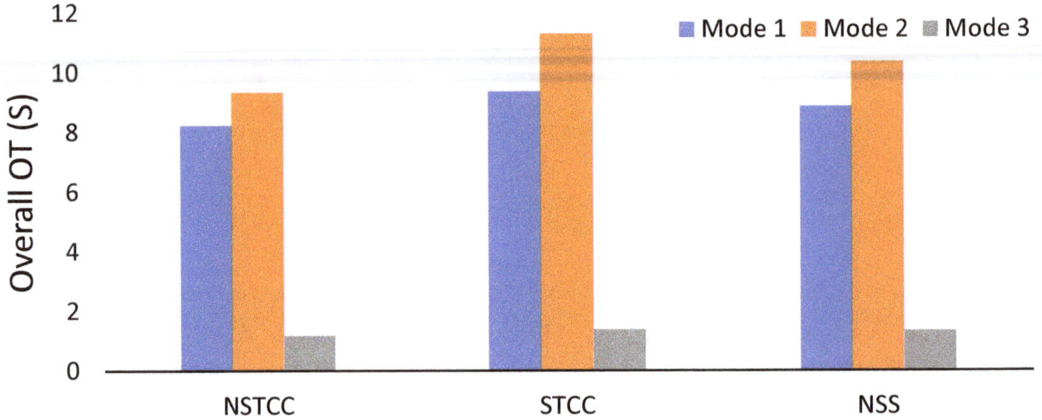

Figure 9. The overall OT in Modes 1, 2, and 3.

4.1.2. The Radial IEC MG Test System

IEC MG benchmark connected to various DG technology types has been used to evaluate the NSTCC in this section. As shown in Figure 10, it has 4 DGs (two wind turbines and two synchronous generators), 5 transformers, and uses 15 OCRs as well as 5 DOCRs which are R1, R3, R5, R8, and R9, Refs. [49,50] give all details about IEC MG. The PS and CTR for each OCR and DOCR are described as follows: for R1, R2, R3, R4, R5, R6, R8, R9, R10, R12, and R15, the PS and CTR are 0.5 and 400/1, respectively. Whereas for R11 and R14, the PS is 0.65 and CTR is the same as the previous OCRs, which is 400/1. The PS and CTR values of the R13 are 0.88 and 400/1. Finally, for R7, the PS and CTR are 1 and 1200/1, respectively [25]. A short circuit on different lines was calculated for each mode in this study. The three-phase faults at different transmission lines in the MG are illustrated in Figure 10. In Figure 10, the DG is considered as a PMSG system where the contribution of fault current is low, similar to our case with the PV system, compared to the DFIG system, which is considered as high fault: around 7 times the full load. However, the DFIG condition will be protected by the first relay after the generator in this work.

These fault cases are: a fault on the line DL-5, named F1; a fault on the line DL-4, named F2; a fault on the line DL-2, named F3; a fault on the line DL-3, named F4; and a fault on the line DL-1, named F5. To optimize the TMS for the coordination of OCRs, GA optimization processes have been carried out to test the NSTCC using MATLAB simulations. For fault conditions, the primary OCR should isolate the fault firstly. If it fails to trip the fault, after allowable CTI, the backup OCR must be operated; it is assumed to be operated between 0.2 s and 0.5 s. Three operational modes have been implemented in this IEC MG.

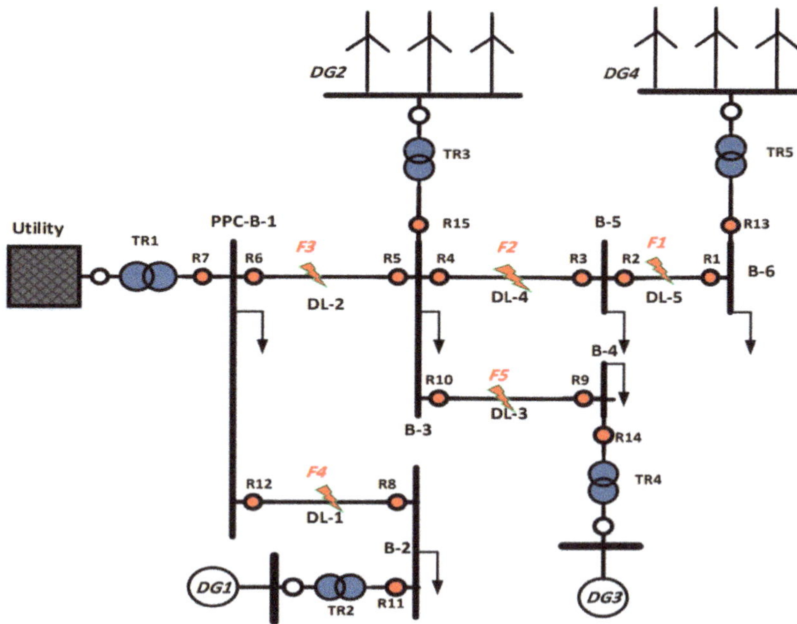

Figure 10. IEC MG Benchmark (Large-Scale Network).

- The radial IEC MG Simulation Results in Mode 1

In this case, the main source is only connected to the IEC MG; however, all the DGs are off. For comparative purposes, the results of the TMS and the operating times for all relays that were obtained from the literature [29,36,51,52] are presented in Table 6. The authors in ref. [25] did not evaluate the IEC MG as a radial system. It can be noticed that the proposed obtained OT equals 2.42 s for the proposed NSTCC, which is a considerable reduction from those reported in the Table 6.

Table 6. The Overall OT for the radial IEC MG—Mode 1.

	[36]	[29]	[51]	[52]	NSTCC	
Relay	**TMS**	**TMS**	**TMS**	**TMS**	**Coefficient A**	**TMS Proposed**
R1						
R2	0.128	0.0500	0.05	0.0500	4.800	0.0501
R3						
R4	0.259	0.1787	4.58	0.2306	4.800	0.3977
R5						
R6	0.402	0.3223	2.16	0.7715	4.886	0.9223
R7	0.290	0.2060	0.05	0.055	4.826	0.1117
R8						
R9						
R10	0.128	0.0500	0.05	0.0500	4.801	0.0500
R11						
R12	0.010	0.0500	0.05	0.0500	4.800	0.0504
R13, R14, R15						
OT(s)	6.64	4.99	4.4	4.19	-	2.42

- The radial IEC MG Simulation Results in Mode 2

In this operational mode, the MG is connected to the main grid and the DG units. The results of the TMS and the operating times for all relays that were obtained from the literature [29,36,51,52] are presented in Table 7. It can be guaranteed that the value of the coordination between the primary and backup relays is achieved by obtaining the lowest tripping time for the NSTCC among the approaches reported in Table 7. The total OT was equal to 4.69 s for NSTCC compared to 11.6 s as the lowest OT value obtained from literature, which equals a 60% reduction for using NSTCC. The coefficient A's optimized value equals approximately 4.8 too according to the maximum fault current in this mode.

Table 7. The Overall OT for the radial IEC MG—Mode 2.

	[36]	[29]	[51]	[52]	NSTCC	
Relay	TMS	TMS	TMS	TMS	Coefficient A	TMS
R1	0.174	0.137	0.217	0.3893	4.801	0.188
R2	0.139	0.050	0.050	0.050	4.800	0.251
R3	0.087	0.050	0.050	0.050	4.801	0.050
R4	0.278	0.189	0.235	0.421	4.801	0.066
R5	0.172	0.115	0.197	0.394	4.801	0.050
R6	0.401	0.320	1.550	0.819	4.853	3.000
R7	0.284	0.244	0.050	0.273	4.820	0.103
R8	0.223	0.191	0.567	0.606	4.801	0.050
R9	0.069	0.050	0.050	0.050	4.808	0.050
R10	0.141	0.050	0.050	0.050	4.801	0.050
R11	0.244	0.218	0.470	0.452	4.802	0.144
R12	0.150	0.050	0.050	0.050	4.840	0.050
R13	0.194	0.167	0.325	0.746	4.800	0.245
R14	0.116	0.100	0.104	0.172	4.801	0.145
R15	0.170	0.136	0.280	0.297	4.801	0.115
OT(s)	17.48	13.66	11.6	12.48	-	4.69

- The radial IEC MG Simulation Results in Mode 3

In mode 3, the MG operates in islanding mode, in which the main grid is in off-grid mode and the load is supplied by all the DG units. The results of the OCR coordination obtained from the literature [29,36,51,52] are shown in Table 8. In this case, too, the proposed approach NSTCC outperformed the other approaches that are reported in Table 8. The proposed NSTCC achieved an OT equal to 4.055 s as the lowest value among others in Table 8. Consequently, the coordination problems may not exist, which means the selectivity is guaranteed. The optimized value of coefficient A has been chosen as 6 in this mode to obtain the lowest OT value in Table 8.

Table 8. The Overall OT for the radial IEC MG—Mode 3.

	[36]	[29]	[51]	[52]	NSTCC	
Relay	TMS	TMS	TMS	TMS	Coefficient A	TMS
R1	0.173	0.137	0.548	0.389	6.000	0.154
R2	0.105	0.050	0.050	0.050	6.000	0.050
R3	0.086	0.050	0.050	0.050	6.000	0.050
R4	0.211	0.157	0.938	0.529	6.000	0.365
R5	0.209	0.155	0.862	0.862	6.000	3.000
R6	0.181	0.151	0.685	0.243	6.000	3.000
R7						
R8	0.248	0.218	0.519	0.499	6.000	0.376
R9	0.069	0.050	0.050	0.050	6.000	0.050
R10	0.112	0.050	0.050	0.050	6.000	0.050
R11	0.265	0.240	0.490	0.431	6.000	0.415
R12	0.105	0.050	0.050	0.050	6.000	0.050
R13	0.193	0.167	0.778	0.100	6.000	0.200
R14	0.116	0.099	0.372	0.104	6.000	0.117
R15	0.177	0.148	0.670	0.630	6.000	1.827
OT(s)	15.56	12.63	9.99	8.96	-	4.055

- Discussion of the radial IEC MG results

In this section, the performance of the proposed NSTCC and approaches from the literature [29,36,51,52] on the radial IEC MG over the different operation modes is presented. The overall operation time, OT, for relays was obtained by using the GA algorithm and is shown in Figure 11. The previous subsections show that the NSTCC approach reduced the overall OT of OCRs for all modes compared to approaches in the literature [29,36,51,52]. For example, compared to the literature approaches [29,36,51,52], the NSTCC reduced the overall OT in Mode 1 by 63.55%, 55.5%, 45.0%, and 47.7%, respectively.

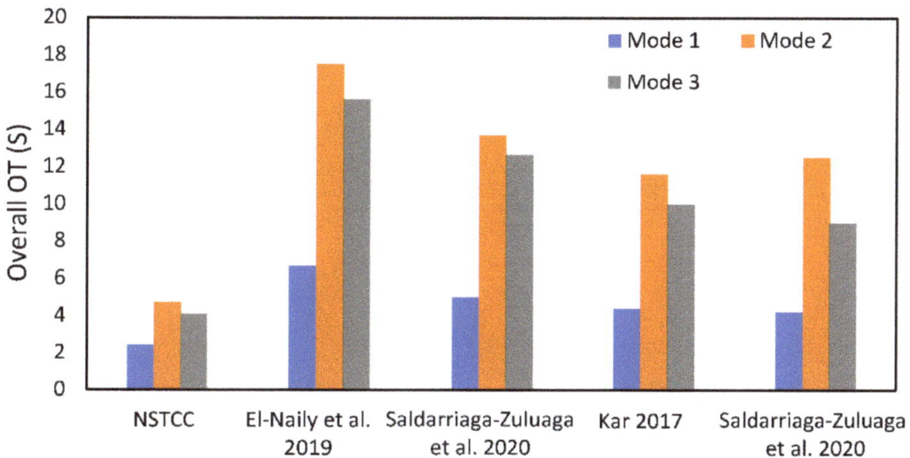

Figure 11. The overall OT in Modes 1, 2, and 3 for the radial IEC MG by different approaches from the literature (Saldarriaga-Zuluaga et al. 2020 [29], El-Naily et al. 2019 [36], Kar 2017 [51], Saldarriaga-Zuluaga et al. 2020 [52]).

4.2. Meshed Networks

In this section, IEEE 9- and 30-bus meshed networks have been implemented to evaluate the proposed NSTCC approach. Figure 12 shows the flowchart for the implementation of NSTCC in radial networks. Afterwards, the short-circuit calculations in various modes and locations were performed. The MATLAB software is used to implement the proposed equation with the constant coefficient 5.8 and with the variable coefficient A for obtaining the short-circuit calculations. The optimal setting, TMS, the coefficient A, and Ip for OCRs have been obtained by using the MATLAB software and applying the hybrid GSA–SQP algorithm for the NSTCC approach. Finally, the NSTCC with a constant coefficient 5.8 and a variable constant A has been compared with the STCC to show the effectiveness of the proposed approach in terms of reducing the OT considerably and ensuring the CTI selectivity.

The main steps of using the hybrid GSA–SQP by applying it in the MATLAB software can be shown as follows:

- The objective function for the IEEE 9-bus system is run on the MATLAB software; it includes the short-circuit calculation values for all points that have been illustrated in Figure 13 from point A to L.
- For example, when the fault occurs at the point A: for primary R1, backup R15, and R17, the fault current was 24779 A, 9150 A, and 15632 A, respectively. The OT of primary and backup relays for faults in A have been calculated based on Equation (4). The best value of the A parameter in Equation (4) will be determined by solving the cost function in Equation (6) by using GSA–SQP in the MATLAB platform.

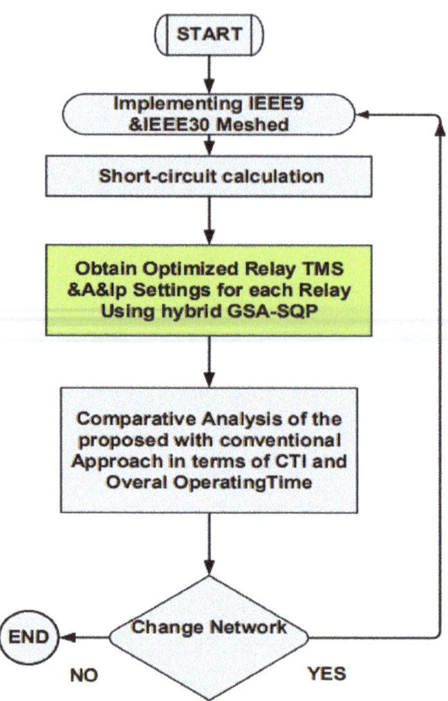

Figure 12. Meshed Networks Flowchart of the Proposed NSTCC.

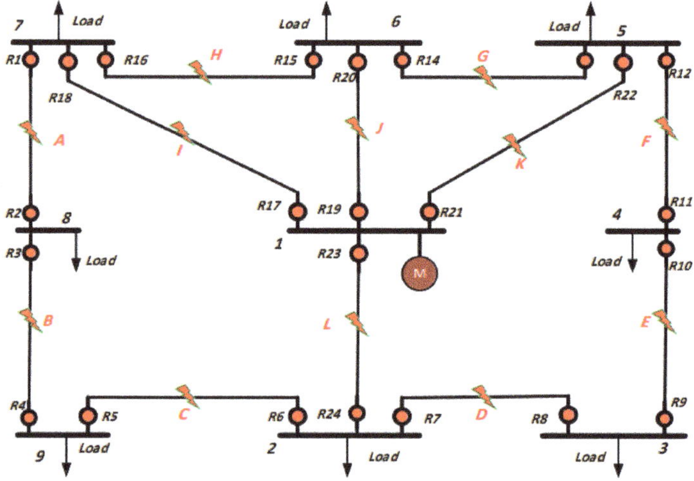

Figure 13. IEEE 9-Bus MG system, meshed network.

4.2.1. Description of the Meshed 9-Bus Test Systems under Study

For solving the OCR coordination problem, the NSTCC is applied to this 9-bus test system by using the hybrid GSA–SQP algorithm. This grid consists of 12 lines and 24 OCRs, and every line has two relays at both ends as illustrated in Figure 13. The power is received via bus 1, which is represented by a source of 100 MVA, 33 KV and more details about this grid are given in [48]. Twelve fault points have been considered, indicated from A to L (one on each line) as shown in Figure 13. For these fault points, Table 9 shows the primary–backup relationship of relays and the CTI is taken at a minimum of 0.2 s. In case of fault at different points, the short-circuit analysis has been conducted to find the current seen by the relays.

Table 9. CTI for the Meshed 9-bus Test System.

Backup Relay	Primary Relay	CTI (s)	Backup Relay	Primary Relay	CTI (s)
15	1	0.200	11	13	0.200
17	1	0.218	21	13	0.200
4	2	0.200	16	14	0.200
1	3	0.200	19	14	0.200
6	4	0.200	13	15	0.200
3	5	0.200	19	15	0.201
8	6	0.200	2	16	0.200
23	6	0.203	17	16	0.200
5	7	0.200	2	18	0.309
23	7	0.200	15	18	0.317
10	8	0.200	13	20	0.322
7	9	0.200	16	20	0.322
12	10	0.200	11	22	0.309
9	11	0.200	14	22	0.320
14	12	0.203	5	24	0.318
21	12	0.235	8	24	0.315

The optimized TMS and Ip values and OTs based on a hybrid GSA–SQP algorithm are compared with obtained results in ref. [8]. The NSTCC reduces the overall operational time of primary OCRs to 1.869 s compared to the results that are illustrated in Table 10. The coefficient A's optimized value is selected to be approximately 6.25 by MATLAB software operations in this case to obtain the lowest OT. The corresponding values of CTI are shown in the Table 9. The optimum results ensure the coordination between primary and backup relays. Further, the CTI is improved by using the NSTCC approach; the sum of CTI values equals 7.392, which is reduced compared with the sum of CTI values in ref. [8], which equals 8.892. The NSTCC scheme results in the best settings.

4.2.2. Description of the Meshed 30-Bus Test Systems under Study

The IEEE 30-bus system is considered as a meshed test system in this study to evaluate the proposed approach's efficiency in solving large, meshed networks. As shown in Figure 14, it has 19 lines and 38 DOCRs and it receives power from three distribution substations by bus 1, 6, and 13; each one is represented by a source of 132 MVA, 33 KV as well as with two DG units. The nineteen fault points are labeled as L1 to L19; for each line, there is one fault point as shown in Figure 14. The primary–backup coordination of relays for these fault points and CTI values are illustrated in Table 11 and the optimal TMS, Ip, and A for the meshed 30-bus test system is presented in Table 12. The configurations of the short circuits' current values and more information can be found in ref. [53]. The TMS range is between 1.1 to 1 and for Ip, from 1.4 to 5.9. The minimum value of the CTI is set as 0.3 s.

Table 10. Optimal TMS, Ip, and A for the Meshed 9-bus Test System.

Relay	STCC [8]			NSTCC with Constant Coefficient A				NSTCC			
	TMS	Ip	$TO_{primary}$	TMS	A	Ip	$TO_{primary}$	TMS	A	Ip	$TO_{primary}$
R1	0.276	2.500	0.629	0.680	5.800	0.500	0.201	0.645	6.250	0.500	0.029
R2	0.091	2.500	0.332	0.107	5.800	0.500	0.157	0.102	6.250	0.500	0.155
R3	0.197	2.500	0.524	0.264	5.800	0.500	0.146	0.267	6.250	0.500	0.161
R4	0147	2.500	0.408	0.243	5.800	0.500	0.171	0.234	6.250	0.500	0.176
R5	0.139	2.500	0.470	0.110	5.800	0.500	0.143	0.118	6.250	0.500	0.159
R6	0.219	2.500	0.509	0.528	5.800	0.500	0.042	0.500	6.250	0.050	0.646
R7	0.221	2.500	0.515	0.553	5.800	0.500	0.044	0.521	6.250	0.050	0.674
R8	0.138	2.500	0.467	0.109	5.800	0.500	0.141	0.117	6.250	0.050	0.157
R9	0.148	2.500	0.402	0.244	5.800	0.500	0.157	0.234	6.250	0.500	0.163
R10	0.312	3.525	0.912	0.263	5.800	0.500	0.143	0.265	6.250	0.500	0.158
R11	0.148	2.500	0.402	0.107	5.800	0.500	0.157	0.102	6.250	0.500	0.155
R12	0.197	2.500	0.766	0.630	5.800	0.500	0.186	0.602	6.250	0.500	0.027
R13	0.092	2.500	0.334	0.175	5.800	0.500	0.101	0.145	6.250	0.500	0.091
R14	0.272	2.500	0.619	0.288	5.800	0.500	0.118	0.166	6.250	0.500	0.076
R15	0.190	2.500	0.508	0.299	5.800	0.500	0.123	0.165	6.250	0.500	0.075
R16	0.189	2.500	0.508	0.172	5.800	0.500	0.100	0.146	6.301	0.500	0.092
R17	0.500	0.542		0.287	5.800	0.890	0.000	0.179	6.250	0.813	0
R18	0.072	1.979	0.232	0.034	5.800	1.553	0.103	0.010	6.364	0.500	0.015
R19	0.315	1.693		0.302	5.800	0.537	0.000	0.130	6.250	1.035	0
R20	0.064	2.201	0.205	0.017	5.800	0.785	0.033	0.010	6.250	0.500	0.014
R21	0.482	0.597		0.188	5.800	1.445	0.000	0.116	6.250	1.632	0
R22	0.092	1.902	0.229	0.049	5.800	0.668	0.093	0.010	6.250	0.500	0.015
R23	0.424	0.682		0.214	5.800	0.871	0.000	0.222	6.354	0.817	0
R24	0.101	0.964	0.271	0.010	5.800	0.500	0.019	0.010	6.250	0.500	0.019
OT(s)	8.892			2.377				1.870			

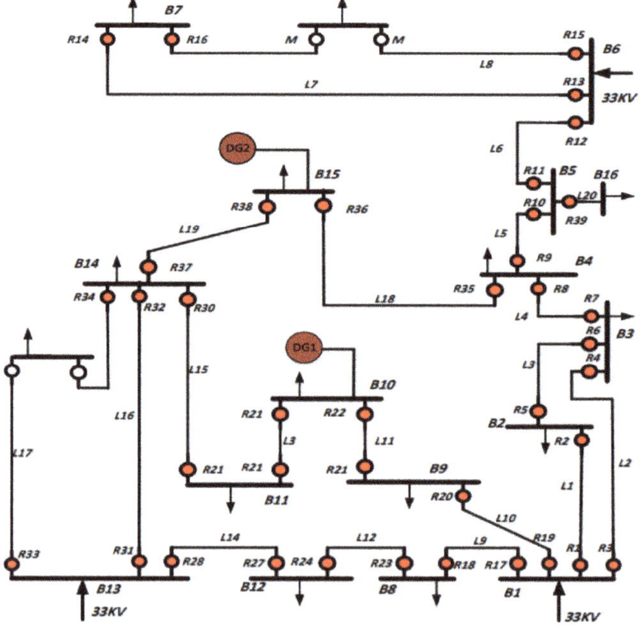

Figure 14. IEEE 30-Bus MG system (Large-Scale Network).

Table 11. CTI for the Meshed 30-bus Test System.

Backup Relay	Primary Relay	CTI (s)	Backup Relay	Primary Relay	CTI (s)	Backup Relay	Primary Relay	CTI (s)
1	3	0.300	15	19	0.300	25	24	0.300
2	4	0.514	15	36	0.463	28	1	0.301
2	22	0.300	16	19	0.310	28	2	0.413
3	4	0.300	16	34	0.301	28	10	0.300
3	21	0.436	16	36	0.488	29	1	0.319
4	5	0.300	17	19	0.301	29	2	0.431
4	18	0.305	17	34	0.305	29	9	0.301
5	6	0.300	17	35	0.349	30	29	0.300
6	7	0.331	18	38	0.300	31	28	0.300
6	8	0.300	19	37	0.301	32	30	0.300
7	27	0.302	20	2	0.443	33	31	0.300
8	26	0.300	20	9	0.300	34	32	0.300
9	12	0.300	20	10	0.317	35	17	0.606
10	11	0.300	21	1	0.327	35	33	0.317
11	13	0.300	21	9	0.301	36	16	0.507
12	14	0.300	21	10	0.317	36	33	0.302
13	15	0.300	22	20	0.300	37	5	0.453
14	16	0.301	23	21	0.647	37	23	0.300
14	17	0.378	23	22	0.300	38	34	0.302
15	19	0.300	24	18	0.458	38	35	0.342
15	35	0.332	24	23	0.300	38	36	0.474

Table 12. Optimal TMS, Ip, and A for the Meshed 30-bus Test System.

Relay	STCC [8]			NSTCC with Constant Coefficient A				NSTCC			
	TMS	Ip	$TO_{primary}$	TMS	A	Ip	$TO_{primary}$	TMS	A	Ip	$TO_{primary}$
R1	0.304	3.816	0.901	0.647	5.800	1.595	0.458	0.550	4.301	2.294	0.274
R2	0.346	1.500	0.713	0.3223	5.800	2.207	0.352	0.331	4.185	2.594	0.164
R3	0.276	2.756	0.946	0.395	5.800	1.895	0.707	0.323	4.199	2.501	0.441
R4	0.258	2.693	0.766	0.424	5.800	1.669	0.525	0.477	4.089	1.892	0.237
R5	0.263	1.659	0.661	0.317	5.800	1.690	0.436	0.2932	4.212	2.438	0.317
R6	0.160	1.981	0.607	0.1801	5.800	1.831	0.443	0.1756	4.123	2.028	0.301
R7	0.173	1.824	0.430	0.203	5.800	1.529	0.215	0.2582	4.024	1.620	0.041
R8	0.157	1.587	0.372	0.190	5.800	1.500	0.199	0.227	4.121	1.689	0.074
R9	0.313	3.353	0.901	0.622	5.800	1.656	0.523	0.535	4.131	2.537	0.293
R10	0.312	3.525	0.912	0.713	5.800	1.500	0.487	0.746	4.188	2.174	0.278
R11	0.280	2.784	1.053	0.910	5.800	2.214	0.910	0.410	4.388	2.538	0.736
R12	0.217	3.594	0.765	0.440	5.800	1.500	0.525	0.392	4.099	1.946	0.252
R13	0.249	2.618	0.872	0.431	5.800	1.525	0.712	0.3734	4.141	2.523	0.538
R14	0.216	2.140	0.772	0.285	5.800	1.500	0.550	0.188	4.181	1.706	0.243
R15	0.233	1.606	0.757	0.298	5.800	1.500	0.612	0.283	4.116	1.836	0.408
R16	0.268	3.071	0.793	0.507	5.800	1.500	0.463	0.5937	4.089	1.660	0.083
R17	0.147	1.629	0.318	0.246	5.800	1.620	0.187	0.207	4.058	1.924	0.010
R18	0.241	2.493	0.735	0.439	5.800	1.500	0.562	0.416	4.119	1.946	0.312
R19	0.241	2.839	0.728	0.447	5.800	1.627	0.535	0.468	4.116	2.142	0.320
R20	0.144	2.398	0.499	0.254	5.800	1.500	0.430	0.217	4.143	1.943	0.256
R21	0.1604	1.500	0.377	0.201	5.800	1.855	0.277	0.179	4.194	1.874	0.106
R22	0.1702	4.008	0.694	0.358	5.800	1.945	0.624	0.363	4.144	2.518	0.449
R23	0.210	2.568	0.731	0.401	5.800	1.500	0.646	0.405	4.092	2.096	0.468
R24	0.202	2.236	0.772	0.302	5.800	2.158	0.764	0.346	4.084	2.168	0.596
R25	0.252	2.663		0.365	5.800	2.885	0.000	0.329	4.599	3.105	0.000
R26	0.111	1.510	0.625	0.100	5.800	1.500	0.334	0.116	4.089	1.623	0.293
R27	0.122	1.636	0.557	0.100	5.800	1.500	0.285	0.107	4.021	1.597	0.210
R28	0.193	2.549	0.963	0.259	5.800	1.808	0.690	0.235	4.174	2.157	0.572
R29	0.164	2.673	0.731	0.2801	5.800	1.500	0.597	0.269	4.064	1.742	0.375
R30	0.184	3.672	0.901	0.378	5.800	1.797	0.808	0.366	4.101	2.371	0.589
R31	0.219	3.069	0.903	0.401	5.800	1.559	0.734	0.390	4.119	2.159	0.541
R32	0.209	3.217	0.953	0.473	5.800	1.500	0.910	0.487	4.159	1.906	0.685
R33	0.286	3.177	0.843	0.582	5.800	1.500	0.582	0.604	4.092	2.209	0.285
R34	0.252	3.661	0.834	0.629	5.800	1.500	0.631	0.712	4.098	1.962	0.329
R35	0.229	2.453	0.729	0.301	5.800	1.718	0.301	0.312	4.072	2.089	0.302
R36	0.100	1.500	0.222	0.140	5.800	4.168	0.315	0.123	4.636	2.725	0.161
R37	0.214	2.355	0.703	0.385	5.800	1.500	0.606	0.408	4.084	1.930	0.408
R38	0.194	2.888	0.790	0.352	5.800	1.644	0.690	0.217	4.108	2.046	0.418
OT(s)	26.826			19.727				12.231			

The optimized values of TMS, Ip, and OTs utilizing NSTCC with the hybrid GSA–SQP algorithm are illustrated in Table 12. It can be noticed from these results that NSTCC is the best approach and achieved the minimum overall OT of all OCRs of 12.231 s compared to 26.826 s and 19.727 s, for the STCC and NSTCC with constant coefficient A, respectively, as it is shown in Table 12. The CTI values calculated from the optimized TMS and Ip are shown in Table 11 and the coordination between primary and backup relays are ensured by the optimal results. Therefore, this signifies that the NSTCC can be efficiently used for solving the OCR coordination problem for the meshed and large-scale power systems.

4.2.3. Discussion of the Meshed Network Results

In this section, the performance of proposed NSTCC and approaches from the literature (STCC and NSTCC with constant coefficient A) on the IEEE 9- and 30-bus meshed networks are presented. The overall OTs for operation modes are shown in Figure 15 and were obtained by using the hybrid GSA–SQP algorithm. The previous subsections show that the NSTCC approach reduced the overall OT of OCRs for all modes compared to approaches from the literature. For example, the NSTCC reduced the overall OT in the meshed 30-bus test system by 54.4% and 37.9% compared to the literature approaches STCC and NSTCC with constant coefficient A, respectively.

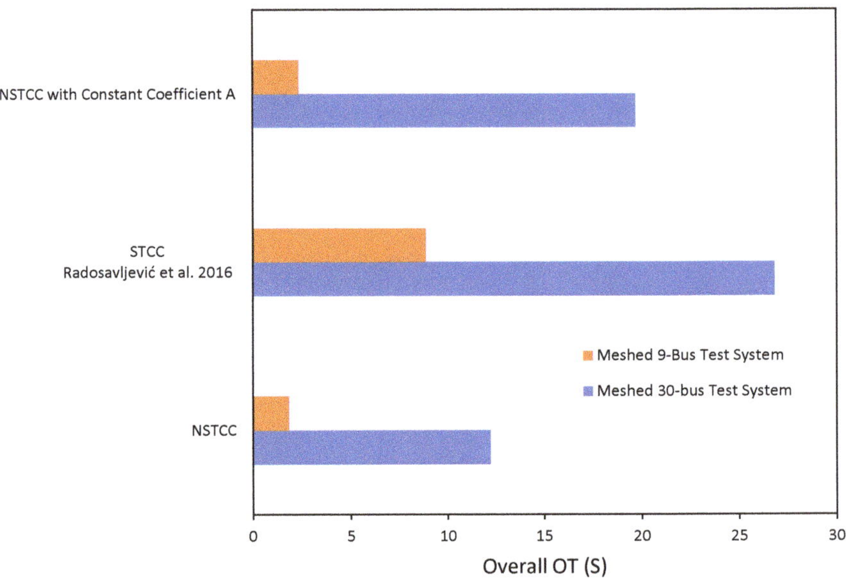

Figure 15. The overall OT in NSTCC with constant coefficient A, STCC by Radosavljević et al., 2016 [8], and NSTCC for the meshed 9- and 30-bus MGs.

5. Conclusions

An approach for the optimal coordination of OCRs in MGs that integrate DGs has been presented in this article. The approach has been successful in sustaining constant CTI between the primary and backup relay pairs and offering lower tripping times than the recent NSS introduced in the literature for different fault currents. The proposed NSTCC applies to long radial feeders and large meshed grids. The NSTCC is implementable when there is sufficient variance in the maximum and minimum fault currents. The GA algorithm was applied with several tests in radial networks, including the IEEE 9-bus test system and benchmark IEC MG that both integrate DGs under various operational modes. In all operational modes (grid-connected and islanding mode), better total operational times have been obtained by the proposed approach. Additionally, the hybrid GSA–

SQP algorithm was carried out based on the proposed scheme in the 9- and 30-bus IEEE standard meshed power system to solve the OCR coordination optimization problem. This illustrates the superiority of the NSTCC in reducing operational time in meshed networks; hence, it is effective in both radial structures and meshed systems. These results broaden our understanding of the concept of using nonstandard curves in industrial relays to obtain optimal, flexible, and reliable outcomes. Regardless, future research could continue to explore the extent of applicability this proposed scheme has in the coordination of overcurrent relays with distance relays in radial networks. In addition, applying the proposed approach to hardware in the loop to confirm its reliability is possible.

Author Contributions: Conceptualization, S.A. and Y.H.; methodology, S.A. and Y.H.; software S.A. and Y.H.; validation, F.A., N.E.-N. and S.A.; formal analysis, F.A., N.E.-N. and S.A.; investigation, S.A. and Y.H.; resources, all authors; data curation, all authors; writing—original draft preparation, F.A., N.E.-N. and S.A.; writing—review and editing, all authors; visualization, all authors; supervision, all authors; project administration, S.A. All authors have read and agreed to the published version of the manuscript.

Funding: This research received no external funding.

Data Availability Statement: Not available.

Conflicts of Interest: The authors declare no conflict of interest.

References

1. Alasali, F.; Haben, S.; Holderbaum, W. Energy management systems for a network of electrified cranes with energy storage. *Int. J. Electr. Power Energy Syst.* **2019**, *106*, 210–222. [CrossRef]
2. Gomes, M.; Coelho, P.; Moreira, C. Microgrid protection schemes. In *Microgrids Design and Implementation*; Springer: Berlin/Heidelberg, Germany, 2019; pp. 311–336.
3. Abeid, S.; Hu, Y. Overcurrent relays coordination optimisation methods in distribution systems for microgrids: A review. In Proceedings of the 15th International Conference on Developments in Power System Protection (DPSP 2020), Liverpool, UK, 9–12 March 2020.
4. Jamali, S.; Borhani-Bahabadi, H. Non-communication protection method for meshed and radial distribution networks with synchronous-based DG. *Int. J. Electr. Power Energy Syst.* **2017**, *93*, 468–478. [CrossRef]
5. So, C.; Li, K.; Lai, K.; Fung, K. Application of genetic algorithm for overcurrent relay coordination. In Proceedings of the Sixth International Conference on Developments in Power System Protection (Conf. Publ. No. 434), Nottingham, UK, 25–27 March 1997. [CrossRef]
6. Zeineldin, H.; El-Saadany, E.; Salama, M. Optimal coordination of overcurrent relays using a modified particle swarm optimization. *Electr. Power Syst. Res.* **2006**, *76*, 988–995. [CrossRef]
7. Bhesdadiya, R.; Trivedi, I.N.; Jangir, P.; Kumar, A. A novel hybrid approach particle swarm optimizer with moth-flame optimizer algorithm. In *Advances in Computer and Computational Sciences*; Springer: Berlin/Heidelberg, Germany, 2017; pp. 569–577.
8. Radosavljević, J.; Jevtić, M. Hybrid GSA-SQP algorithm for optimal coordination of directional overcurrent relays. *IET Gener. Transm. Distrib.* **2016**, *10*, 1928–1937. [CrossRef]
9. Purwar, E.; Vishwakarma, D.N.; Singh, S.P. A Novel Constraints Reduction-Based Optimal Relay Coordination Method Considering Variable Operational Status of Distribution System with DGs. *IEEE Trans. Smart Grid* **2019**, *10*, 889–898. [CrossRef]
10. Saberi, N.A.; Sadeh, J.; Rajabi, M.H. A New Index for Constraint Reduction in Relay Coordination Problem Considering Uncertainty. *J. Iran. Assoc. Electr. Electron. Eng.* **2011**, *8*, 59–67.
11. Zeineldin, H.H.; Sharaf, H.M.; Ibrahim, D.K.; El-Zahab, E.E.-D.A. Optimal Protection Coordination for Meshed Distribution Systems With DG Using Dual Setting Directional Over-Current Relays. *IEEE Trans. Smart Grid* **2016**, *7*, 1757. [CrossRef]
12. Beder, H.; Mohandes, B.; El Moursi, M.S.; Badran, E.A.; El Saadawi, M.M. A New Communication-Free Dual Setting Protection Coordination of Microgrid. *IEEE Trans. Power Deliv.* **2021**, *36*, 2446–2458. [CrossRef]
13. Alam, M.N.; Das, B.; Pant, V. An interior point method based protection coordination scheme for directional overcurrent relays in meshed networks. *Int. J. Electr. Power Energy Syst.* **2016**, *81*, 153–164. [CrossRef]
14. Yazdaninejadi, A. Protection coordination of directional overcurrent relays: New time current characteristic and objective function. *IET Gener. Transm. Distrib.* **2018**, *12*, 190–199. [CrossRef]
15. Henville, C. Combined use of definite and inverse time overcurrent elements assists in transmission line ground relay coordination. *IEEE Trans. Power Deliv.* **1993**, *8*, 925–932. [CrossRef]
16. Saleh, K.A.; Zeineldin, H.H.; Al-Hinai, A.; El-Saadany, E.F. Optimal coordination of directional overcurrent relays using a new time–current–voltage characteristic. *IEEE Trans. Power Deliv.* **2014**, *30*, 537–544. [CrossRef]
17. Jamali, S.; Borhani-Bahabadi, H. Recloser time–current–voltage characteristic for fuse saving in distribution networks with DG. *IET Gener. Transm. Distrib.* **2017**, *11*, 272–279. [CrossRef]

18. Kılıçkıran, H.C.; Akdemir, H.; Şengör, I.; Kekezoğlu, B. A non-standard characteristic based protection scheme for distribution networks. *Energies* **2018**, *11*, 1241. [CrossRef]
19. Keil, T.; Jager, J. Advanced coordination method for overcurrent protection relays using nonstandard tripping characteristics. *IEEE Trans. Power Deliv.* **2007**, *23*, 52–57. [CrossRef]
20. Yazdaninejadi, A.; Nazarpour, D.; Golshannavaz, S. Dual-setting directional over-current relays: An optimal coordination in multiple source meshed distribution networks. *Int. J. Electr. Power Energy Syst.* **2017**, *86*, 163–176. [CrossRef]
21. Ojaghi, M.; Ghahremani, R. Piece-wise Linear Characteristic for Coordinating Numerical Overcurrent Relays. *IEEE Trans. Power Deliv.* **2017**, *32*, 145–151. [CrossRef]
22. Azari, M.; Mazlumi, K.; Ojaghi, M. Efficient non-standard tripping characteristic-based coordination method for overcurrent relays in meshed power networks. *Electr. Eng.* **2022**, 1–18. [CrossRef]
23. Sharaf, H.M.; Zeineldin, H.H.; Ibrahim, D.K.; EL-Zahab, E.E.D.A. A proposed coordination strategy for meshed distribution systems with DG considering user-defined characteristics of directional inverse time overcurrent relays. *Int. J. Electr. Power Energy Syst.* **2015**, *65*, 49–58. [CrossRef]
24. Karegar, H.K. Relay curve selection approach for microgrid optimal protection. *Int. J. Renew. Energy Res. (IJRER)* **2017**, *7*, 636–642.
25. Alasali, F.; El-Naily, N.; Zarour, E.; Saad, S.M. Highly sensitive and fast microgrid protection using optimal coordination scheme and nonstandard tripping characteristics. *Int. J. Electr. Power Energy Syst.* **2021**, *128*, 106756. [CrossRef]
26. Alasali, F.; Zarour, E.; Holderbaum, W.; Nusair, K.N. Highly Fast Innovative Overcurrent Protection Scheme for Microgrid Using Metaheuristic Optimization Algorithms and Nonstandard Tripping Characteristics. *IEEE Access* **2022**, *10*, 42208–42231. [CrossRef]
27. Kida, A.A.; Rivas, A.E.L.; Gallego, L.A. An improved simulated annealing–linear programming hybrid algorithm applied to the optimal coordination of directional overcurrent relays. *Electr. Power Syst. Res.* **2020**, *181*, 106197. [CrossRef]
28. Darabi, A.; Bagheri, M.; Gharehpetian, G.B. Highly sensitive microgrid protection using overcurrent relays with a novel relay characteristic. *IET Renew. Power Gener.* **2020**, *14*, 1201–1209. [CrossRef]
29. Saldarriaga-Zuluaga, S.D.; López-Lezama, J.M.; Muñoz-Galeano, N. Optimal coordination of overcurrent relays in microgrids considering a non-Standard characteristic. *Energies* **2020**, *13*, 922. [CrossRef]
30. Hussain, N.; Nasir, M.; Vasquez, J.C.; Guerrero, J.M. Recent developments and challenges on AC microgrids fault detection and protection systems–a review. *Energies* **2020**, *13*, 2149. [CrossRef]
31. Elmitwally, A.; Kandil, M.S.; Gouda, E.; Amer, A. Mitigation of DGs impact on variable-topology meshed network protection system by optimal fault current limiters considering overcurrent relay coordination. *Electr. Power Syst. Res.* **2020**, *186*, 106417. [CrossRef]
32. Manditereza, P.T.; Bansal, R. Renewable distributed generation: The hidden challenges–A review from the protection perspective. *Renew. Sustain. Energy Rev.* **2016**, *58*, 1457–1465. [CrossRef]
33. Telukunta, V.; Pradhan, J.; Agrawal, A.; Singh, M.; Srivani, S.G. Protection challenges under bulk penetration of renewable energy resources in power systems: A review. *CSEE J. Power Energy Syst.* **2017**, *3*, 365–379. [CrossRef]
34. El-Naily, N.; Saad, S.M.; Mohamed, F.A. Novel approach for optimum coordination of overcurrent relays to enhance microgrid earth fault protection scheme. *Sustain. Cities Soc.* **2020**, *54*, 102006. [CrossRef]
35. Din, Z.; Zhang, J.; Xu, Z.; Zhang, Y.; Zhao, J. Low voltage and high voltage ride-through technologies for doubly fed induction generator system: Comprehensive review and future trends. *IET Renew. Power Gener.* **2021**, *15*, 614–630. [CrossRef]
36. El-Naily, N.; Saad, S.M.; Hussein, T.; Mohamed, F.A. A novel constraint and non-standard characteristics for optimal over-current relays coordination to enhance microgrid protection scheme. *IET Gener. Transm. Distrib.* **2019**, *13*, 780–793. [CrossRef]
37. Chabanloo, R.M.; Maleki, M.G.; Agah, S.M.M.; Habashi, E.M. Comprehensive coordination of radial distribution network protection in the presence of synchronous distributed generation using fault current limiter. *Int. J. Electr. Power Energy Syst.* **2018**, *99*, 214–224. [CrossRef]
38. Aghdam, T.S.; Karegar, H.K.; Zeineldin, H.H. Optimal coordination of double-inverse overcurrent relays for stable operation of DGs. *IEEE Trans. Ind. Inform.* **2018**, *15*, 183–192. [CrossRef]
39. Meskin, M.; Domijan, A.; Grinberg, I. Optimal co-ordination of overcurrent relays in the interconnected power systems using break points. *Electr. Power Syst. Res.* **2015**, *127*, 53–63. [CrossRef]
40. Adelnia, F.; Moravej, Z.; Farzinfar, M. A new formulation for coordination of directional overcurrent relays in interconnected networks. *Int. Trans. Electr. Energy Syst.* **2015**, *25*, 120–137. [CrossRef]
41. Coffele, F.; Booth, C.; Dyśko, A. An adaptive overcurrent protection scheme for distribution networks. *IEEE Trans. Power Deliv.* **2014**, *30*, 561–568. [CrossRef]
42. Benmouyal, G.; Meisinger, M.; Burnworth, J.; Elmore, W.A.; Freirich, K.; Kotos, P.A.; Leblanc, P.R.; Lerley, P.J.; McConnell, J.E.; Mizene, J.; et al. IEEE standard inverse-time characteristic equations for overcurrent relays. *IEEE Trans. Power Deliv.* **1999**, *14*, 868–872. [CrossRef]
43. Darabi, A.; Bagheri, M.; Gharehpetian, G. Dual feasible direction-finding nonlinear programming combined with metaheuristic approaches for exact overcurrent relay coordination. *Int. J. Electr. Power Energy Syst.* **2020**, *114*, 105420. [CrossRef]
44. Protection, O. 7SJ62 SIPROTEC 4 7SJ62 multifunction protection relay. *Siemens* **2006**, *38*, 44.
45. Tirumala Pallerlamudi Srinivas, S.; Swarup, K.S. Optimal Protection Coordination of Nonstandard Overcurrent Relays Using Hybrid QCQP Method. *Electr. Power Compon. Syst.* **2020**, *48*, 1327–1338. [CrossRef]

46. Wadood, A.; Gholami Farkoush, S.; Khurshaid, T.; Kim, C.-H.; Yu, J.; Geem, Z.W.; Rhee, S.-B. An optimized protection coordination scheme for the optimal coordination of overcurrent relays using a nature-inspired root tree algorithm. *Appl. Sci.* **2018**, *8*, 1664. [CrossRef]

47. Alasali, F.; Haben, S.; Holderbaum, W. Stochastic optimal energy management system for RTG cranes network using genetic algorithm and ensemble forecasts. *J. Energy Storage* **2019**, *24*, 100759. [CrossRef]

48. Bedekar, P.P.; Bhide, S.R. Optimum coordination of directional overcurrent relays using the hybrid GA-NLP approach. *IEEE Trans. Power Deliv.* **2010**, *26*, 109–119. [CrossRef]

49. Kennedy, J.; Eberhart, R. Particle swarm optimization. In Proceedings of the ICNN'95-International Conference on Neural Networks, Perth, Australia, 27 November–1 December 1995.

50. Kar, S. A comprehensive protection scheme for micro-grid using fuzzy rule base approach. *Energy Syst.* **2017**, *8*, 449–464. [CrossRef]

51. Saldarriaga-Zuluaga, S.D.; López-Lezama, J.M.; Muñoz-Galeano, N. Optimal coordination of over-current relays in microgrids considering multiple characteristic curves. *Alex. Eng. J.* **2021**, *60*, 2093–2113. [CrossRef]

52. Saldarriaga-Zuluaga, S.D.; López-Lezama, J.M.; Muñoz-Galeano, N. An approach for optimal coordination of over-current Relays in Microgrids with distributed generation. *Electronics* **2020**, *9*, 1740. [CrossRef]

53. Christie, R. Power Systems Test Case Archive: 30 Bus Power Flow Test Case. 1993. Available online: https://labs.ece.uw.edu/pstca/pf30/pg_tca30bus.htm (accessed on 4 July 2022).

Review

Battery Energy Storage for Photovoltaic Application in South Africa: A Review

Bonginkosi A. Thango * and Pitshou N. Bokoro

Department of Electrical and Electronic Engineering Technology, University of Johannesburg, Johannesburg 2028, South Africa
* Correspondence: bonginkosit@uj.ac.za; Tel.: +27-65-564-7287

Abstract: Despite the significant slowdown of economic activity in South Africa by virtue of the COVID-19 outbreak, load shedding or scheduled power outages remained at a high level. The trend of rising load-shedding hours has persisted throughout most of the year 2022. Operational issues within the South African power utility inflamed the unpredictable nature of generation capacity, resulting in unscheduled outages at several generating units, mostly due to multiple breakdowns. To forestall substantial spikes in energy costs, an increasing number of enterprises and homeowners have started to gradually adopt renewable energy technologies to sustain their operational demand. Therefore, there is an increase in the exploration and investment of battery energy storage systems (BESS) to exploit South Africa's high solar photovoltaic (PV) energy and help alleviate production losses related to load-shedding-induced downtime. As a result, the current work presents a comprehensive and consequential review conducted on the BESS specifically for solar PV application and in the South African context. The research investigations carried out on BESS for PV application are crucially examined, drawing attention to their capacities, shortcomings, constraints, and prospects for advancement. This investigation probed several areas of interest where the BESS-PV scheme is adopted, viz., choice of battery technology, mitigating miscellaneous power quality problems, optimal power system control, peak load shaving, South African BESS market and status of some Real BESS-PV projects. The techno-economic case scenario has been proposed in the current research and results yield that lithium-ion batteries are more viable than Lead–acid batteries.

Keywords: South Africa; load shedding; battery energy storage systems (BESS); photovoltaic (PV)

Citation: Thango, B.A.; Bokoro, P.N. Battery Energy Storage for Photovoltaic Application in South Africa: A Review. *Energies* **2022**, *15*, 5962. https://doi.org/10.3390/en15165962

Academic Editors: Najib El Ouanjli, Saad Motahhir and Mustapha Errouha

Received: 12 July 2022
Accepted: 14 August 2022
Published: 17 August 2022

Publisher's Note: MDPI stays neutral with regard to jurisdictional claims in published maps and institutional affiliations.

1. Introduction

The aging power plant infrastructure of the South African national electric utility, coupled with unscheduled blackouts, have hampered the power producer's generating capacity for years, and these bottlenecks continue to be a major hurdle to future expansion. The utility issues, which include a congested electric grid, an outdated, unreliable, and inadequately maintained generating fleet, and a yearning for new generation capacity, have been accentuated by the increasing load-shedding days since 2018, as shown in Figure 1 [1]. Until significant supplemental wattage capacity is invested, their vulnerability to load shedding will be incessant. Increasing levies by national power utilities and municipalities, in addition to load shedding, have augmented the investment case for industries in renewable energy generation and power efficiency initiatives [2,3]. To circumvent hefty increases in electricity costs, an influx of major corporations is considering implementing alternative energy sources to support their daily operations. PV grid-tied systems are playing a central role in this shift in the South African energy sector on account of their environmental merits and attenuated carbon emissions [4,5]. Since the enactment of the Integrated Resource Plan (IRP), in March 2011, by the Department of Energy (DoE), there has been a gradual increase in the deployment of solar PV. The IRP 2019 steers the energy sector, with a transition from coal power generation, increasing the adoption of renewables and thereby reducing South

African dependence on coal. The IRP advocates for 7958 MW of solar PV to be generated by end of 2030 [6,7].

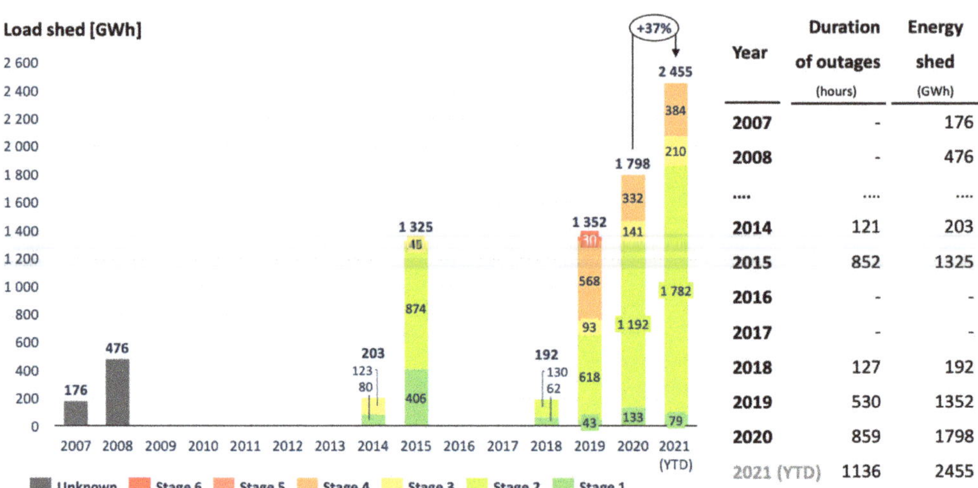

Figure 1. The number of days of load shedding in South Africa.

This load shedding concurs with South Africa's power utility decaying Energy Availability Factor (EAF), which estimates the performance of electricity-generating stations in accordance with the electrical energy they supply to the national grid. In 2021, 2020 and 2019, the EAF was estimated at about 61.8%, 65% and 66.9%, respectively. The EAF had come down rapidly since 2018, which was estimated at about 71.9%, just below the power utility's 74% target. In 2021, a low of about 53.3% was reported on a weekly average EAF. Figure 2 demonstrates interest or progress in terms of renewable energy in South Africa in the context of installed generation capacity. The planned capacity by 2030 is expected to contribute about 10.5% of South Africa's generation capacity [6,7].

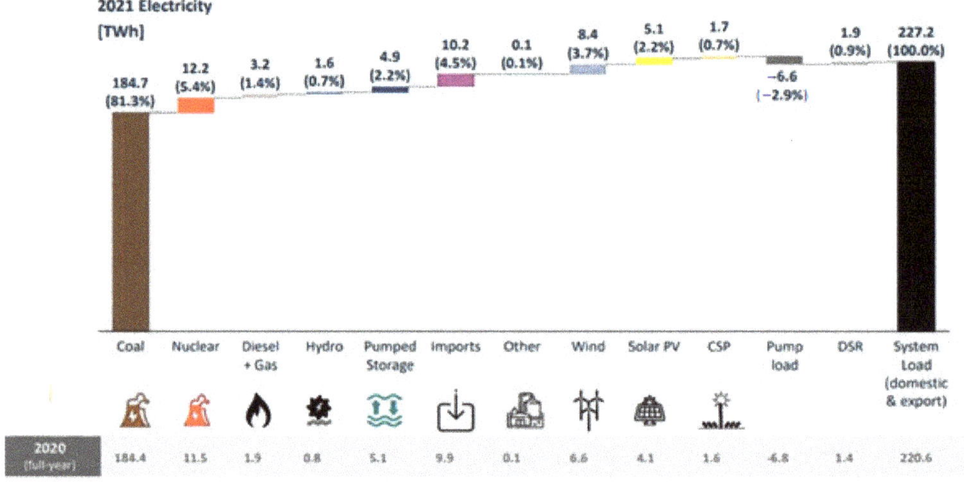

Figure 2. Installed generation capacity in South Africa [1].

It can be observed that coal generation is yet the main electrical power source, responsible for approximately 81.4% of the power fed to the grid, succeeded by renewable energy sources, viz., solar and wind—responsible for approximately 6.7%. The increase in the deployment of renewable energies will aid in curbing not only the emission of greenhouse gases but also the dependency on coal generation and reduction in load shedding. Commercial activities will be able to continue their operations with investments in BESS-PV scheme.

Although solar PV is proliferating, it also poses operational challenges attributable to its unpredictability and investment cost. Subsequently, solar PV exhibits variations inherent to renewable energy. These variations can emerge at any interval from seconds to minutes, necessitating the deployment of supplemental energy management devices. PV systems are additionally confronted by the cost differential during peak hours and the power quality given to the power grid. As a result, energy storage technologies are integral parts that can support PV systems to be able to provide energy for longer hours in the absence of sunlight.

In the literature [8], energy storage systems have been suggested as a mechanism to feasibly alleviate faults [8]. These issues are connected to energy penetration levels and can equip solar PV with highly desired adaptability and robustness [9,10]. Solar PV storage systems, in this regard, endorse power management, such as load levelling or peak demand reduction, for power balancing and quality upgrades [11]. Consequently, these systems provide on-site load flow control, allowing for power storage during low-demand times to be utilized during peak hours [12]. As a result, South Africans can lower rates associated with electricity consumption while also enhancing the quality of the power grid [13].

The fundamental issue with solar energy is the availability of sunlight, which does not correlate to the demand. This is particularly troublesome for residential or business users, who are frequently prohibited from returning surplus PV power to the grid that is generated when there is a plethora of individual load demand. Solar PV-battery systems permit these customers to store surplus energy for later usage, as demonstrated in Figure 3, thereby enhancing their solar energy consumption [14].

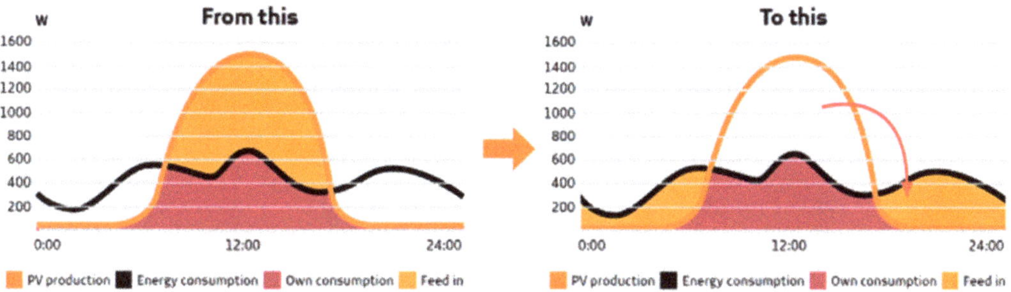

Figure 3. Solar PV self-consumption in South Africa.

A summary of recent related research works is tabulated in Table 1 to outline the contribution that can be made by the current study and the main focus thereof.

This work has been organized as follows:

A brief account of solar PV and battery energy storage system technologies with their crucial information is covered in Section 2. Research on battery storage systems applications is comprehensively detailed with supporting arguments and opposing arguments in Section 3. Current status and some real PV-battery projects are discussed briefly in Section 4. A simulation case scenario with a techno-economic analysis of two different BESS-PV systems is performed to assess the economic performance in Section 5. Feasible future courses and suggestions for the research on BESS-PV systems are also presented. This review research is generally concluded in Section 6 by describing the importance of the findings.

Table 1. Summary of recent related research works.

Ref. No	Year	Addressed Challenges	Limitations	Outcomes
[15]	2021	Peak shaving	Isolated microgrid system	An algorithm was proposed and tested in a microgrid under different load conditions and PV generations.
[16]	2022	BESS-PV sizing	Hourly based energy generation and consumption profiles of 128 residents	An energy management tool that suggests the BESS-PV sizes in accordance with answering a few straightforward energy consumption questions.
[17]	2021	Techno-economic analysis	HOMER Grid software	Provides behind-the-meter application of BESS-PV to efficaciously use renewable energy under the conditions of various portions of renewable energy.
[18]	2021	Inverter Control	MATLAB software	Reactive power control of smart inverters for BESS-PV to enhance the PV hosting capacity of distribution networks.
[19]	2021	Scheduling		BESS-PV under uncertainty using model predictive control.
Current Study	2022	Peak shaving PV-BESS sizing Techno-economic analysis BESS-PV market BESS-PV Policies	Limited to South Africa BESS market	The study aims to crucially examine BESS-PV capacities, shortcomings, constraints, and prospects for advancement. Probed areas of interest: choice of battery technology, mitigating miscellaneous power quality problems, optimal power system control, peak load shaving, South African BESS market and status of some Real BESS-PV projects.

2. Solar PV and Battery Energy Storage System

The rooftop solar PV systems convert solar radiation into electrical energy that may be consumed by South African residents, as shown in Figure 4 [20]. Any power that is not utilized is fed into the main grid. To conserve energy generated throughout the day, large-scale batteries can be coupled to solar PV systems. When the system is not producing enough power, particularly at night or in adverse weather conditions, this energy may be consumed. Using the power generated by a solar PV system throughout the day alleviates the amount of power purchased from the grid, lowering the energy costs.

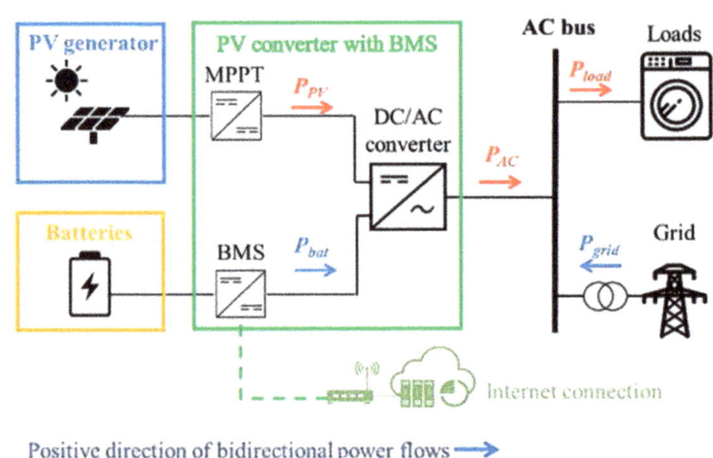

Figure 4. Solar PV-Battery Energy Storage System.

2.1. Selection and Deploying a Solar PV-Battery System

A variety of factors will significantly affect which solar PV-battery system is ideal for a household, including:

- Amount of power and time of consumption;
- Dimensions of available rooftop space;
- Positioning and direction of solar PV panels.

Understanding the amount of energy consumption in a household may facilitate the evaluation of the impact of a solar PV system on the energy costs and establish whether battery storage is a cost-effective option. To realize the best options, licensed solar installers certified by South African PV GreenCard may further be consulted.

2.2. Emplacement of the Solar PV-Battery System

When solar PV panels are oriented directly toward effective solar irradiation between 09:00 a.m. and 15:00 p.m., they achieve the greatest power generation:

- The PV panels must not be exposed to any shade. Even if a single PV cell is obscured by objects such as branches, roof vents, or satellite dishes, many other PV cells will lose power. Due to variations in the flow of energy through the panel, the latter will have a significant influence on the output of the panel.
- The efficacy of batteries can be affected by the temperature in the surrounding environment.
- Batteries necessitate a setting or housing that is well-insulated and well-ventilated. A battery enclosure should ideally be placed on the south or west side of a South African building.

2.3. Connecting Solar PV-Battery System to the Power Grid

Under its jurisdiction to administer electricity tariffs in South Africa, the National Energy Regulator of South Africa (NERSA) approved the establishment of a Renewable Energy Feed-in Tariff (REFIT) for the country. The feed-in tariff obligates the Renewable Energy Purchasing Agency (REPA), to purchase renewable energy from eligible producers at preset rates [21,22].

3. Battery Technologies

3.1. Lead–Acid Battery

The lead–acid battery, created in 1859, is the first kind of rechargeable battery ever developed [23]. Lead–acid batteries have a suboptimal energy density when compared to contemporary rechargeable batteries.

Lead–acid cells are composed of lead alloy grids (solid electrodes) that operate as current collectors and mechanically support the positive and negative active elements. The grids are interlaced with a permeable, electrically isolator and arrayed as positive and negative plates. The plate stack is embedded into an adequately contoured polymer housing to embody the cell elements and the electrolyte with the coupled positive and negative plates, terminals, a lid and venting arrangements. The construction of a lead–acid battery is shown in Figure 5.

The operating voltage of the lead–acid cell is reasonably high at approximately 2.05 V. The positive active material (PAM) is considerably permeable lead dioxide (PbO_2) and the negative active material (NAM) is delicately isolated lead. The electrolyte utilized in the discharge process is thinned liquefied sulfuric acid (HSO_4). HSO_4 ions move to the negative electrode during discharge, producing H+ ions and lead sulfate (SO_4^{2-}). Lead dioxide reacts with the electrolyte at the positive electrode to yield lead sulfate particles and water (H_2O), as shown in Figure 6. Both electrodes are discharged to a feeble conductor, lead sulfate ($PbSO_4$), and the electrolyte is incrementally diluted as the discharge progresses. On charging, the reactions are reversible [24,25].

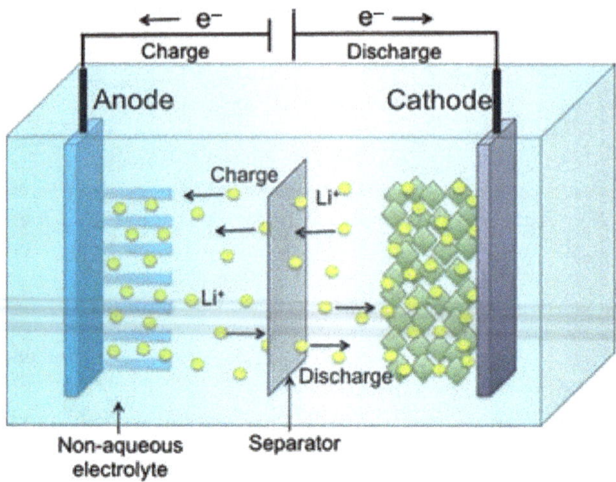

Figure 5. Lead–acid battery construction.

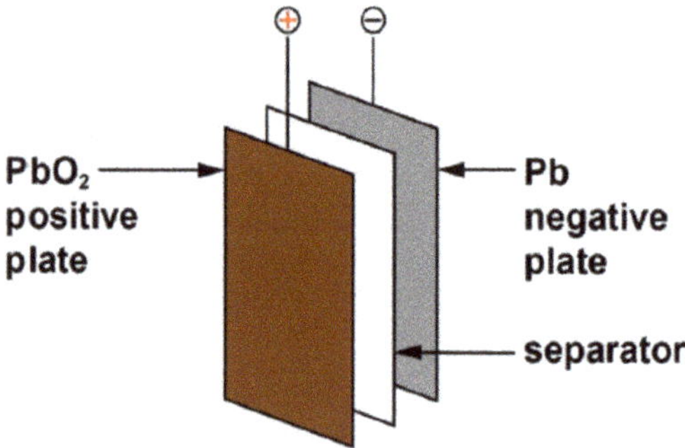

$$Pb + PbO_2 + 2H_2SO_4 \rightleftharpoons 2PbSO_4 + 2H_2O$$

Figure 6. Chemistry of lead–acid battery.

The most recent innovations [26,27] have used enhanced lead batteries in a variety of grid-related plans as well as smaller-scale industrial and domestic energy storage applications. In recent years, systems with integrated super-capacitors have been described in addition to conventional lead–acid batteries; they are commonly referred to as carbon-enhanced (LC) lead batteries. These could have a negative electrode made of a mix of lead–acid and supercapacitor negatives made of carbon. The positive electrode is exactly like the one in a typical lead–acid battery in every way. The current tendency in operating renewable energy sources, especially solar PV sources, is for periodic discharges rather than a continuous restoration of the battery to a full state of charge (SOC). This partial state-of-charge (PSoC) behavior can be detrimental to lead–acid batteries since it induces permanent corrosion of the negative electrode, and sustainable development strategies are still being investigated [28,29].

3.2. Lithium-Ion Battery

Lithium-ion (Li-ion) batteries store energy in positive electrode materials composed of lithium extracts adequate for reversible physical adsorption of Li-ions and negative electrode materials made of carbon and can properly support Li in the solid state [30]. Since

Li interacts severely with water, non-aqueous electrolytes are employed [31]. These are ionizable organic diluents, such as propylene carbonate, in solution with adequate lithium salts. To improve the safety of the cells, the separators are microporous plastic strips that may be covered with ceramic particles, as shown in Figure 7.

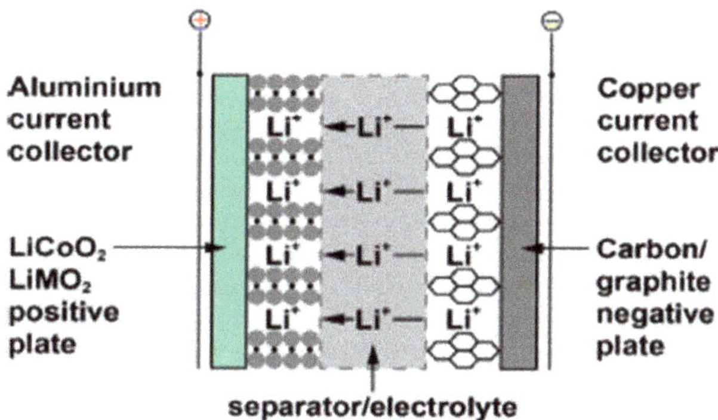

Figure 7. Chemistry and principal components of a lithium-ion battery.

Li-ion cells must be meticulously assessed in terms of safety. They have a high energy density and a volatile organic electrolyte.

3.3. Sodium–Sulfur Battery

The anodes of sodium–sulfur (Na-S) batteries are viscous liquid sodium and sulfur, and they run at hot temperatures, around 300 and 350 °C, to maintain the electrode's liquid and to provide strong ionic conductivity in the electrolyte, which is a ceramic material [32,33]. At processing temperature, the electrolyte is beta-alumina (b-Al$_2$O$_3$), which transmits sodium ions. When sodium and sulfur are released, sodium polysulfide is produced [34]. They have a far superior energy density and durability over lead–acid batteries. To preclude cell defects from proliferating, security is essential, and careful design is essential. The chemistry and principal components of a sodium–sulfur battery are shown in Figure 8.

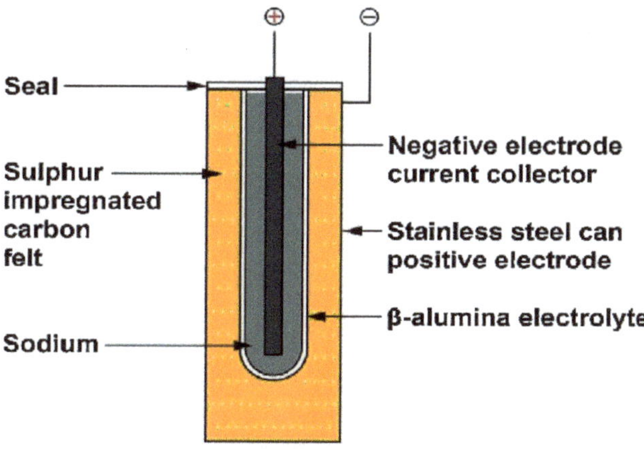

$$2Na + xS \rightleftharpoons Na_2 S_x$$

Figure 8. Chemistry and principal components of a sodium–sulfur battery.

Although Na-S batteries are made from abundant and inexpensive raw items, the production processes, as well as the necessity for insulation, cooling, and temperature control, make them fairly pricey. They are more cost-effective in large units since the thermal management of smaller batteries contributes to the cost relative to the battery's volume. As a consequence, they are primarily employed for utility load levelling in big substations.

3.4. Sodium–Nickel Chloride Battery

In Na-NiCl$_2$ batteries, a beta-alumina electrolyte is employed; however, the cathode is nickel chloride submerged in sodium aluminum chloride (NaAlCl$_4$), a molten salt that conducts sodium ions, as opposed to a sulfur electrode [35]. Nickel metal and sodium chloride are created when sodium reacts with nickel chloride during discharge [35,36]. The system still needs heating, insulation, and temperature control even though it operates at a lower temperature than Na-S batteries (300 °C) [37]. Compared to Na-S batteries, the energy density is higher, but the battery life is longer. Figure 9 depicts the chemistry and main parts of a sodium–nickel chloride battery.

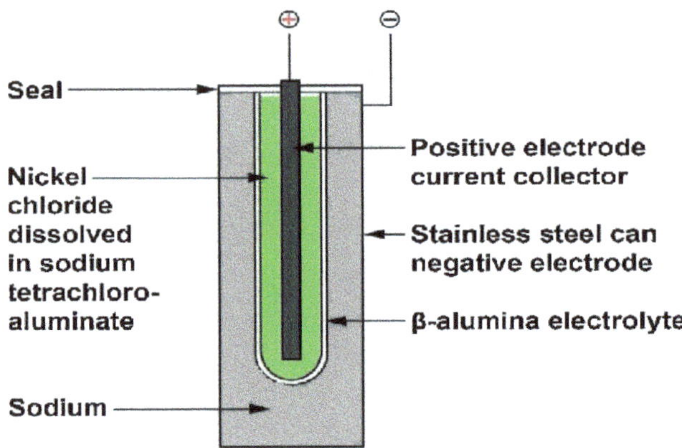

Figure 9. Chemistry and principal components of a sodium–nickel chloride battery.

3.5. Flow Batteries

Utility-scale energy storage has some promise thanks to flow batteries. There are many different compositions, but they all have energy-producing cells with electrode material stored remotely, making it possible for very large storage batteries to be made [38,39]. Vanadium redox batteries (VRB) are made up of cells with carbon composite electrodes submerged in a fluid containing aqueous acid and vanadium sulfate, with different valence states separated by an ion-selective membrane. At the positive electrode during discharge, V^{5+} is converted to V^{4+}, while V^{2+} is converted to V^{3+} at the negative electrode. The volume of the vanadium sulfate solution, and hence the battery's capacity, is potentially limitless because it is kept in a storage tank. Recharging causes reverse reactions, which replenish the materials. The batteries are complicated to use and made of heavy materials, but their expected lifespan is very lengthy. Only a few prototype systems have been implemented so far, and given the size of the battery, VRB batteries are only practical for utility energy storage. Figure 10 depicts the chemistry and main parts of a vanadium redox flow battery.

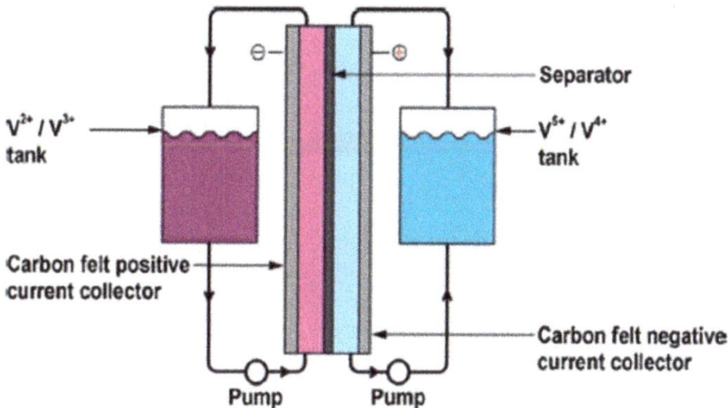

$$VO_2^+ + V^{2+} + 2H^+ \rightleftharpoons VO^{2+} + V^{3+} + H_2O$$

Figure 10. Chemistry and principal components of a vanadium redox flow battery.

Another kind of flow battery is the zinc–bromine (Zn-Br_2) battery, as shown in Figure 11. Zinc bromide is synthesized when Zn reacts with Br_2 within the cell. Br_2 is injected into cells with carbon electrodes and a microporous plastic separator in an aqueous solution as an organic complexing agent [40,41].

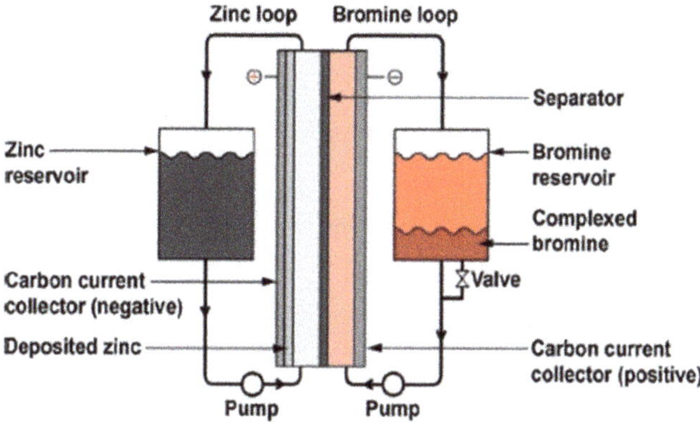

$$Zn + Br_2 \rightleftharpoons ZnBr_2$$

Figure 11. Chemistry and principal components of a zinc–bromine battery.

Metallic Zn is formed on charging, and while Br_2 is housed in tanks, the Zn electrode enforces a limit on the capacity for any specific design. The price is cheaper than VRB batteries, but the average lifetime is less. The discharge of bromine is a perceived threat that must be avoided. Zn-Br_2 batteries, like other flow batteries, have only been employed in moderate numbers for utility usage. There are a few similar types of flow batteries, such as iron–chromium batteries; however, they are not broadly utilized.

The technical comparison of the aforementioned battery technologies has been tabulated as demonstrated in Table 2.

Table 2. Technical comparison of battery technology in South Africa [23–37].

Battery	Lead–Acid [23]	Lithium-Ion [24]	Sodium–Sulfur [25]	Sodium–Nickel Chloride [26]	Zinc–BROMINE [27]	Vanadium Redox [28]
Energy density (Wh/L)	80–90	250–693	110	100–120	15–65	15–25
Nominal cell voltage (V)	2.1	3.6/3.7/3.8/3.85, LiFePO$_4$ 3.2	1.78–2.208	2.58	1.8	1.15–1.55
Specific energy (Wh/kg)	35–40	100–265	150	350	60–85	10–20
Self-discharge rate	1%/day	5%/day	20%/day	5–20%/day	10%/day	
Cycle durability	<350	400–1200	4500	4500	>2000	>12,000–14,000
Charge/discharge efficiency	50–95%	80–90%	80%	85–95%	75.9%	75–80%<
Key Challenges and Limitations	- Low cycle and calendar life - Self-discharge - Lead toxicity	- Fire protection - Large-scale controls - Self-discharge - Temp sensitivity	- Long life - High number of charge and discharge cycles - Ability to discharge fully with no effects on the performance. - Low energy to size ratio	- Operate at −20 °C to 60 °C - Recyclable - No fire hazards - −80% Depth of discharge - Expensive to use	- Utility-scale to be proven - Low energy density (More Storage Space) - Electrolyte Leak and Mechanical Pumps	

4. Battery Storage Systems Applications

The fixed resolutions comprising deconcentrated generation, management, and control tactics for power supervision and storage assisted to establish the concept of smart networks. The intellect of these networks is not just about the reduction the technical restrictions but also manufacturing the electric system to be greener, more competent, compliant to the customer desires, and consequently cost-effective. These networks aggregate the employment of information technology personnel, allowing two-party communication between the energy network and the construction customers, which results in detection on both sides, making the network intelligent as they are more competent and elastic than the conventional energy network [42,43]. Consequently, intelligent networks unlock the industry to new utilization with comprehensive interdisciplinary influences due to their capability to safely provide and integrate more sustainable power sources, grid-based generators, and elegant buildings [43]. Hence, extraordinarily dependable communication will be imperative to transfer a substantial volume of information. Thus, communication and system engineering will play a substantial function in the integration of elegant buildings and energy systems [43]. Numerous alterations in the electric network have been taking place since the development of the smart network concepts, resulting from a reorganization in the sector and technological developments.

These amendments also produce regulative alterations and can generate numerous intrinsic advantages to energy generation networks through the amenities given by these networks. Nonetheless, energy storage is unconstrained in gathering long-term alterations in energy output generated by short-term irregular intrinsic renewable sources. These networks submit significant characteristics to the electrical network and consumers. Energy storage networks may have very different applications and capabilities and, thus, have a slow or fast response [43]; a few of these services as explained below.

The applications of BESS are highlighted below in summary format. The economic value proposition for some applications is presented, with formulas.

4.1. Regulation

Regulation involves handling energy flow with other command areas to merge planned flow and instant fluctuations in demand. The primary motive for regulation is to preserve frequency and voltage inside industry-conventional standards [44]. In realistic conditions, this application is marked by the constant balance between the provision and need of electricity, concerning the frequency or load, and the regulation of operating (low) and responsive powers (high) [45]. Voltage regulation is a necessity in the electric energy system [45]. This application comprises the supervision of reactivity, generated by grid-connected apparatus that produces, sends, or employs electricity and frequently has or presents features for instance inductors and capacitors in an electric circuit [44]. Hence, these energy plants (reactive energy—VAR) might either be substituted by energy storage tactically located within the network at central positions or across the supplied method, embedding various VAR support storage networks closely to large loads [44].

Furthermore, the frequency response operation, which is like regulation, excludes the fact that it reacts to system requirements in an even shorter time, in the order of seconds to below one minute, when there is an unexpected loss of frequency response [45].

4.2. Integration of Renewable Power Generation

Energy storage completes load smoothing produced by the intermittence of wind and PV networks [46]. In this operation, within minutes, there is a load ramp support to react to a fast or at random floating load outline [46]. Consequently, renewable energy production can be controlled, smoothed, and expeditious, particularly in distant areas [46]. Additionally, diverse modes of process must merge to reach feasibility, for instance, energy quality control, load supervision, and others are considered next [46].

4.3. Energy Arbitrage

Energy arbitration is where energy is kept throughout low production expense/tariff time for transmission at high production cost/tariff time. Accordingly, the network's incompetence does not surpass the cost distinction, so the business case could be positive. The general formulae for value realization over the life of the asset are shown in Equation (1).

$$Value = Energy\ Supplied \times Peak\ Rate - Energy\ Consumed \times Off - Peak\ rate \qquad (1)$$

where,

$$Energy\ Supplied = Energy\ Consumed \times Efficiency\ of\ BESS$$

4.4. Peak Clipping

The release of energy at a certain volume factor during peak times can be employed to lessen peaking volume thus saving costs. The general formulae for value realization over the life of the asset to the generator are expressed as follows in Equation (2).

$$Value = Demand\ Reduction \times Differential\ Cost\ of\ Peaking\ Generation \qquad (2)$$

The general formulae for value realization over the life of the asset to the end consumer are expressed as follows in Equation (3).

$$Value = Demand\ Reduction \times Demand\ Charge\ (R/kWh) \qquad (3)$$

The energy arbitrage and peak shaving are demonstrated in Figure 12.

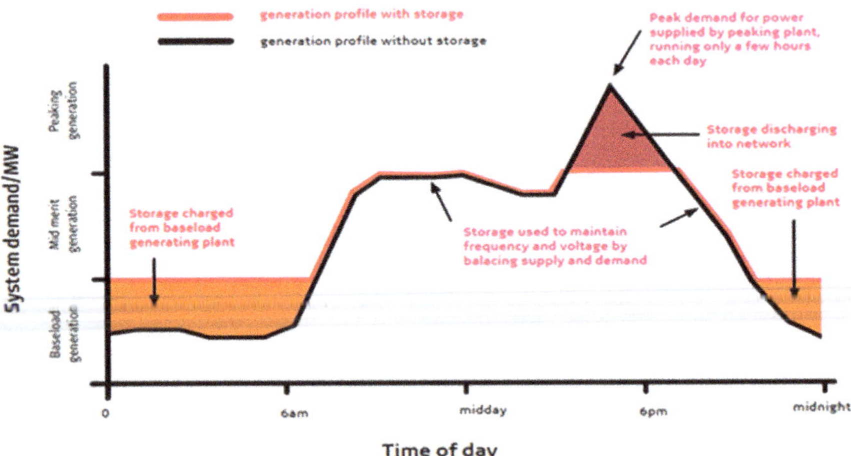

Figure 12. Peak shaving application for BESS.

4.5. Spinning/Instantaneous Reserves

BESS techniques have an extremely quick response, far faster than synchronous networks [47]. Fast distribution can have extra profits, mainly when linked with traditional non-instantaneous spinning resources, for example, gas turbines that may well take minutes to achieve rated output.

4.6. Frequency Support

BESS volume can be employed in energy mode (short duration), as a source or load, to supply fast volume support that can capture frequency excursions [48]. Frequency must function under the rate of change in the frequency curve. The general formula for value realization is shown in Equation (4).

$$Value = Investment\ in\ Traditional\ Spinning\ Reserves - Equivalent\ BESS\ lifecycle\ cost \tag{4}$$

4.7. Voltage Support

Equally, BESS volume can be employed in energy mode (short duration), as a generator or load, to supply fast voltage aid that can restrain frequency excursions. This can be battery-operated volume energy or non-battery-reactive energy in static VAR compensator mode.

4.8. Quality of Supply/Critical Power

BESS can be employed to ride short-duration quality of provision phenomena for instance voltage dips and flicker.

4.9. Capex Deferral

BESS can be employed to increase volume and support the network at crucial periods of the day, thus evading what can be an expensive investment in infrastructure development. The general formula for value realization is shown in Equation (5).

$$Value = investment\ in\ traditional\ infrastructure - Equivalent\ BESS\ lifecycle\ cost \tag{5}$$

4.10. PV Smoothing

PV smoothing refers to the practice of storing solar energy throughout the day for release at peak demand periods. With the help of this operation, services can maximize the amount of solar energy that the network will allow, allowing for both provision and storage. The non-violation of network inertia constraints is the main emphasis of everything here [49].

4.11. Wind Energy Firming

It is possible to size battery storage to accommodate varying wind production. The BESS is charged when the wind exceeds the set limit. When production falls below a certain threshold, it is released. In a manner similar to PV, adequate storage can render wind dispatchable.

4.12. Backup Supply

In UPS mode, a BESS solution can be employed as a backup energy source. There are numerous engineering production inroads in the SA market in the backup energy and energy security market [50]:

- Lithium-ion (Li-ion) and lead–acid battery techniques, which are the most attempted and verified, remain the leaders in this market;
- There are other storage skills available, but they either do not have present pilot projects in SA, or they have not exhibited promise in medium-to-large-measure storage applications when compared with their direct competitors.

A look into the BESS market in South Africa is illustrated in Table 3.

Table 3. BESS in South African Market [50].

Battery Technology	Application	Major Advantage	Current Limitations	Cost Range (R/kWh)	Replacement Cost (R/kWh)	Operation and Maintenance Cost (R/kWh/Year)	Installation and Other Charges (EUR/kW)
Lead–Acid	Backup power, UPS	- Time tested - Economical - Advanced technology	- Low energy density - Heavy - Damage the environment - Confined full discharge cycles	200–R1000	1773	143.59	426
Li-ion	Industrial-scale storage, Backup, UPS	- High energy density - Minimal maintenance - Renowned in the market - Continually evolving	- Expensive - Transport constraints - Advancing chemical combinations and developments	4000–R10,000	7000	0	287.17
Vanadium Redox	Industrial-scale storage, Backup power	- High depth of discharge - Adjustable electrolyte tanks - Unlimited storage potential	Accessing markets	21,793–25,146	23,000	-	-

Due to the fact that networks are typically developed specifically for each application, hydrogen storage and vanadium redox flow batteries have not gained the necessary access to markets. These situations often involve substantial capital-scale applications, and the viability of the project is determined throughout its life cycle by the levelized cost of storage (LCOS). Initial analysis shows that lithium iron technology may have the most significant commercial presence. Lithium iron phosphate ($LiFePO_4$) is currently the dominant technology, mostly because of its low production costs, established performance rankings, and evaluated effective stability. The majority of Li-ion-linked base systems are used in off-network applications where the end-user has limited or no access to service energy or where energy security is crucial for the continuity of business operations. These outcomes often take the shape of a hybrid mini-network with integrated renewable production (mostly solar PV), diesel production, and battery storage (see this case study). Energy storage system installation has increased in high-end homes and businesses as a way to mitigate the effects of load shedding.

Some of the technical challenges can be circumvented as tabulated in Table 4 using the relevant application, control algorithm and duration, which is critical.

Table 4. BESS control algorithms and application.

Battery Technology	Application	Control Algorithm	Duration
Li-ion	Output power smoothing [51]	ANN and grid-exchanged power profile	8760 h
	Peak generation/load shaving [52]	Stochastic optimization-based battery operation framework	24 h
	Frequency regulation [53]	State-machine-based coordinated control	24 h
	Voltage and frequency regulation [53]	Fuzzy logic-based intelligent control technique	18 s
	PV plant dispatchability [54]	Optimal power control strategy	72 h
	Fault-ride-through [55]	Master–slave control mode	18 s
	Black start [56]	Stratified optimization strategy	60 min
	Energy arbitrage [57]	Classification-based scheme	21 months
Lead–Acid	Output power smoothing [58]	Simple moving average	10 s
	Frequency regulation [53]	Step-wise inertial control method	100 s
	Fault-ride-through [55]	Supervisory control system	240 s
	Black start [56]	A copula selection and goodness-of-fit-based method	80 min
	Dynamic program approach [59]	Dynamic program approach	720 h

It can be observed that by using a Li-ion battery over a Pb acid battery, the issues of power smoothing, load shaving, frequency regulation, PV plant dispatchability and energy arbitration can be circumvented.

5. Current Status and Some Real PV-Battery Projects

This segment examines some South African situations wherein energy storage systems have been used conjointly with PV generation, highlighting their modes of operation, energy storage forms, and current outcomes.

5.1. Canadian Farm

The Canadian farm, located in Lephalale, Limpopo, South Africa has a System size (kW + kWh) of about 200–1200 kWh and is equipped with a BESS as described in Table 5.

Table 5. BESS Canadian farm in Limpopo, South Africa.

Technology	Description	Quantity
Batteries	7.4 kWh Solar Md Li-ion	156
Inverters	8 kVA inverters SMA	21
	50 kW grid-tied inverter	2
Dimension	40-foot containerized solution	
Annual energy stored (kWh)	2200	
Electricity tariff reduction (%)	100	

5.2. Botha Huis

Botha huis, located in Mosselbay, South Africa has a capacity of 13.2 kWp (kW + kWh) and is equipped with a battery energy storage system as described in Table 6.

Table 6. BESS Botha Huis, Mosselbay, South Africa.

Technology	Description	Quantity
PV modules	270 W × 60 cells of polycrystalline	49
Batteries	BYD B-Box	2 × 2.56 kWh
Inverters	8 kVA	1
Annual energy yield (kWh)	15,018.1 kWh	
Annual energy stored (kWh)	3312.2 kWh	
Electricity tariff reduction (%)	70	

5.3. Matjhabeng Solar PV with Battery Energy Storage Systems Project

The Matjhabeng 400 MW Solar Photovoltaic Power Plant with 80 MW (320 MWh) battery energy storage systems (henceforth referred to as the "Project"), which is situated north and south of the town of Odendaalsrus in the Free State Province, has been proposed by SunElex Energy (Pty) Ltd. (the Applicant). The planned project will be designed to meet the energy needs of the Matjhabeng Local Municipality and will produce electricity for delivery to the regional or global grid. Locality map of the project's Phase 1 and Phase 2 sites show in Figure 13. The two (2) phases listed below will be used to develop the proposed utility-scale project:

- Phase 1: A 200 MW solar photovoltaic system with a 40 MW (160 MWh) battery energy storage system (referred to as the "Phase 1 Site"), which is located on a site south of Odendaalsrus;
- Phase 2: 200 MW of solar photovoltaic capacity with a 40 MW (160 MWh) battery energy storage system (on the site north of Odendaalsrus, referred to as the "Phase 2 Site" in the following). The project's electrical output will be fed into Eskom's pre-existing 132 kV distribution network.

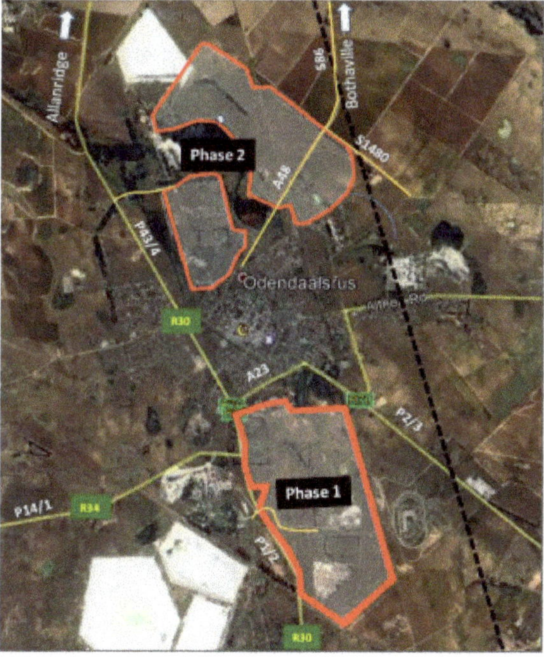

Figure 13. Locality map of the project's Phase 1 and Phase 2 sites.

5.4. Planned BESS Projects

Eskom, a state-owned enterprise has recognized 24 sites in the Western Cape Province, South Africa, where the planned BESS projects will be realized with a capacity of about 148.5 MW. The following criteria were contemplated in the choice of fitting sites:

- Vicinity of electricity clienteles to existing PV generators;
- Decrease in energy supply losses;
- Peak load abatement on severely loaded network components;
- Abatement in congestion of upstream high-voltage networks;
- Enhancement of local network characteristics and quality of supply;
- Peak load abatement where the peak load is coincident with the national system peak (i.e., winter evenings);
- Accessibility of sufficient MV connection capacity for the BESS;
- Accessibility of sufficient space at the substation for the deployment of BESS containers.

A view of one of the sites planned for BESS implementation, Eskom Paleisheuwel, is shown in Figure 14.

Figure 14. Eskom Paleisheuwel, Western Cape province.

6. Deployment of Utility-Scale Battery Energy Storage

The Eskom BESS project involves implementing outcomes at several locations in various operating units (OUs). Sizes of the results range from 1 MW to 60 MW. The standard size of an installation is 4 MW/16 MWh, which equals an estimated total of 90 installations. To optimize the usefulness of the BESS, all results will have a primary function and supporting roles that are "stacked benefits" in nature. As an illustration, a unit designed primarily for capex deferral during peak times in the winter will be available for operations such as frequency support at any time and peak clipping in the summer. The maximum discharge period will be 4 h. Figure 15 illustrates the steps used to appear in a technical investigation.

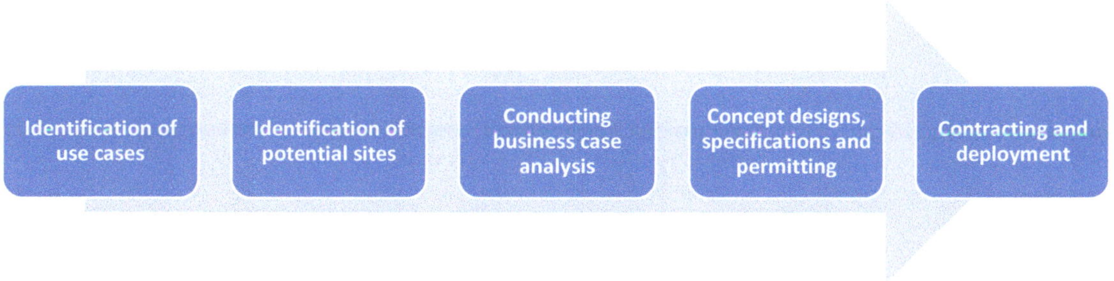

Figure 15. Process flow for the BESS project.

Step 1:

Wherever possible, supply-delayed investment and congestion supervision were given priority because these are the most enticing applications.

Step 2:

Active OU recognized potential sites, where appropriate. Substations located in electrically distant operational units were chosen in cases where local improvements could not be immediately realized. Prioritizing areas that relieve congestion upstream and reduce failures was one endeavor. In addition to replacing peak energy (kWh) and demand, these locations will supply gains and losses (kW).

Step 3:

In this stage, the business benefit of installing BESS is compared to other viable options (such as adding new supply/transmission substations and feeders, installing voltage regulators, adding more peaking power generators, etc.), as well as whether the investment will be recouped within a reasonable amount of time. The case for the project is made on the basis of both direct and indirect benefits, such as lower distribution costs for bulk purchases and lower production costs overall.

Step 4:

Conducting technical due diligence on potential locations is the first stage of this cycle. This entails showcasing a number of system planning studies, such as the worst case (maximum charging and discharging) load flows, dynamic time-series training, and quality of distribution studies. Conceptual plans for efficient locations have been carried out. These are based on standard BESS yard and station yard layouts, but the winning engineering, procurement, and construction (EPC) contractor are responsible for finishing them at the detail design stage. This phase also saw the creation and approval of the following technical specifications:

BESS equipment;
AC equipment;
General BESS and substation yard;
Protection and control;
Distributed energy resources management system (DERMS);
Application performance monitoring (APM) tool.

In order to ensure a smooth transition between system circumstances and BESS stationing, the DERMS will be implemented into the SCADA. A "BESS fleet" will be successfully run by it. The life management of the BESS divisions is important, since some interactions tend to diminish over time, making the technical advancements less fully implementable. Therefore, it is crucial to check the unit's longevity, especially the chemical storage unit. The APM tool is used to achieve this. Authorizing involves locating land and directing environmental impact assessments in accordance with the relevant laws.

Step 5:

The preparation of bid documents (bills of quantities and assessment criteria) in accordance with the provided templates is part of the contracting phase. Eskom has provided the necessary technical paperwork and validated the possibility for each location. The list of required documentation includes, among other things:

1. A planning report showing the use case;
2. Business case and system limits;
3. System diagrams for the anticipated BESS and substation yards;
4. Several service technical conditions.

Primary assessment, technical assessment, and finally economic development assessment make up the evaluation criteria. The development stage takes into account energy and capital costs.

7. Results and Discussion

In this section, to provide a significant innovation and contribution in the field of implementing battery energy storage for photovoltaic applications, a techno-economic analysis of two battery technologies incorporated with the Photovoltaic Grid-Connected System is carried out by adopting the HOMER-Pro-software with contemplation of actual load profiles and resource data. Consequently, the BESS-PV incorporated with a Li-ion battery brought forth a LCOE of 5.46 R/kWh in comparison with the BESS-PV system embedded with a Pb-acid battery postulating a LCOE of 5.8 R/kWh. Conversely, a total present cost (TPC) of the BESS-PV system with Li-ion batteries turned out to have a total of about R245,774 in comparison to the BESS-PV system with Pb-acid battery yielded a TPC of R257,841. The levelized cost of electricity (LCOE) outcome of the current study is proved to be favorable. The comparative analysis of Pb-acid and Li-ion battery technology in reference to various measure of effectiveness is tabulated in Table 7.

Table 7. Pb-acid vs. Li-ion battery technology in reference to various measure of effectiveness.

Battery Technology	PV (kW)	Number of Battery (Units)	Converter (kW)	Total TPC (R)	LCOE (R)	Operating Cost	PV Fraction (%)
Li-ion	10	6	5	24,577	5.46	13,757	90
Pb-acid	10	10	5	257,841	5.8	27,157	91

As observed above, for each type of BESS with similar input PV, the number of batteries, converter parameters postulated, the state of charge (%), battery capacity (Ah), and lifetime (years) feature an output of Li-ion batteries (100%, 167, 11) is discovered to be enhanced compared to a Pb-acid battery (100%, 83, 4). Moreover, as shown in Table 8, it could be absorbed as evidence for Li-ion batteries to be exploited in solar PV generation due to their enhanced energy capability.

Table 8. Pb-acid vs. Li-ion battery technology in reference to techno-economic analysis results.

Battery Technology	Energy in (kWh/Year)	Energy Out (kWh/Year)	Storage Depletion (kWh/Year)	Losses (kWh/Year)	Annual Throughput (kWh/Year)	Estimated Life (Year)
Li-ion acid	1898	1712	3.7	192	1804	11.2
Pb-ion acid	2129	1707	4.1	427	1908	4.1

By this investigation, the results lead to the conclusion that the BESS-PV system with Li-ion batteries necessitates about 41% fewer batteries in comparison to Pb-acid batteries and is supplementary in the establishment of an unswerving power source with lower expenditure. Furthermore, Li-ion battery technology delivered lower TPC and LCOE, and the BESS-PV system that has a higher solar PV fraction necessitates a greater number of batteries reciprocally. Generally, considering the standard application scenario investigated, Li-ion batteries are ascertained to be lucrative in both technical and economic countenances, and thus, they are advisable as a fill-in workable solution in combating the problem of load shedding in South Africa.

8. Conclusions

In the South African context, as well as in many other countries, electricity supply capacity could be best increased by promoting the diversity of energy sources in the generation. In this generation mix, renewable energies and particularly PV solar are one of the leading renewable sources of energy despite challenges related to their inability to meet the base load demand of electricity. Therefore, large-scale PV solar projects for reliable electricity supply require both in-depth knowledge pursuit as well as financial investment in energy storage technologies. This work discusses the knowledge gap in the three critical areas concerning the implementation of large-scale electrical energy storage in the South African context.

Based on the proposed case scenario, Li-ion batteries are ascertained to be lucrative in both technical and economic countenances, and thus, they are advisable as a fill-in workable solution in combating the problem of load shedding in South Africa. Some of the technical challenges, i.e., output power smoothing, load shaving, frequency regulation, PV plant dispatchability and energy arbitration can be circumvented using the control algorithms furnished and their corresponding duration thereof.

As a proposal, further investigations should be conducted in order to crack the problem of economic viability under distinctive application set-ups.

Author Contributions: B.A.T., conceptualized, carried out the investigation, wrote and prepared the article. P.N.B. was responsible for reviewing, editing, and drafting closing remarks of the article. The published version of the article has been reviewed and approved by all authors. All authors have read and agreed to the published version of the manuscript.

Funding: This research received no external funding.

Institutional Review Board Statement: Not applicable.

Informed Consent Statement: Not applicable.

Data Availability Statement: Not applicable.

Conflicts of Interest: The authors declare no conflict of interest.

References

1. CSIR. Load Shedding Statistics. December 2021. Available online: https://www.csir.co.za/load-shedding-statistics (accessed on 17 July 2022).
2. Bashe, M.; Shuma-Iwisi, M.; van Wyk, M.A. Assessing the Costs and Risks of the South African Electricity Portfolio: A Portfolio Theory Approach. *J. Energy S. Afr.* **2016**, *27*, 91–100. [CrossRef]
3. Bruwer, J.P. South African Residential Electricity Price Increase between 2013 and 2020: An Online Desktop Study. *Bus. Re-Solut.* **2021**. [CrossRef]
4. Dauda, L.; Long, X.; Mensah, C.N.; Salman, M.; Boamah, K.B.; Ampon-Wireko, S.; Dogbe, C.S.K. Innovation, trade openness and CO_2 emissions in selected countries in Africa. *J. Clean. Prod.* **2021**, *281*, 125143. [CrossRef]
5. Friedrich, E.; Trois, C. GHG emission factors developed for the collection, transport and landfilling of municipal waste in South African municipalities. *Waste Manag.* **2013**, *33*, 1013–1026. [CrossRef]
6. Department of Mineral Resources and Energy. Integrated Resource Plan (IRP2019). Government Notice; October 2019. Available online: http://www.energy.gov.za/IRP/2019/IRP-2019.pdf (accessed on 17 July 2022).
7. Department of Mineral Resources and Energy. IRP Update Draft Report 2018. Government Notice; August 2018. Available online: http://www.energy.gov.za/IRP/irp-update-draft-report2018/IRP-Update-2018-Draft-for-Comments.pdf (accessed on 17 July 2022).
8. Gür, T.M. Review of electrical energy storage technologies, materials and systems: Challenges and prospects for large-scale grid storage. *Energy Environ. Sci.* **2018**, *11*, 2696–2767. [CrossRef]
9. Jiang, J.; Li, Y.; Liu, J.; Huang, X.; Yuan, C.; Wen, X.; Lou, D. Recent Advances in Metal Oxide-based Electrode Architecture Design for Electrochemical Energy Storage. *Adv. Mater.* **2012**, *24*, 5166–5180. [CrossRef]

10. Bruce, P.; Freunberger, S.; Hardwick, L.; Tarascan, J.M. Li–O$_2$ and Li–S batteries with high energy storage. *Nat. Mater.* **2012**, *11*, 19–29. [CrossRef]

11. Sabihuddin, S.; Kiprakis, A.E.; Mueller, M. A Numerical and Graphical Review of Energy Storage Technologies. *Energies* **2015**, *8*, 172–216. [CrossRef]

12. Agyenim, F.; Hewitt, N.; Eames, P.; Smyth, M. A review of materials, heat transfer and phase change problem formulation for latent heat thermal energy storage systems (LHTESS). *Renew. Sustain. Energy Rev.* **2010**, *14*, 615–628. [CrossRef]

13. Zahedi, A. Maximizing solar PV energy penetration using energy storage technology. *Renew. Sustain. Energy Rev.* **2011**, *15*, 866–870. [CrossRef]

14. Mararakanye, N.; Bekker, B. Renewable energy integration impacts within the context of generator type, penetration level and grid characteristics. *Renew. Sustain. Energy Rev.* **2019**, *108*, 441–451. [CrossRef]

15. Rana, M.M.; Romlie, M.F.; Abdullah, M.F.; Uddin, M.; Sarkar, M.R. A novel peak load shaving algorithm for isolated microgrid using hybrid PV-BESS system. *Energy* **2021**, *234*, 121157. [CrossRef]

16. Korjani, S.; Casu, F.; Damiano, A.; Pilloni, V.; Serpi, A. An online energy management tool for sizing integrated PV-BESS systems for residential prosumers. *Appl. Energy* **2022**, *313*, 118765. [CrossRef]

17. Peng, C.-Y.; Kuo, C.-C.; Tsai, C.-T. Optimal Configuration with Capacity Analysis of PV-Plus-BESS for Behind-the-Meter Application. *Appl. Sci.* **2021**, *11*, 7851. [CrossRef]

18. Gush, T.; Kim, C.-H.; Admasie, S.; Kim, J.-S.; Song, J.-S. Optimal Smart Inverter Control for PV and BESS to Improve PV Hosting Capacity of Distribution Networks Using Slime Mould Algorithm. *IEEE Access* **2021**, *9*, 52164–52176. [CrossRef]

19. Nair, U.R.; Sandelic, M.; Sangwongwanich, A.; Dragičević, T.; Costa-Castelló, R.; Blaabjerg, F. An Analysis of Multi Objective Energy Scheduling in PV-BESS System under Prediction Uncertainty. *IEEE Trans. Energy Convers.* **2021**, *36*, 2276–2286. [CrossRef]

20. Thirugnanam, K.; Kerk, S.K.; Yuen, C.; Liu, N.; Zhang, M. Energy Management for Renewable Microgrid in Reducing Diesel Generators Usage with Multiple Types of Battery. *IEEE Trans. Ind. Electron.* **2018**, *65*, 6772–6786. [CrossRef]

21. Venu, C.; Riffonneau, Y.; Bacha, S.; Baghzouz, Y. Battery Storage System sizing in distribution feeders with distributed photovoltaic systems. In Proceedings of the 2009 IEEE Bucharest PowerTech, Bucharest, Romania, 28 June–2 July 2009; pp. 1–5. [CrossRef]

22. Wang, Z.; Gu, C.; Li, F.; Bale, P.; Sun, H. Active Demand Response Using Shared Energy Storage for Household Energy Management. *IEEE Trans. Smart Grid* **2013**, *4*, 1888–1897. [CrossRef]

23. Kumar, D.; Zare, F.; Ghosh, A. DC Microgrid Technology: System Architectures, AC Grid Interfaces, Grounding Schemes, Power Quality, Communication Networks, Applications, and Standardizations Aspects. *IEEE Access* **2017**, *5*, 12230–12256. [CrossRef]

24. Lezhniuk, P.; Komar, V.; Rubanenko, O. Information Support for the Task of Estimation the Quality of Functioning of the Electricity Distribution Power Grids with Renewable Energy Source. In Proceedings of the 2020 IEEE 7th International Conference on Energy Smart Systems (ESS), Kyiv, Ukraine, 12–14 May 2020; pp. 168–171. [CrossRef]

25. *450-2020*; IEEE Recommended Practice for Maintenance, Testing, and Replacement of Vented Lead-Acid Batteries for Stationary Applications—Redline. IEEE: Manhattan, NY, USA, 2021; pp. 1–115.

26. *P1361/D4*; IEEE Draft Guide for Selection, Charging, Test and Evaluation of Lead-Acid Batteries Used in Stand-Alone Photovoltaic (PV) Systems. IEEE: Manhattan, NY, USA, 2013; pp. 1–41.

27. *1188-2005*; IEEE Recommended Practice for Maintenance, Testing, and Replacement of Valve-Regulated Lead-Acid (VRLA) Batteries for Stationary Applications—Redline. IEEE: Manhattan, NY, USA, 2006; pp. 1–49.

28. Li, T.; Chen, Y.; Gou, H.Y.; Chen, X.Y.; Tang, M.G.; Lei, Y. A DC Voltage Swell Compensator Based on SMES Emulator and Lead-Acid Battery. *IEEE Trans. Appl. Supercond.* **2019**, *29*, 1–4. [CrossRef]

29. Freitas, D.C.C.; de Moraes, J.L.; Neto, E.C.; Sousa, J.R.B. Battery Charger Lead-Acid using IC BQ2031. *IEEE Lat. Am. Trans.* **2016**, *14*, 32–37. [CrossRef]

30. McKeon, B.B.; Furukawa, J.; Fenstermacher, S. Advanced Lead–Acid Batteries and the Development of Grid-Scale Energy Storage Systems. *Proc. IEEE* **2014**, *102*, 951–963. [CrossRef]

31. Chiu, H.; Lin, L.; Pan, P.; Tseng, M. A novel rapid charger for lead-acid batteries with energy recovery. *IEEE Trans. Power Electron.* **2006**, *21*, 640–647. [CrossRef]

32. Hannan, M.A.; Hoque, M.M.; Hussain, A.; Yusof, Y.; Ker, P.J. State-of-the-Art and Energy Management System of Lithium-Ion Batteries in Electric Vehicle Applications: Issues and Recommendations. *IEEE Access* **2018**, *6*, 19362–19378. [CrossRef]

33. Zhang, W.; Wang, L.; Wang, L.; Liao, C.; Zhang, Y. Joint State-of-Charge and State-of-Available-Power Estimation Based on the Online Parameter Identification of Lithium-Ion Battery Model. *IEEE Trans. Ind. Electron.* **2022**, *69*, 3677–3688. [CrossRef]

34. Al-Humaid, Y.M.; Khan, K.A.; Abdulgalil, M.A.; Khalid, M. Two-Stage Stochastic Optimization of Sodium-Sulfur Energy Storage Technology in Hybrid Renewable Power Systems. *IEEE Access* **2021**, *9*, 162962–162972. [CrossRef]

35. Saruta, K. Long-term performance of sodium sulfur battery. In Proceedings of the 2016 IEEE 2nd Annual Southern Power Electronics Conference (SPEC), Auckland, New Zealand, 5–8 December 2016; pp. 1–5. [CrossRef]

36. Hatta, T. Applications of sodium-sulfur batteries. In *PES T&D 2012*; IEEE: Manhattan, NY, USA, 2012; pp. 1–3. [CrossRef]

37. Manzoni, R. Sodium Nickel Chloride batteries in transportation applications. In Proceedings of the 2015 International Conference on Electrical Systems for Aircraft, Railway, Ship Propulsion and Road Vehicles (ESARS), Aachen, Germany, 3–5 March 2015; pp. 1–6. [CrossRef]

38. Marcondes, A.; Scherer, H.F.; Salgado, J.R.C.; de Freitas, R.L.B. Sodium-nickel chloride single cell battery electrical model—Discharge voltage behavior. In Proceedings of the 2019 Workshop on Communication Networks and Power Systems (WCNPS), Brasilia, Brazil, 3–4 October 2019; pp. 1–4. [CrossRef]

39. Benato, R.; Sessa, S.D.; Necci, A.; Palone, F. A general electric model of sodium-nickel chloride battery. In Proceedings of the 2016 AEIT International Annual Conference (AEIT), Capri, Italy, 5–7 October 2016; pp. 1–6. [CrossRef]

40. Challapuram, Y.R.; Quintero, G.M.; Bayne, S.B.; Subburaj, A.S.; Harral, M.A. Electrical Equivalent Model of Vanadium Redox Flow Battery. In Proceedings of the 2019 IEEE Green Technologies Conference (GreenTech), Lafayette, LA, USA, 3–6 April 2019; pp. 1–4. [CrossRef]

41. Vins, M.; Sirovy, M. Assessing Suitability of Various Battery Technologies for Energy Storages: Lithium-ion, Sodium-sulfur and Vanadium Redox Flow Batteries. In Proceedings of the 2020 International Conference on Applied Electronics (AE), Pilsen, Czech Republic, 8–9 September 2020; pp. 1–5. [CrossRef]

42. Lim, J.-U.; Lee, S.-J.; Kang, K.-P.; Cho, Y.; Choe, G.-H. A modular power conversion system for zinc-bromine flow battery based energy storage system. In Proceedings of the 2015 IEEE 2nd International Future Energy Electronics Conference (IFEEC), Taipei, Taiwan, 1–4 November 2015; pp. 1–5. [CrossRef]

43. Nakatsuji-Mather, M.; Saha, T.K. Zinc-bromine flow batteries in residential electricity supply: Two case studies. In Proceedings of the 2012 IEEE Power and Energy Society General Meeting, San Diego, CA, USA, 22–26 July 2012; pp. 1–8. [CrossRef]

44. Lawder, M.T.; Suthar, B.; Northrop, P.W.C.; De, S.; Hoff, C.M.; Leitermann, O.; Crow, M.L.; Santhanagopalan, S.; Subramanian, V.R. Battery Energy Storage System (BESS) and Battery Management System (BMS) for Grid-Scale Applications. *Proc. IEEE* **2014**, *102*, 1014–1030. [CrossRef]

45. *2030.2.1-2019*; IEEE Guide for Design, Operation, and Maintenance of Battery Energy Storage Systems, Both Stationary and Mobile, and Applications Integrated with Electric Power Systems. IEEE: Manhattan, NY, USA, 2019; pp. 1–45. [CrossRef]

46. Wong, Y.S.; Lai, L.L.; Gao, S.; Chau, K.T. Stationary and mobile battery energy storage systems for smart grids. In Proceedings of the 2011 4th International Conference on Electric Utility Deregulation and Restructuring and Power Technologies (DRPT), Weihai, China, 6–9 July 2011; pp. 1–6. [CrossRef]

47. Shi, M.; Hu, J.; Han, H.; Yuan, X. Design of Battery Energy Storage System based on Ragone Curve. In Proceedings of the 2020 4th International Conference on HVDC (HVDC), Xi'an, China, 6–9 November 2020; pp. 37–40. [CrossRef]

48. Akhil, A. Trends and status of battery energy storage for utility applications. In Proceedings of the Tenth Annual Battery Conference on Applications and Advances, Long Beach, CA, USA, 10–13 January 1995; pp. 273–277. [CrossRef]

49. Helling, F.; Götz, S.; Weyh, T. A battery modular multilevel management system (BM3) for electric vehicles and stationary energy storage systems. In Proceedings of the 2014 16th European Conference on Power Electronics and Applications, Lappeenranta, Finland, 26–28 August 2014; pp. 1–10. [CrossRef]

50. Akhil, A.; Kraft, S.; Symons, P.C. Market feasibility study of utility battery applications: Penetration of battery energy storage into regulated electric utilities. In Proceedings of the Twelfth Annual Battery Conference on Applications and Advances, Long Beach, CA, USA, 14–17 January 1997; pp. 195–200. [CrossRef]

51. Li, X.; Hui, D.; Lai, X. Battery Energy Storage Station (BESS)-Based Smoothing Control of Photovoltaic (PV) and Wind Power Generation Fluctuations. *IEEE Trans. Sustain. Energy* **2013**, *4*, 464–473. [CrossRef]

52. Oudalov, A.; Cherkaoui, R.; Beguin, A. Sizing and Optimal Operation of Battery Energy Storage System for Peak Shaving Application. In Proceedings of the 2007 IEEE Lausanne Power Tech, Lausanne, Switzerland, 1–5 July 2007; pp. 621–625. [CrossRef]

53. Mejía-Giraldo, D.; Velásquez-Gomez, G.; Muñoz-Galeano, N.; Cano-Quintero, J.B.; Lemos-Cano, S. A BESS Sizing Strategy for Primary Frequency Regulation Support of Solar Photovoltaic Plants. *Energies* **2019**, *12*, 317. [CrossRef]

54. Gilmanova, A.; Wang, Z.; Gosens, J.; Lilliestam, J. Building an internationally competitive concentrating solar power industry in China: Lessons from wind power and photovoltaics. *Energy Sources Part B Econ. Plan. Policy* **2021**, *16*, 515–541. [CrossRef]

55. Berger, M.; Kocar, I.; Farantatos, E.; Haddadi, A. Dual Control Strategy for Grid-tied Battery Energy Storage Systems to Comply with Emerging Grid Codes and Fault Ride through Requirements. *J. Mod. Power Syst. Clean Energy* **2022**, *10*, 977–988. [CrossRef]

56. Li, J.; You, H.; Qi, J.; Kong, M.; Zhang, S.; Zhang, H. Stratified Optimization Strategy Used for Restoration with Photovoltaic-Battery Energy Storage Systems as Black-Start Resources. *IEEE Access* **2019**, *7*, 127339–127352. [CrossRef]

57. Abdullah, W.S.W.; Osman, M.; Ab Kadir, M.Z.A.; Verayiah, R. Battery energy storage system (BESS) design for peak demand reduction, energy arbitrage and grid ancillary services. *Int. J. Power Electron. Drive Syst.* **2020**, *11*, 398–408. [CrossRef]

58. Shi, L.; Fa, L.; Zhu, H.; Shi, J.; Wu, F.; He, W.; Wang, C.; Lee, K.Y.; Lin, K. Photovoltaic active power control based on BESS smoothing. *IFAC-PapersOnLine* **2019**, *52*, 443–448. [CrossRef]

59. Zhuo, W. An Approximate Dynamic Programming Approach for Wind Power Dispatch in Wind Farms. In Proceedings of the 2018 37th Chinese Control Conference (CCC), Wuhan, China, 25–27 July 2018; pp. 7502–7508. [CrossRef]

Article

Robust Voltage Control of a Buck DC-DC Converter: A Sliding Mode Approach

Salah Beni Hamed [1,*], Mouna Ben Hamed [2] and Lassaad Sbita [2]

[1] Physic Department, High School of Engineers of Tunis, Tunis 1008, Tunisia
[2] Electrical Department, National Engineering School of Gabes, Zrig Eddakhlania 6029, Tunisia
* Correspondence: salahbenihamed@yahoo.fr

Abstract: This paper deals with voltage control in a buck DC-DC converter. In fact, dynamic mathematical equations describing the principle behavior of the above system have been derived. Due to the nonlinearity of the established model, a nonlinear control algorithm is adopted. It is based on the sliding mode control approach. To highlight the performance of the latter, a comparative study with four control algorithms is carried out. The validity of the model and the performance of the conceived algorithms are verified in simulation. Both the system and the algorithm controls are implemented in the Matlab/Simulink environment. Extensive results under different operational conditions are presented and discussed.

Keywords: buck DC-DC converter; parameter variation; Matlab/Simulink environment; sliding mode control

Citation: Hamed, S.B.; Hamed, M.B.; Sbita, L. Robust Voltage Control of a Buck DC-DC Converter: A Sliding Mode Approach. *Energies* **2022**, *15*, 6128. https://doi.org/10.3390/en15176128

Academic Editors: Saad Motahhir, Najib El Ouanjli and Mustapha Errouha

Received: 15 June 2022
Accepted: 28 July 2022
Published: 24 August 2022

1. Introduction

Energy production is one of the most important development priorities. On the one hand, it has been observed that the world's electrical energy consumption is rapidly increasing [1–9]. On the other hand, energy production uses fossil fuels [10–13] such as oil, coal, and natural gas. They are all burnt and used as energy sources for production. The use of fossil fuels can help to offset the energy demand. However, there is too much carbon dioxide (CO_2) in the atmosphere, which can lead to big problems both for people and the earth. This forces humans to seek out alternative energy sources that may be capable of saving both people and the planet.

Many alternative sources are suggested. Among these sources, renewable energy is the most commonly used one. Nowadays, photovoltaic solar energy [14–16] and wind energy have become the most used alternative energy sources. The operating power of both the photovoltaic generator and the wind energy sources depends on metrological conditions such as temperature, irradiation, wind speed, etc. [17–21]. The optimal use of the produced energy can be assumed only if the produced wind and photovoltaic sources' maximum power are extracted and tracked for any change in the metrological conditions. Tools to track these specific points are required [22–27].

Most of the proposed power point tracking algorithms are based on varying the generator characteristics in such a way as to be adapted to the latter of the load. The generator characteristic changes are assumed by using a DC-DC converter and a maximum power point tracking (MPPT) algorithm [28–30]. Different DC-DC converters are used as buck converters, boost converters, buck–boost converters, etc. [31–34]. The boost and buck converters are the most commonly used. The buck DC-DC converter is used especially in DC link control, as the DC voltage of the photovoltaic/wind conversion system output depends on the metrological conditions. DC-DC converters are naturally classified as nonlinear systems due to their commuting properties. They are the most commonly used circuits in power electronics, especially in DC link voltage stabilization.

In general, the DC link voltage must be fixed at a desired value despite input voltage and load variations [35]. To regulate the DC voltage magnitude, and obtain a constant and stable output voltage and fast response, many types of controllers are used [27], such as fuzzy logic, PI controllers, PID controllers, sliding mode control, etc. Conventional controllers are in general conceived by using small signal state equations obtained at a specific operation point. The conceived control algorithm remains efficient only around the specified operating point. To avoid the drawbacks of conventional controllers, nonlinear control approaches are investigated and used in the control voltage loop of the buck DC-DC converter. Among these nonlinear control algorithms, the sliding mode approach is the most used. Different sliding mode algorithms, both for continuous and discrete times, are proposed in the literature [36–43]. In [36], a cascade loop control is proposed. The voltage control loop is based on a classical PID controller, and the current loop control is based on zero order sliding mode control over a continuous time. The validity of the conceived algorithm is confirmed under different working conditions, including target voltage variation, load variation and input voltage variations. In [37], a sliding mode controlled pole is conceived both in voltage and current control loops. An integral switching surface is used here. Simulation results are given under target voltage variation and load variation. The fractional order sliding mode is also used in the literature [38]. Different sliding mode approaches for discrete time are used. In fact, in [39], discrete time sliding mode control is investigated for the output voltage control. The performance of the used algorithm is tested only for a fixed target output voltage. Discrete time fast terminal sliding mode control with mismatched disturbance is conceived for DC-DC buck converters, and the control strategy is investigated in [40]. The validity of the proposed algorithm, both in simulation and experimentation, is assessed under both load variations and fluctuations in the input voltage. A discrete repetitive adaptive sliding mode control for the DC-DC buck converter under only variable target output voltage perturbed with a Gaussian noise is conceived in [41]. Simulation results under fixed target voltage and load variations are given. An Adaptive Global Sliding Mode Controller based on the Lyapunov approach is designed for perturbed DC-DC buck converters [42]. In this work, the external disturbances and dynamic uncertainties are modeled with a sinusoidal function. In [43,44], to cope with the chattering problem, a high-order sliding mode control of the DC-DC buck converter is conceived. Practical results under target output voltage variation are presented. In most of the mentioned studies, the performance of the conceived algorithms is validated against external disturbances as input voltage and load variations. The internal disturbances in terms of DC-DC buck converter parameter variations are omitted. Thus, in this paper, we are interested in the performance of the first-order sliding mode in DC-DC buck output voltage control with both external and internal disturbances. Besides this, to highlight the good performance of the investigated control algorithm, a comparison study with four control algorithms is carried out.

The paper is organized as follows. Section 2 presents the modeling of the buck DC-DC converter feeding a resistive load. Section 3 describes the sliding mode controller principle and its application in buck DC-DC converter output voltage control. Section 4 presents the internal model controller. The fuzzy logic controller is given in Section 5. Section 6 illustrates the obtained results in simulation, both for internal and external disturbances. Section 5 concludes the work and presents some suggested prospects.

2. Modelling of DC-DC Buck Converter Mathematical

The synoptic scheme of the DC-DC buck converter is depicted in Figure 1. It consists of an on and off controlled semiconductor (Transistor IGBT, T), a natural commuted semiconductor (Diode, D), a smoothing current system (inductor, L), and a smoothing voltage system (capacitor, C). It is powered by a direct current voltage source and feeds a resistive load.

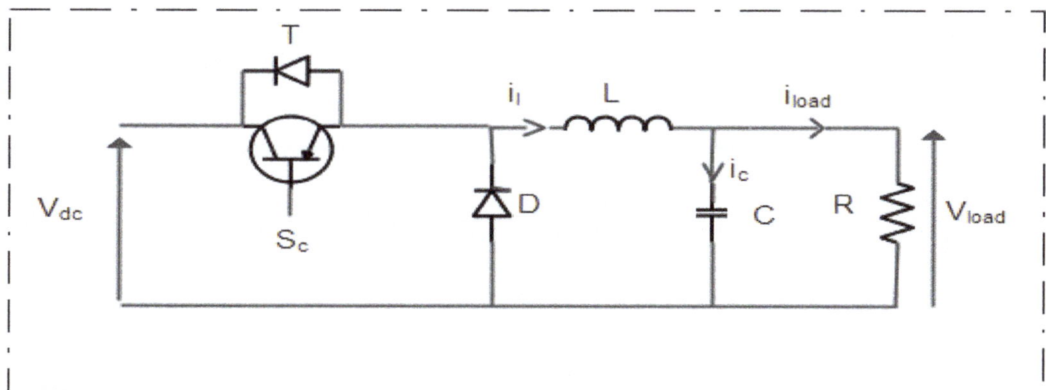

Figure 1. Synoptic scheme of the DC-DC buck converter.

2.1. Bilinear Switching Model of DC-DC Buck Converter

The working principle is based on two alternative phenomena: charging and discharging, based on the control signal state, Sc. When considering the continuous conduction mode and the control signal state levels, two modes are to be considered [45–53].

Mode 1:

For the ON mode in which $S_c = 1$, the transistor T is closed and the diode D is open. Based on Kirchhoff's current and voltage laws, we can write:

$$\begin{cases} \frac{di_l}{dt} = \frac{V_{dc}}{L} - \frac{V_{load}}{L} \\ \\ \frac{dV_{load}}{dt} = \frac{i_l}{C} - \frac{i_{load}}{C} \end{cases} \tag{1}$$

Mode2:

$S_c = 0$, the transistor T becomes open and the diode D begins closed. The state equations describing the inductor current and the output voltage dynamics are given in (2).

$$\begin{cases} \frac{di_l}{dt} = -\frac{V_{load}}{L} \\ \\ \frac{dV_{load}}{dt} = \frac{i_l}{C} - \frac{i_{load}}{C} \end{cases} \tag{2}$$

The combination of the two sub models leads to the general buck DC-DC converter model, as illustrated in (3).

$$\begin{cases} \frac{di_l}{dt} = S_c\frac{V_{dc}}{L} - \frac{V_{load}}{L} \\ \\ \frac{dV_{load}}{dt} = \frac{i_l}{C} - \frac{i_{load}}{C} \end{cases} \tag{3}$$

For a resistive load, (3) begins

$$\begin{cases} \frac{di_l}{dt} = S_c\frac{V_{dc}}{L} - \frac{V_{load}}{L} \\ \\ \frac{dV_{load}}{dt} = \frac{i_l}{C} - \frac{V_{load}}{RC} \end{cases} \tag{4}$$

2.2. Averaged Dynamic Model of DC-DC Buck Converter

By using the state-space averaging method [46], Equation (4) can be written as illustrated with Equation (5).

$$\begin{cases} \dfrac{d\langle i_l \rangle_{T_s}}{dt} = \langle S_c \rangle_{T_s} \dfrac{\langle V_{dc} \rangle_{T_s}}{L} - \dfrac{\langle V_{load} \rangle_{T_s}}{L} \\[2ex] \dfrac{d\langle V_{load} \rangle_{T_s}}{dt} = \dfrac{\langle i_l \rangle_{T_s}}{C} - \dfrac{\langle V_{load} \rangle_{T_s}}{RC} \end{cases} \tag{5}$$

where $\langle i_l \rangle_{T_s}$, $\langle V_{load} \rangle_{T_s}$, $\langle S_c \rangle_{T_s}$ and $\langle V_{dc} \rangle_{T_s}$ are the averaged values of inductor current, output voltage, control signal, and input voltage, respectively, in a switching period T_s.

This can be put into the more compact form of an uncertain nonlinear system, as indicated by (6).

$$\dot{X} = f(X) + g(X)\alpha \tag{6}$$

The nonlinear equations $f(X)$ and $g(X)$, and averaged state vector X, are defined as follows:

$$g(X) = \begin{pmatrix} \dfrac{V_{dc}}{L} \\ 0 \end{pmatrix} \tag{7}$$

$$f(X) = \begin{pmatrix} 0 & -\dfrac{V_{load}}{L} \\ \dfrac{i_l}{C} & -\dfrac{V_{load}}{RC} \end{pmatrix} \tag{8}$$

$$X = \left[\langle i_l \rangle_{T_s} \ \langle V_{load} \rangle_{T_s} \right]^T \tag{9}$$

Here, α is the duty cycle.

2.3. Small Signal Dynamic Model of DC-DC Buck Converter

The small signal DC-DC buck converter model is obtained by linearizing the averaged model around an operating point. Thus, the input control and the output signals' expressions are to be represented by the sum of their quiescent values and a small alternative current (AC) variation. So, we can write,

$$\langle i_l \rangle_{T_s} = I_l + \tilde{i}_l \tag{10}$$

$$\langle V_{load} \rangle_{T_s} = U_{load} + \tilde{V}_{load} \tag{11}$$

$$\langle V_{dc} \rangle_{T_s} = U_{dc} + \tilde{V}_{dc} \tag{12}$$

$$\langle S_c \rangle_{T_s} = \alpha_e + \tilde{\alpha} \tag{13}$$

where I_l, U_{load}, U_{dc} and α_e are the inductor current, the load voltage, the input voltage, and the duty cycle at the operating point, respectively.

Replacing $\langle i_l \rangle_{T_s}$, $\langle V_{dc} \rangle_{T_s}$, $\langle V_{load} \rangle_{T_s}$ and $\langle S_c \rangle_{T_s}$ with their expressions, Equation (5) begins

$$\begin{cases} \dfrac{d(I_l + \tilde{i}_l)}{dt} = (\alpha_e + \tilde{\alpha}) \dfrac{(U_{dc} + \tilde{V}_{dc})}{L} - \dfrac{(U_{load} + \tilde{V}_{load})}{L} \\[2ex] \dfrac{d(U_{load} + \tilde{V}_{load})}{dt} = \dfrac{(I_l + \tilde{i}_l)}{C} - \dfrac{(U_{load} + \tilde{V}_{load})}{RC} \end{cases} \tag{14}$$

By separating steady state terms and small-signal terms, we obtain

$$\begin{cases} \dfrac{d\tilde{i}}{dt} + \dfrac{dI_l}{dt} = \underbrace{\dfrac{\alpha_e U_{dc}}{L} - \dfrac{U_{load}}{L}}_{DC\ term} + \underbrace{\dfrac{\alpha \tilde{V}_{dc}}{L} + \dfrac{\tilde{\alpha} U_{dc}}{L} - \dfrac{\tilde{V}_{load}}{L}}_{1^{st}\ order\ AC\ term} + \underbrace{\dfrac{\tilde{\alpha} \tilde{V}_{dc}}{L}}_{2^{nd}\ order\ AC\ term} \\[3ex] \dfrac{d\tilde{V}_{load}}{dt} + \dfrac{dU_{load}}{dt} = \underbrace{\dfrac{I_l}{C} - \dfrac{U_{load}}{RC}}_{DC\ term} + \underbrace{\dfrac{\tilde{i}_l}{C} - \dfrac{\tilde{V}_{load}}{RC}}_{1^{st}\ order\ AC\ term} \end{cases} \tag{15}$$

In the system of Equation (15), the DC terms contain the DC terms only, the first-order AC term contains a product of a DC term with an AC term, and the second order AC term contains a product between two AC terms.

The second-order AC terms are much smaller in magnitude than the firs- order AC terms. Therefore, the second small AC quantity is neglected. Moreover, U_{dc}, I_l and U_{load} are constant DC terms. As a result, the sum of the DC term and its derivative are zero. Consequently, only the linear term remains, and the small signal dynamic DC-DC buck converter is defined with (16).

$$\begin{cases} \frac{d\tilde{i_l}}{dt} = \frac{\tilde{\alpha}U_{dc}}{L} - \frac{\tilde{V}_{load}}{L} \\[2mm] \frac{d\tilde{V}_{load}}{dt} = \frac{\tilde{i_l}}{C} - \frac{\tilde{V}_{load}}{RC} \end{cases} \tag{16}$$

2.4. Comparative Study

To assess the performance of the used mathematical model of the DC-DC buck converter, the three developed models are implemented in Matlab/Simulink platform and compared to one that was established using the predefined electronic components in matlab/Simulink packages. The parameters of the used DC-DC buck converter are grouped in Table 1.

Table 1. DC-DC buck converter parameters.

Parameter	Value
Input voltage (v)	400
Capacitor (µF)	5
Inductor (mH)	20
Switching frequency (Khz)	10
Resistive load (Ω)	10

In order to assess the validity of the established models, various operating points are considered. The most significant simulation results are displayed and commented on. Figure 2a shows the evolution of the duty cycle used as a control signal for the averaged and small-signal models. The switching signal obtained at the pulse width modulation bloc, as a control signal for the bilinear switching model and established using the predefined electronic components, is depicted in Figure 2b. Output voltage waveforms are given in Figure 2c. Both dynamic and steady-state working modes are considered. As shown in Figure 2c, output voltage ripples are omitted both for the small signal model and the averaged model, and appear for the switching model, as established using the predefined electronic components in the matlab/Simulink packages. The accuracy of the established model is proven as the modeling error is a few percentage points lower at high duty cycle values, and increases at low duty cycle values, according to the losses in the used power semi-conductors in the DC-DC buck (Figure 2d). Consequently, to cope with the nonlinearities and modeling error, a robust nonlinear control law is to be used. This allows us to use the sliding model approach for DC-DC buck converter control.

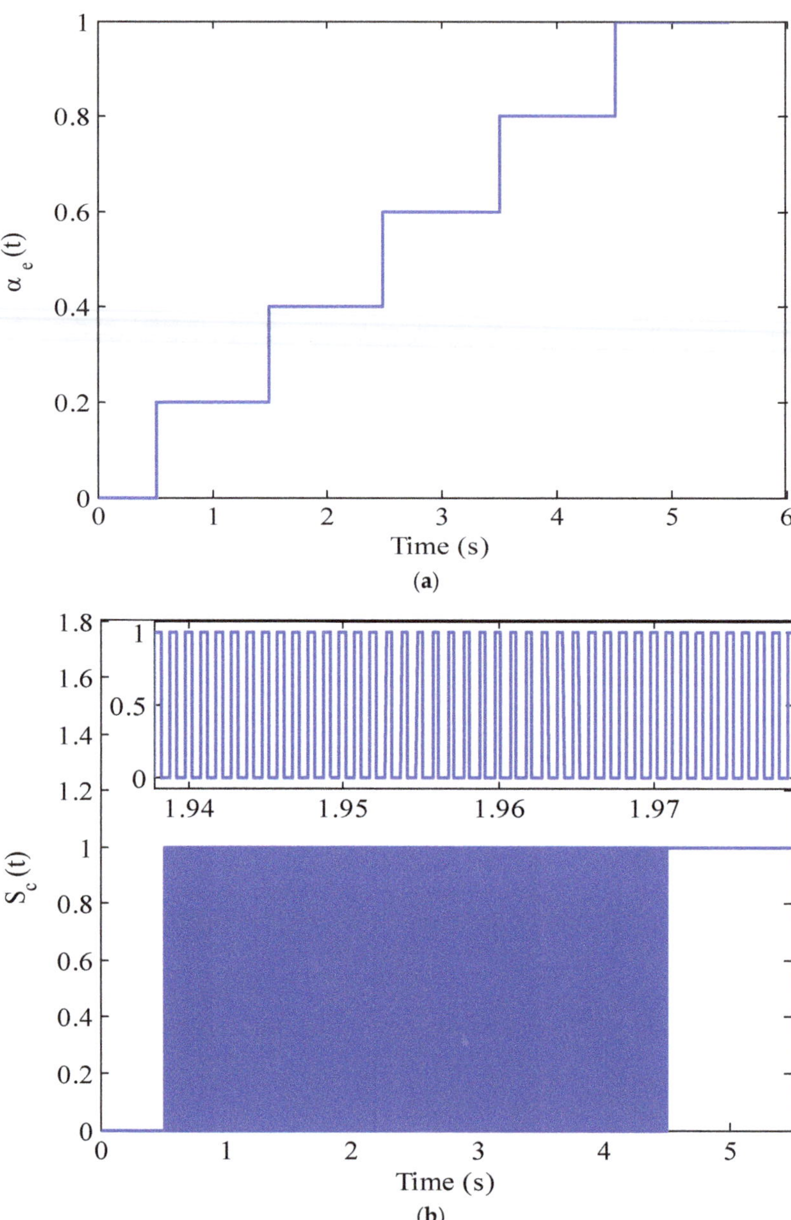

(a)

(b)

Figure 2. *Cont.*

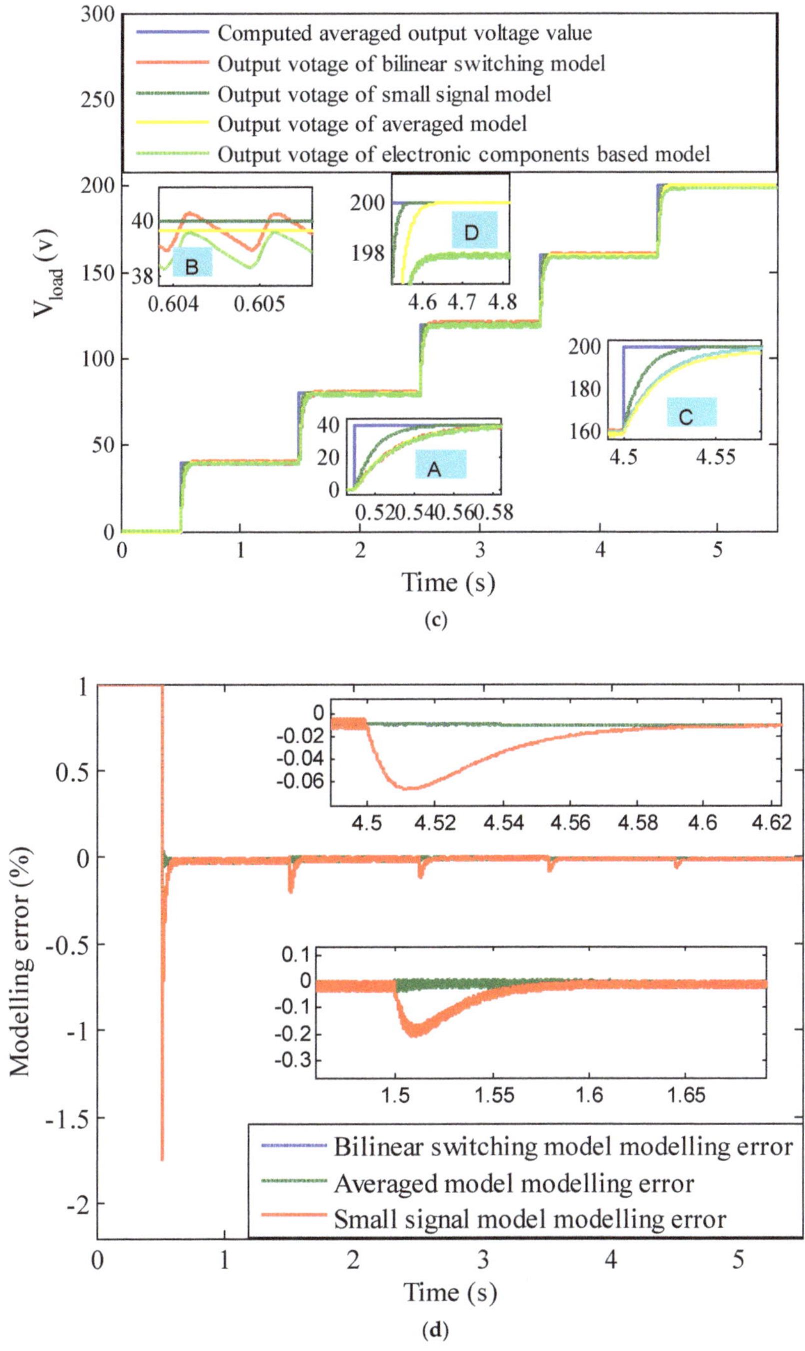

Figure 2. Comparative study simulation results: (**a**) duty cycle, (**b**) control signal, (**c**) output voltages of DC-DC buck converter models and (**d**) model modeling error.

3. Sliding Mode Control Approach for Buck Converter Voltage Control

Control of nonlinear systems using a sliding mode approach was conceived in 1992 by Vadim Ulkin [54] in order to solve the conventional controller's problems. Typical sliding-mode control operates in the form of these two modes. The first is named the "approaching mode". When this mode is reached, the convergence of the system state to a predefined manifold called the sliding mode surface in finite time is assumed. The second, designed with sliding mode, follows the sliding surface and returns to the origin. Many approaches to sliding mode control have been conceived. The equivalent control approach [54–59] is the most commonly used. Let us denote with S the sliding mode function. In this case, the output's voltage is controlled. A linear sliding mode surface is adopted. It is defined for the first sliding mode control as follows:

$$S = C_1 e + \dot{e} \tag{17}$$

Here e is the output voltage error and $\dot{e}$ is its derivative value.
The output voltage error is defined by (13).

$$e = V_{loadref} - V_{load} \tag{18}$$

$$\dot{e} = -\frac{1}{C}\left[i_l - \frac{V_{load}}{R}\right] \tag{19}$$

Substituting e and $\dot{e}$ with their expressions, the sliding mode surface becomes:

$$S = -\frac{1}{C}i_l + \left(\frac{1}{RC} - C_1\right)V_{load} + C_1 V_{loadref} \tag{20}$$

Its derivative is defined by (21).

$$\dot{S} = \left(\frac{1 - C_1 RC}{RC^2}\right)i_l - \left(\frac{L - R^2 C - C_1 RCL}{R^2 C^2 L}\right)V_{load} - \frac{\alpha V_{dc}}{LC} \tag{21}$$

The equivalent control α_{eq} is deduced from the following equality.

$$\dot{S} = 0 \tag{22}$$

It is defined as:

$$\alpha_{eq} = \left(\frac{L - C_1 RCL}{RCV_{dc}}\right)i_l - \left(\frac{L - R^2 C - C_1 RCL}{R^2 CV_{dc}}\right)V_{load} \tag{23}$$

Since the duty cycle must be in $\begin{bmatrix} 0 & 1 \end{bmatrix}$, the real control signal is given by (24).

$$\alpha(t) = \begin{cases} 1 & si & \alpha(t) > 1 \\ \alpha_{eq} + Msign(S) & & 0 \leq \alpha(t) \leq 1 \\ 0 & & \alpha(t) < 0 \end{cases} \tag{24}$$

4. Internal Model Control Approach for Buck Converter Voltage Control

The general block diagram of the internal model control (IMC) loop is given in Figure 3. $G(s)$ is the real system's open loop transfer function, $Gi(s)$ is the system model's open loop transfer function, and $Q(s)$ is the IMC controller's transfer function [60].

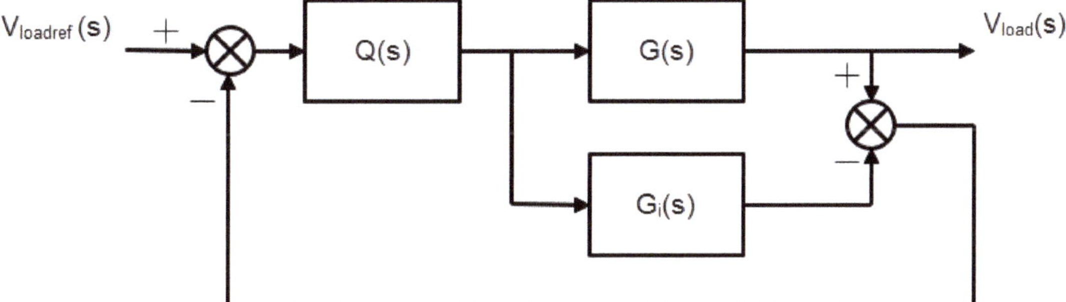

Figure 3. The internal model control of the feedback scheme.

Referring to Figure 3, we can write

$$V_{load} = G(s)[1 + Q(s)(G(s) - G_i(s))]^{-1}Q(s)V_{loadref} \tag{25}$$

When the modeling system is perfect, we can write

$$V_{load} = G(s)Q(s)V_{loadref} \tag{26}$$

The closed loop is stable if and only if $G(s)$ and $Q(s)$ are stable. However, if $G(s)$ is in the non-minimum phase, $G^{-1}(s)$ is not stable. Besides this, if the $G^{-1}(s)$ numerator degree is higher than the denominator degree, then $G^{-1}(s)$ cannot be implemented. Referring to the H_2 optimization leads to choosing $Q(s) = G^{-1}(s)$. To assume the feasibility condition, a low-pass filter is added. The final expression of the IMC controller is given in (27).

$$Q(s) = \frac{a}{s+a}G^{-1}(s) \tag{27}$$

where a is the filter parameter.

5. Fuzzy Logic for Buck Converter Voltage Control

Fuzzy logic is a computational approach based on degrees of truth. It was first discovered in the 1960s by Lotfi Zadeh [61]. Since the above approach does not require a mathematical model, and is based on human decision-making, this approach can present high efficiency. A mamdani-type fuzzy logic is used for the voltage control. In this fuzzy logic controller (FLC), the voltage error $ev(k)$ and the change in voltage error $cev(k)$ are the inputs of the fuzzy system, while the change in the duty cycle $c\alpha(k)$ is considered as the output of this system.

The equations for $ev(k)$ and $cev(k)$ are as follows:

$$e_v(k) = V_{load}(k) - V_{loadref}(k) \tag{28}$$

When the modeling system is perfect, we can write

$$ce_v(k) = \frac{e_v(k) - e_v(k-1)}{T_e} \tag{29}$$

where T_e is the sample time.

According to Figure 4, three essential steps are to be followed in the mamdani system's conception. At the fuzification step, the crisp variables $e_v(k)$, $Ce_v(k)$ and $C\alpha(k)$ are converted to fuzzy sets using triangular membership functions, as can be seen in Figure 5a, b and c, respectively. The linguistic variables GN, PN, Z, PP and GP indicate negative big, negative small, zero, positive small and positive big. The number and the type of the membership function used for the system variables are determined through a trial and error test. The obtained fuzzy output variables are then processed by an inference engine. A sum-prod inference algorithm is adopted in this work. Based on the input membership functions number, the number of rules is obtained. The if–then rules that map input to output are conceived as indicated in Table 2. At the defuzzification step, the inference engine output variable is converted into a crisp value. The centroid defuzzification algorithm is used in this paper. The control signal to be applied in the real system is obtained using a recurrent equation, as indicated in (30).

$$\alpha(k) = \alpha(k+1) + AC_\alpha(k) \tag{30}$$

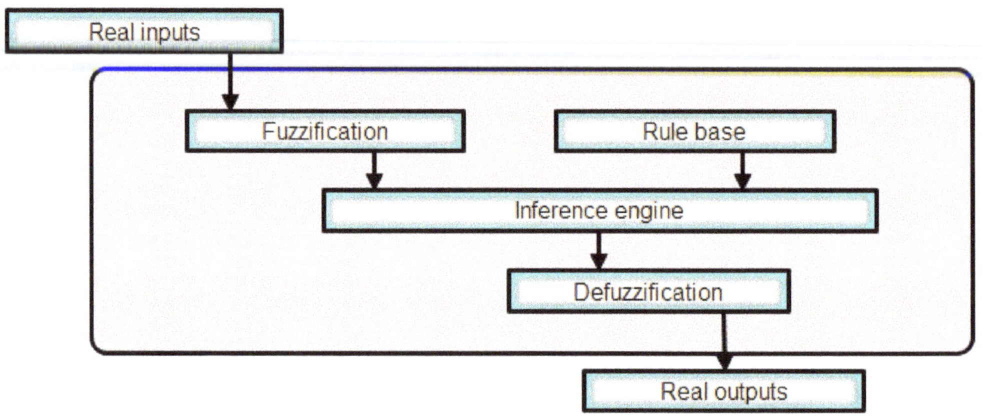

Figure 4. Schematic of the fuzzy logic controller.

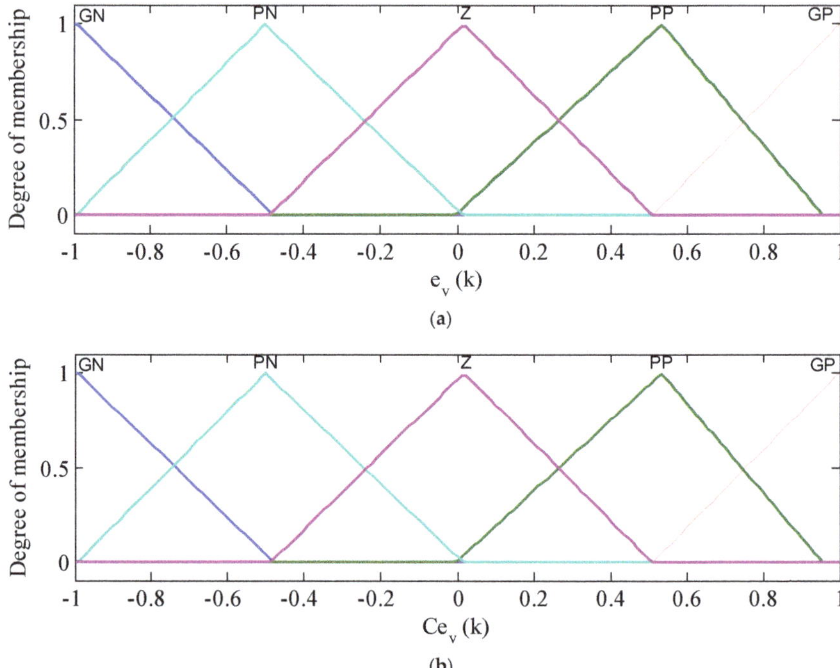

Figure 5. *Cont.*

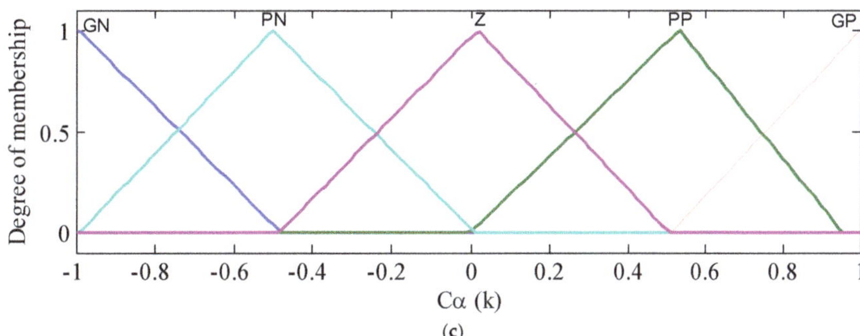

Figure 5. Fuzzy logic controller membership functions: (**a**) error membership function, (**a**) change error membership function and (**b**) evolution of target, and (**c**) change duty cycle membership function.

Table 2. Rule-based table of the fuzzy logic controller.

$C_\alpha(k)$		$Ce_v(k)$				
		GN	**PN**	**Z**	**PP**	**GP**
	GN	GN	GN	PN	PN	Z
	PN	GN	PN	PN	Z	PP
$e_v(k)$	Z	GN		Z	PP	GP
	PP	PN	Z	PP	PP	GP
	GP	Z	PP	PP	GP	GP

A is adjustable positive gain.

6. Simulation Results

To highlight the effectiveness and robustness of the proposed output voltage controller for the DC-DC buck converter, the overall drive scheme illustrated in Figure 6 was implemented in the Matlab/Simulink environment. Different scenarios were simulated in which the conceived algorithm was evaluated in comparison with four control algorithms: PI, IP, FLC, and IMC. Five scenarios, including abrupt target output voltage variation, triangular target output voltage variation, abrupt input voltage variation, abrupt load variation, and DC-DC buck converter parameter variation, are considered.

6.1. Sliding Mode Parameter Choice

The performance of the conceived sliding mode controller depends on the discontinuous term coefficient. A simulation study was carried out to determine the best choice of this factor. Figure 7 shows the obtained results. It should be noted that high sliding mode parameters lead to reduced response times both for the system and the controller outputs as shown in the zones 1 to 4. On the other hand, it increases the magnitude of oscillation both in the controller and the system responses, which may lead to disrupting the system. Thus, in this work, the considered parameters are chosen in such a way that a compromise between rapidity and stability is achieved.

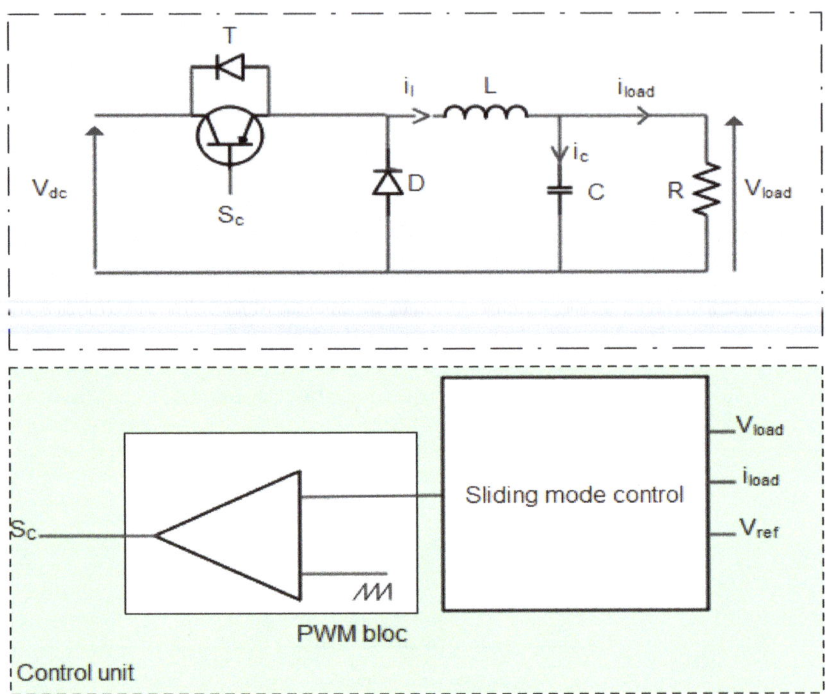

Figure 6. Closed loop control of buck control using sliding mode control approach.

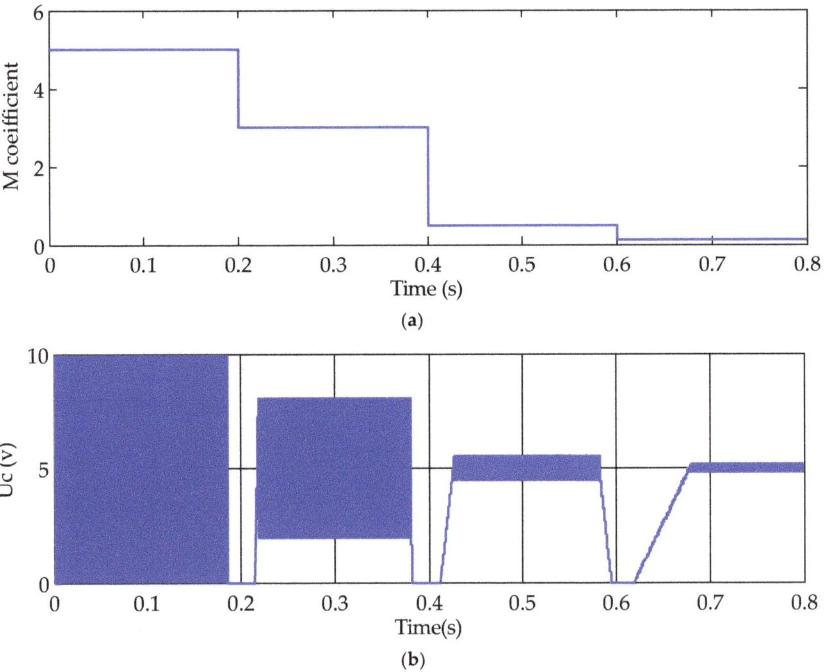

Figure 7. *Cont.*

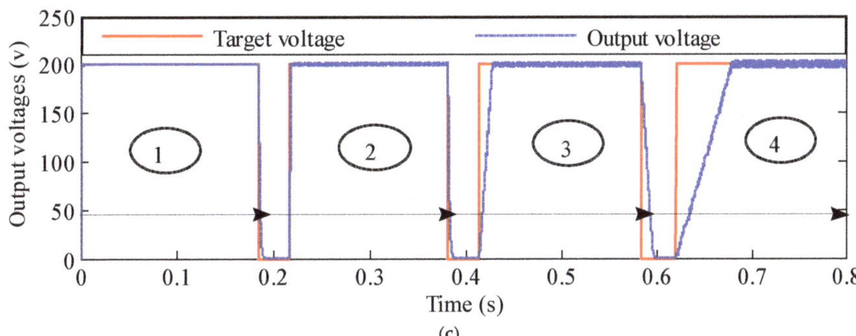

Figure 7. Discontinuous function coefficient effect: (**a**) evolution of M coefficient, (**b**) evolution of the control voltage and (**c**) evolution of the output voltages.

6.2. Controller's Behavior under Abrupt Target Output Voltage Variations

In this case, the aim is to test the tracking behavior of the proposed sliding mode controller against abrupt target output voltage variations. All five algorithms are implemented and simulated in the same test conditions. In fact, the load resistor, the input voltage, and the DC-DC buck converter parameters are all maintained at their nominal values. The target output voltage is first fixed at 150 v. At t = 0.03 s, it changes to 350 v and decreases to 250 v at t = 0.07 s. The five obtained output voltages and control signals are respectively reported in Figure 8a,b. Comparative performances are extracted and summarized in Table 3.

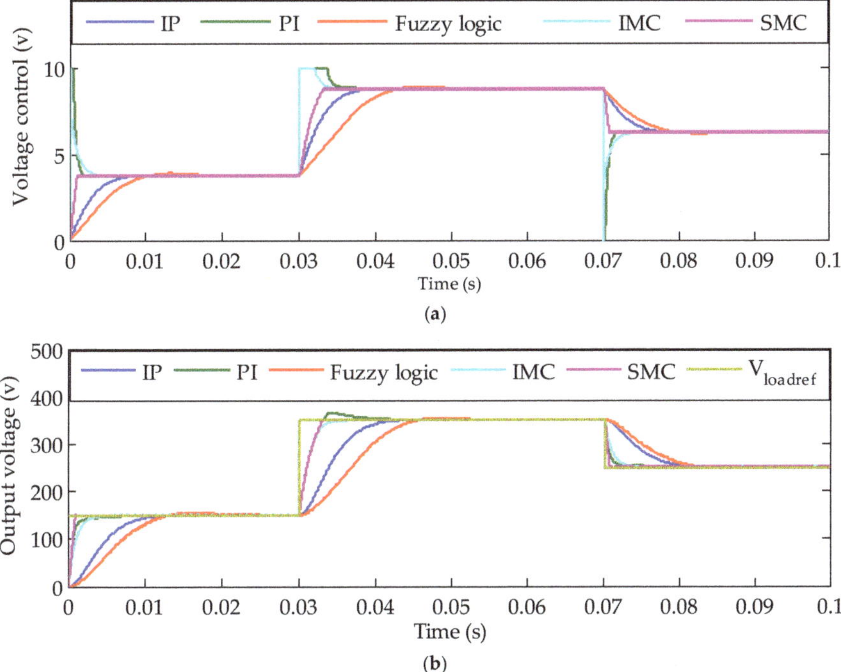

Figure 8. Target output voltage variation: (**a**) evolution of the control voltage and (**b**) evolution of the output voltages.

Table 3. Comparative study between the five algorithms.

Operating Modes	Values	Parameters	Control Algorithms				
			PI	IP	FLC	IMC	SMC
Abrupt target output variation	150 v	Response time (ms)	9.58	2.62	1.5	2.8	0.91
		Tracking error (%)	0.013	0.016	0.13	0.12	0.003
		Overshoot (%)	$\cong 0$	0.266	3.26	$\cong 0$	1.6
	350 v	Response time (ms)	8.5	2.6	2.5	2.8	2.6
		Tracking error (%)	2.8×10^{-3}	2.9×10^{-4}	1.2×10^{-3}	1.4×10^{-3}	7.1×10^{-5}
		Overshoot (%)	4.48	0.17	1.17	0.014	0.002
	250 v	Response time (ms)	1.4	7.8	9.2	2.1	0.7
		Tracking error (%)	0.33	0.01	0.04	0.02	0.001
		Voltage loss (%)	$\cong 0$	0.11	0.44	$\cong 0$	0.8
Triangular target output variation	20×10^3 v/s	Tracking error (v)	71.5	15	77	20	−14.5
Abrupt input voltage variation	400 v	Settling time	1.92	9.58	13.4	2.8	1.3
		Tracking error	0.025	0.04	0.1	0.175	5×10^{-5}
		Overshoot	$\cong 0$	0.25	2	$\cong 0$	0.675
	350 v	Stabilization time (ms)	2.1	9.8	1.3	8	$\cong 0$
		Tracking error (%)	2×10^{-4}	0.001	$\cong 0$	0.1	$\cong 0$
		Voltage loss (%)	2.25	6.65	8.3	3.55	0.001
	300 v	Stabilization time (ms)	18	22	23.8	10.9	$\cong 0$
		Tracking error	0.04	0.01	$\cong 0$	0.1	$\cong 0$
		Voltage loss (%)	3	8	9.75	4.25	$\cong 0$
Abrupt load resistor variation	15 Ω	Settling time (ms)	2.4	3.24	13.6	3.7	0.89
		Tracking error (%)	0.1	1.014	0.2	0.15	0.05
		Overshoot (%)	$\cong 0$	$\cong 0$	0.45	0.1	1.7
	10 Ω	Stabilization time (ms)	3.2	6.8	2	8.77	0.8
		Tracking error (%)	0.15	0.12	0.13	0.05	5×10^{-5}
		Overshoot (%)	7.25	12.4	13.75	10.2	$\cong 0$
		Voltage loss (%)	1.15	26.2	31.4	28.6	26.05
	5 Ω	Settling time (ms)	2.4	9.98	11.5	2.19	1.5
		Tracking error (%)	0.001	0.006	0.125	0.025	$\cong 0$
		Overshoot (%)	7.1	12.15	13.15	10.45	$\cong 0$
		Voltage loss (%)	46.7	48.35	48.6	46.8	46.45
Parameter variations	Abrupt capacitor variation	Stabilization time (ms)	-	-	13.7	13.9	$\cong 0$
		Tracking error (%)	-	-	0.1	0.15	$\cong 0$
		Overshoot/Voltage loss (%)	-	-	12.1	30	$\cong 0$

Table 3. *Cont.*

Operating Modes	Values	Parameters	Control Algorithms				
			PI	IP	FLC	IMC	SMC
	Abrupt inductor variation	Stabilization time (ms)	-	-	0.11	0.12	$\cong 0$
		Tracking error (%)	-	-	5×10^{-5}	0.05	$\cong 0$
		Overshoot/Voltage loss (%)	-	-	0.25	0.23	5×10^{-4}
Number of controller tuning parameters			02	02	03	01	01

6.3. Controller's Behavior under Triangular Target Output Voltage Variations

In order to access the dynamic response of the proposed control algorithm, the target output is rapidly changed as a triangular signal is chosen for the five control algorithms. The obtained results are obtained as recorded in Figure 9. Comparative performances in this case are extracted and grouped in Table 3.

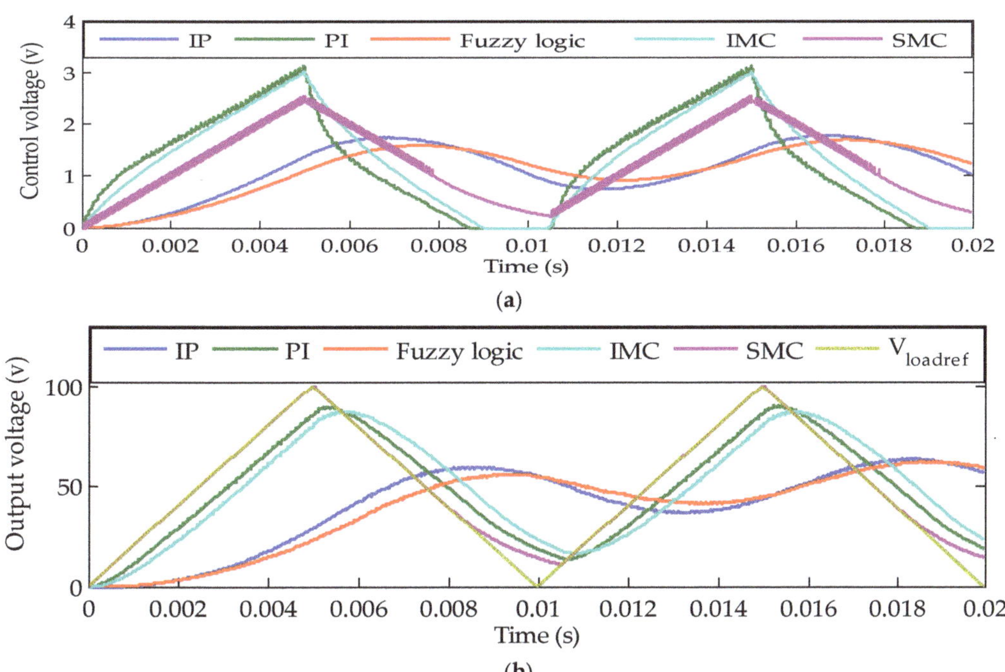

Figure 9. Triangular target output voltage variation: (**a**) evolution of the control voltage and (**b**) evolution of the output voltages.

6.4. Controller's Behavior under Input Voltage Variations

The DC input voltage variations for the five control algorithms are represented in Figure 10. For this test case, the load resistor and the DC-DC buck converter parameters are all maintained at their nominal values. The DC input voltage is fixed at 400 v, and decreased to 350 v and 300 v at t = 0.05 s and t = 0.1 s, respectively. Comparative performances are extracted as represented in Table 3.

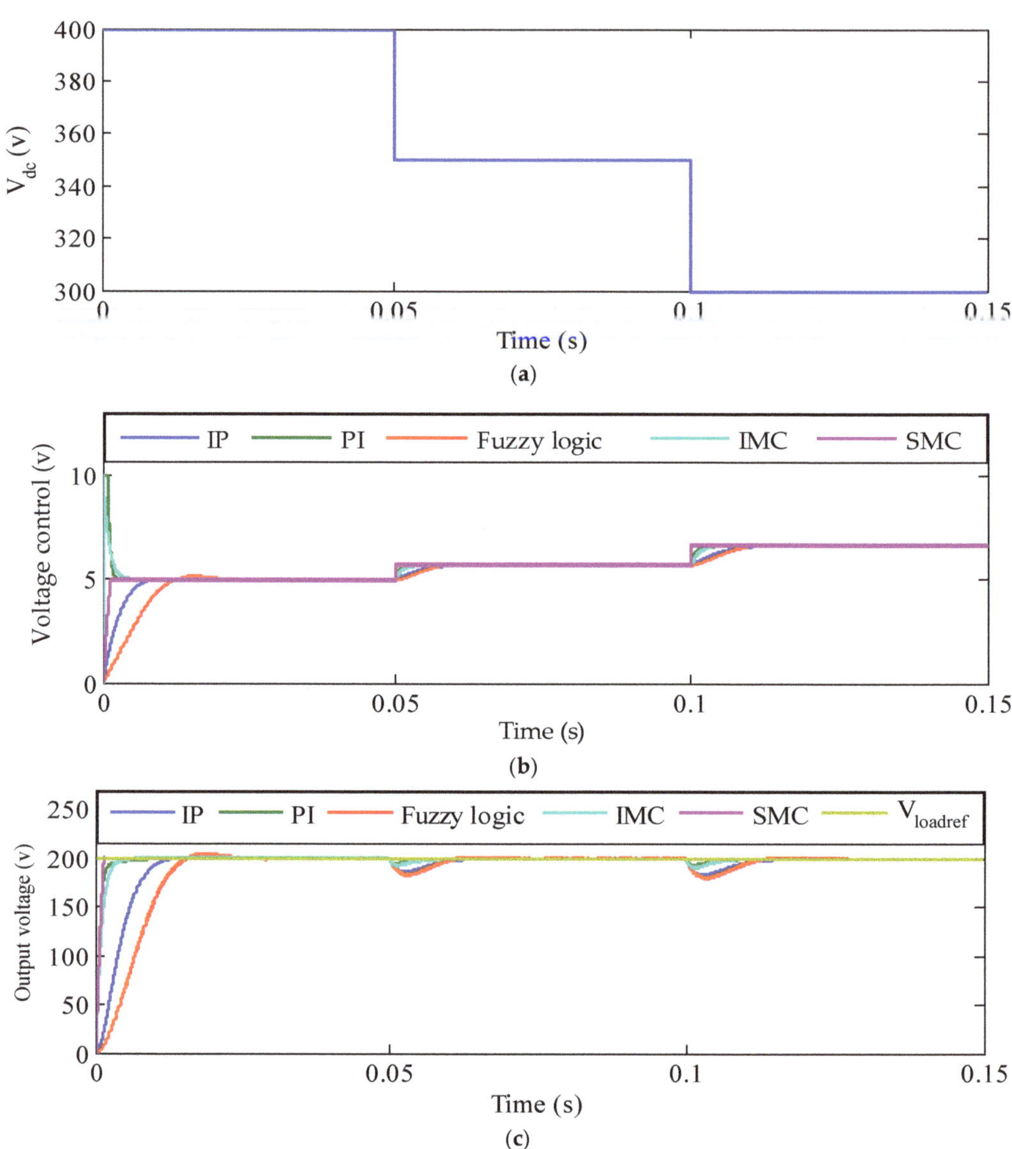

Figure 10. Input voltage variation: (**a**) evolution of the control voltage, (**b**) evolution of the control voltage and (**c**) evolution of the control voltage.

6.5. Controller's Behavior under Resistor Load Variations

In this test, the input DC voltage and the DC-DC buck converter are held constant at their nominal values. The target output voltage is also fixed to a constant value. The load variation trajectory is given in Figure 11a. The obtained responses of the five algorithms are shown in Figure 11b,c. Table 3 gives the comparison performances in this case.

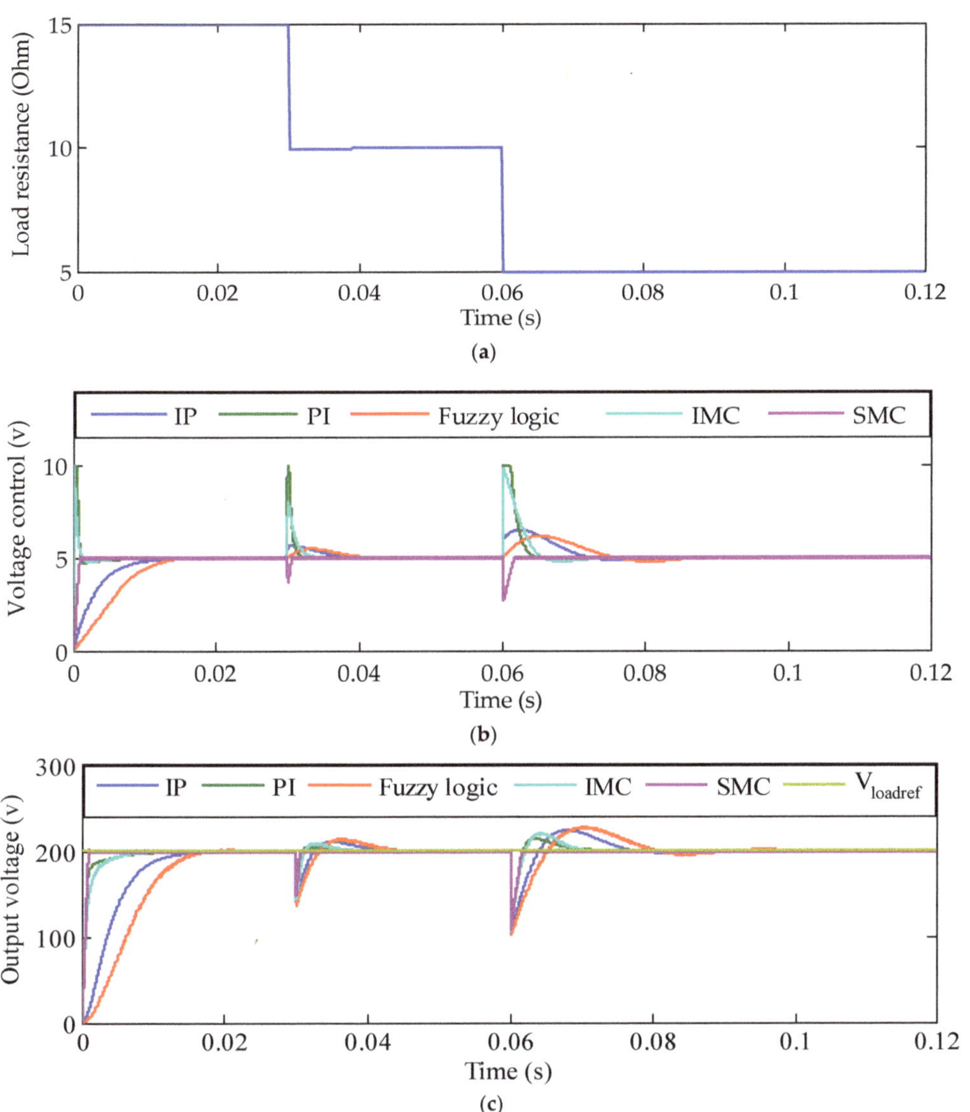

Figure 11. Resistor load variation: (**a**) evolution of the control voltage and (**b**) evolution of target and (**c**) actual output voltages.

6.6. Controller's Behavior under DC-DC Buck Converter Parameter Variations

As is well known, the PI and IP controllers are sensitive to system parameter variations. Thus, in this case, only the behavior of FLC, IMC, and the proposed SMC algorithms are tested. The DC input voltage, the load resistor, and the target output voltage are all held constant. Only the DC-DC buck converter parameter variations are considered in this case. Figure 12a,b show the adopted trajectories for the DC-DC buck converter inductor and capacitor, respectively. The obtained simulation results in this case are given in Figure 12c,d. The comparison performances of the three control algorithms in this case are summarized in Table 3.

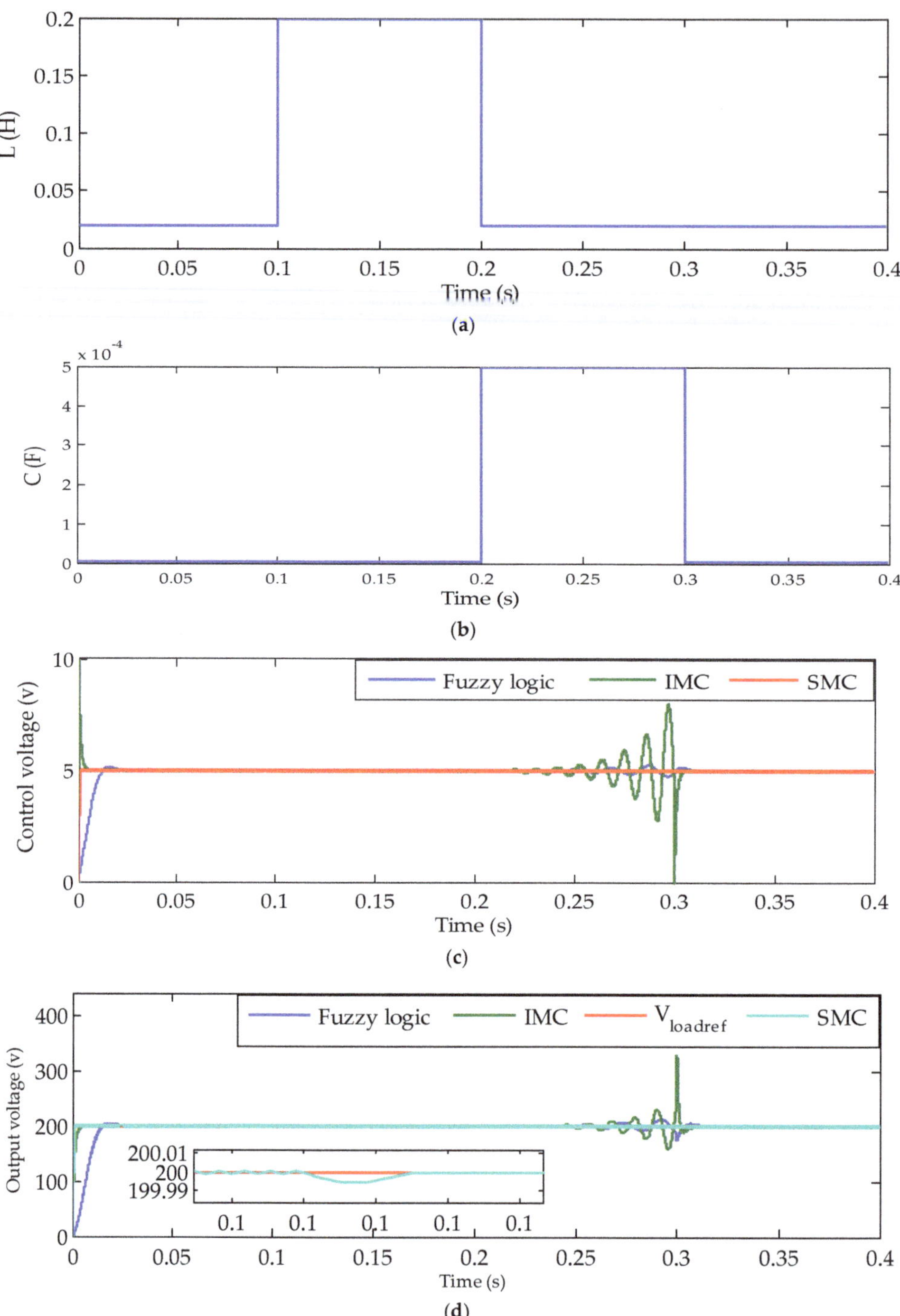

Figure 12. DC-DC buck converter parameters variations: (**a**) trajectory of the inductor, (**b**) trajectory of capacitor, (**c**) control voltage and (**d**) output voltage.

It can be easily noted from both the obtained simulation results for different cases and the comparative study shown in Table 3 that the highest performance is achieved by the conceived SMC control algorithm, compared to the others.

7. Conclusions

In this paper, a mathematical model of the buck DC-DC converter is established. A robust control strategy is adopted for the output control voltage of the system. The validity of the latter is demonstrated using the Matlab/Simulink environment. In a comparative study with four control algorithms, PI, IP, FLC and IMC, the simulation results show that the designed sliding mode controller has robust characteristics and a fast dynamic response in the different studied cases.

Despite all these good performances, the chattering phenomenon remains the major problem of the used sliding mode control algorithm. To avoid this issue, we intend to use high-order sliding mode control for buck DC-DC converter control in future works.

Author Contributions: Supervision, M.B.H.; Validation, L.S.; Writing—review and editing, S.B.H. All authors have read and agreed to the published version of the manuscript.

Funding: This research received no external funding.

Institutional Review Board Statement: Not applicable.

Informed Consent Statement: Not applicable.

Data Availability Statement: Not applicable.

Conflicts of Interest: The authors declare no conflict of interest.

References

1. Roncancio, J.S.; Vuelvas, J.; Patino, D.; Correa-Flórez, C.A. Flower greenhouse energy management to offer local flexibility markets. *Energies* **2022**, *15*, 4572. [CrossRef]
2. Turoń, K.; Kubik, A.; Chen, F. What Car for Car-Sharing? Conventional Electric, Hybrid or Hydrogen Fleet? Analysis of the Vehicle Selection Criteria for Car-Sharing Systems. *Energies* **2022**, *15*, 4344. [CrossRef]
3. Fady, M. Urban Freight Transport Electrification in Westbank, Palestine: Environmental and Economic Benefits. *Energies* **2022**, *15*, 4058.
4. Sun, L.; Zhang, T.; Liu, S.; Wang, K.; Rogers, T.; Yao, L.; Zhao, P. Reducing energy consumption and pollution in the urban transportation sector: A review of policies and regulations in Beijing. *J. Clean. Prod.* **2021**, *285*, 125339. [CrossRef]
5. Streltsov, A.; Malof, J.M.; Huang, B.; Bradbury, K. Estimating residential building energy consumption using overhead imagery. *Appl. Energy* **2020**, *280*, 116018. [CrossRef]
6. Bandara, W.C.; Godaliyadda, G.M.R.I.; Ekanayake, M.P.B.; Ekanayake, J.B. Coordinated photovoltaic re-phasing: A novel method to maximize renewable energy integration in low voltage networks by mitigating network unbalances. *Appl. Energy* **2020**, *285*, 116022. [CrossRef]
7. Guo, S.; Liu, Q.; Sun, J.; Jin, H. A review on the utilization of hybrid renewable energy. *Renew. Sustain. Energy Rev.* **2018**, *91*, 1121–1147. [CrossRef]
8. Nie, P.; Roccotelli, M.; Fanti, M.P.; Ming, Z.; Li, Z. Prediction of home energy consumption based on gradient boosting regression tree. *Energy Rep.* **2021**, *7*, 1246–1255. [CrossRef]
9. Mofidipour, E.; Babaelahi, M. New procedure in solar system dynamic simulation, thermodynamic analysis, and multi-objective optimization of a post-combustion carbon dioxide capture coal-fired power plant. *Energy Convers. Manag.* **2020**, *224*, 113321. [CrossRef]
10. Aboagye, B.; Gyamfi, S.; Ofosu, E.A.; Djordjevic, S. Status of renewable energy resources for electricity supply in Ghana. *Sci. Afr.* **2020**, *11*, e00660. [CrossRef]
11. IEA. *World Energy Outlook 2015*; OECD/IEA: Paris, France, 2015.
12. NREL. *Photovoltaic System Pricing Trends: Historical, Recent, and Near-Term Projections*; National Renewable Energy Laboratory: Golden, CO, Canada, 2015.
13. Panwar, N.L.; Kaushik, S.C.; Kothari, S. Role of renewable energy sources in environmental protection: A review. *Renew. Sustain. Energy Rev.* **2011**, *15*, 1513–1524. [CrossRef]
14. Nassar, Y.F.; Abdunnabi, M.J.; Sbeta, M.N.; Hafez, A.A.; Amer, K.A.; Ahmed, A.Y.; Belgasim, B. Dynamic analysis and sizing optimization of a pumped hydroelectric storage-integrated hybrid PV/Wind system: A case study. *Energy Convers. Manag.* **2021**, *229*, 113744. [CrossRef]

15. Leirpoll, M.E.; Næss, J.S.; Cavalett, O.; Dorber, M.; Hu, X.; Cherubini, F. Optimal combination of bioenergy and solar photovoltaic for renewable energy production on abandoned cropland. *Sol. Energy* **2021**, *168*, 45–56. [CrossRef]
16. Tervo, E.J.; Callahan, W.A.; Toberer, E.S.; Steiner, M.A.; Ferguson, A.J. Solar thermoradiative-photovoltaic energy conversion. *Cell Rep. Phys. Sci.* **2020**, *1*, 100258. [CrossRef]
17. Mellit, A.; Rezzouk, H.; Messai, A.; Mejahed, B. FPGA–based real time implementation of MPPT controller for photovoltaic systems. *Renew. Energy* **2011**, *36*, 1652–1661. [CrossRef]
18. Houssamo, I.; Locment, F.; Sechilariu, M. Maximum power tracking for photovoltaic power system: Development and experimental comparison of two algorithms. *Renew. Energy* **2010**, *35*, 2381–2387. [CrossRef]
19. Goudan, N.A.; Peter, S.A.; Nallandula, H.; Krithiga, S. Fuzzy logic controller with MPPT using line commutated inverter for three phase grid connected photovoltaic systems. *Renew. Energy* **2009**, *34*, 900–915.
20. Ko, S.H.; Chao, R.M. Photovoltaic dynamic MPPT on a moving vehicle. *Sol. Energy* **2012**, *86*, 1750–1760. [CrossRef]
21. Dawoud, S.M. Developing different hybrid renewable sources of residential loads as a reliable method to realize energy sustainability. *Alex. Eng. J.* **2021**, *60*, 2435–2445. [CrossRef]
22. Debastiani, G.; Nogueira, C.E.C.; Acorci, J.M.; Silveira, V.F.; Siqueira, J.A.C.; Baron, L.C. Assessment of the energy efficiency of a hybrid wind-photovoltaic system for Cascavel. *Renew. Sustain. Energy Rev.* **2020**, *131*, 110013. [CrossRef]
23. Loukil, K.; Abbes, H.; Abid, H.; Abid, M.; Toumi, A. Design and implementation of reconfigurable MPPT fuzzy controller for photovoltaic systems. *Ain Shams Eng. J.* **2019**, *11*, 319–328. [CrossRef]
24. Padmanaban, S.; Priyadarshi, N.; Bhaskar, M.S.; Holm-Nielsen, J.B.; Hossain, E.; Azam, F. A hybrid photovoltaic fuel cell for grid integration with Jaya-based maximum power point tracking:experimental performance evaluation. *IEEE Access* **2019**, *7*, 82978–82990. [CrossRef]
25. Priyadarshi, N.; Padmanaban, S.; Maroti, P.K.; Sharma, A. An extensive practical investigation of FPSO-based MPPT for grid integrated PV system under variable operating conditions with anti-islanding protection. *IEEE Syst. J.* **2019**, *13*, 1861–1871. [CrossRef]
26. Shahid, H.; Kamran, M.; Mehmood, Z.; Saleem, M.Y.; Mudassar, M.; Haider, K. Implementation of the novel temperature controller and incremental conductance MPPT algorithm for indoor photovoltaic system. *Sol. Energy* **2018**, *163*, 235–242. [CrossRef]
27. Bouchama, Z.; Khatir, A.; Benaggoune, S.; Harmas, M.N. Design and experimental validation of an intelligent controller for dc–dc buck converters. *J. Frankl. Inst.* **2020**, *357*, 10353–10366. [CrossRef]
28. Batiyah, S.; Sharma, R.; Abdelwahed, S.; Alhosaini, W.; Aldosari, O. Predictive Control of PV/Battery System under Load and Environmental Uncertainty. *Energies* **2022**, *15*, 4100. [CrossRef]
29. Chen, L.; Zheng, P.; Gao, W.; Jiang, J.; Chang, J.; Wu, R.; Ai, C. Frequency Modulation Control of Hydraulic Wind Turbines Based on Ocean Used Wind Turbines and Energy Storage. *Energies* **2022**, *15*, 4086. [CrossRef]
30. Akram, N.; Khan, L.; Agha, S.; Hafeez, K. Global Maximum Power Point Tracking of Partially Shaded PV System Using Advanced Optimization Techniques. *Energies* **2022**, *15*, 4055. [CrossRef]
31. Ge, Q.; Li, Z.; Sun, Z.; Xu, J.; Long, H.; Sun, T. Low Resistance Hot-Spot Diagnosis and Suppression of Photovoltaic Module Based on I-U Characteristic Analysis. *Energies* **2022**, *15*, 3950. [CrossRef]
32. Olalla, C.; Rodriguez, M.; Clement, D.; Maksimovic, D. Architectures and Control of Sub module Integrated DC-DC Converters for Photovoltaic Applications. *IEEE Trans. Power Electron.* **2012**, *28*, 2980–2997. [CrossRef]
33. Pilawa-Podgurski, R.C.N.; Perreault, D.J. Submodule integrated distributed maximum power point tracking for solar photovoltaic applications. *IEEE Trans. Power Electron.* **2013**, *28*, 2957–2967. [CrossRef]
34. Carbone, R.; Borrello, C. Experimenting with a Battery-Based Mitigation Technique for Coping with Predictable Partial Shading. *Energies* **2022**, *15*, 4146. [CrossRef]
35. Aluko, A.; Buraimoh, E.; Oni, O.E.; Davidson, I.E. Advanced Distributed Cooperative Secondary Control of Islanded DC Microgrids. *Energies* **2022**, *15*, 3988. [CrossRef]
36. Guldemir, H. Study of sliding mode control of DC-DC buck converter. *Energy Power Eng.* **2011**, *3*, 401–406. [CrossRef]
37. He, Y.; Luo, F.L. Study of sliding mode control for DC-DC converters. In Proceedings of the 2004 International Conference on Power System Technology, Singapore, 21–24 November 2004.
38. Babes, B.; Boutaghane, A.; Hamouda, N.; Mezaache, M. Design of a Robust Voltage Controller for a DC-DC Buck Converter Using Fractional-Order Terminal Sliding Mode Control Strategy. In Proceedings of the 2019 International Conference on Advanced Electrical Engineering (ICAEE), Algiers, Algeria, 19–21 November 2019; pp. 1–6.
39. Qamar, M.A.; Feng, J.; Rehman, A.U.; Raza, A. Discrete time sliding mode control of DC-DC buck converter. In Proceedings of the 2015 IEEE Conference on Systems, Process and Control (ICSPC), Sunway, Malaysia, 18–20 December 2015; pp. 91–95.
40. Wang, Z.; Shihua, L.; Qi, L. Discrete-Time Fast Terminal Sliding ModeControl Design for DC–DC Buck Converters with Mismatched Disturbances. *IEEE Trans. Ind. Inform.* **2020**, *16*, 1204–1213. [CrossRef]
41. Dehri, K.; Nouri, A.S. A discrete repetitive adaptive sliding mode control for DC-DC buck converter. *Proc. Inst. Mech. Eng. I J.* **2021**, *235*, 1698–1708. [CrossRef]
42. Mobayen, S.; Bayat, F.; Lai, C.-C.; Taheri, A.; Fekih, A. Adaptive Global Sliding Mode Controller Design for Perturbed DC-DC Buck Converters. *Energies* **2021**, *14*, 1249. [CrossRef]
43. Rakhtala, S.M.; Yasoubi, M.; Hosseinnia, H. Design of second order sliding mode and sliding mode algorithms: A practical insight to DC-DC buck converter. *IEEE/CAA J. Autom. Sin.* **2017**, *4*, 483–497. [CrossRef]

44. Shi, S.; Xu, S.; Gu, J.; Min, H. Global High-Order Sliding Mode Controller Design Subject to Mismatched Terms: Application to Buck Converter. *IEEE Trans. Circuits Syst.* **2019**, *66*, 4840–4849. [CrossRef]
45. Chi, X.; Lin, F.; Wang, Y. Disturbance and Uncertainty-Immune Onboard Charging Batteries With Fuel Cell by Using Equivalent Load Fuzzy Logic Estimation-Based Backstepping Sliding-Mode Control. *IEEE Trans. Transp. Electrif.* **2021**, *7*, 1249–1259. [CrossRef]
46. Meng, Z.; Shao, W.; Tang, J.; Zhou, H. Sliding-Mode Control Based on Index control. Law for MPPT in Photovoltaic Systems. *CES Trans. Electr. Mach. Syst.* **2018**, *2*, 303–311. [CrossRef]
47. Roy, T.K.; Faria, F.; Ghosh, S.K.; Pramanik, A.H. Robust Adaptive Backstepping Sliding Mode Controller for a DC-DC Buck Converter Fed DC Motor. In Proceedings of the 2021 Joint 10th International Conference on Informatics, Electronics & Vision (ICIEV) and 2021 5th International Conference on Imaging, Vision & Pattern Recognition (icIVPR), Virtual, 16–20 August 2021; pp. 1–6.
48. Lim, S.; Kim, S.-K.; Kim, Y. Active Damping Injection Output Voltage Control with Dynamic Current Cut-Off Frequency for DC/DC Buck Converters. *Energies* **2021**, *14*, 6848. [CrossRef]
49. Argyrou, M.C.; Marouchos, C.C.; Kalogirou, S.A.; Christodoulides, P. Modeling a residential grid-connected PV system with battery–super capacitor storage: Control design and stability analysis. *Energy Rep.* **2021**, *7*, 4988–5002.
50. Abdelmalek, S.; Dali, A.; Bakdi, A.; Bettayeb, M. Design and experimental implementation of a new robust observer-based nonlinear controller for DC-DC buck converters. *Energy* **2020**, *213*, 118816. [CrossRef]
51. Abedi, A.; Rezaie, B.; Khosravi, A.; Shahabi, M. DC-bus voltage control based on direct Lyapunov method for a converter based stand-alone DC micro-grid. *Electr. Power Syst. Res.* **2020**, *187*, 106451. [CrossRef]
52. Ding, S.; Zheng, W.; Sun, J.; Wang, J. Second-Order Sliding-Mode Controller Design and Its Implementation for Buck Converters. *IEEE Trans. Ind. Inform.* **2018**, *14*, 1990–2000. [CrossRef]
53. Derbeli, M.; Charaabi, A.; Barambones, O.; Sbita, L. Optimal Energy Control of a PEM Fuel Cell/Battery Storage System. In Proceedings of the 10th International Renewable Energy Congress (IREC), Sousse, Tunisia, 26–28 March 2019; pp. 1–5.
54. Utkin, V. Variable structure systems with sliding modes. *IEEE Trans. Autom. Control* **1977**, *22*, 212–222. [CrossRef]
55. Mihoub, M.; Nouri, A.S.; Ben Abdennour, R. Real-time application of discrete second order sliding mode control to a chemical reactor. *Control Eng. Pract.* **2009**, *19*, 1089–1095. [CrossRef]
56. Utkin, V.I. *Sliding Mode in Control and Optimization*; Springer: Berlin/Heidelberg, Germany, 1992.
57. Levant, A. Sliding order and sliding accuracy in sliding mode control. *Int. J. Control* **1993**, *58*, 1247–1263. [CrossRef]
58. Sira-Ramirez, H. Structure at infinity, zero dynamics and normal forms ofsystems undergoing sliding motion. *Int. J. Syst. Sci.* **1988**, *21*, 665–674. [CrossRef]
59. Sira-Ramirez, H. Non-linear discrete variable structure systems in quasi-sliding mode. *Int. J. Control* **1991**, *54*, 1171–1187. [CrossRef]
60. Flah, A.; Sbita, L. A novel IMC controller based on bacterial foraging optimization algorithm applied to a high speed range PMSM drive. *Appl. Intell.* **2013**, *38*, 114–129. [CrossRef]
61. Zadeh, L.A. Fuzzy sets. *Inf. Technol. Control* **1965**, *8*, 338–353. [CrossRef]

Article

Hybrid MLI Topology Using Open-End Windings for Active Power Filter Applications

Abdullah M. Noman *, Abdulaziz Alkuhayli, Abdullrahman A. Al-Shamma'a and Khaled E. Addoweesh

Faculty of Engineering, Electrical Engineering Department, King Saud University, Riyadh 12372, Saudi Arabia
* Correspondence: anoman@ksu.edu.sa

Abstract: Different multilevel converter topologies have been presented for achieving more output voltage steps, hence improving system performance and lowering costs. In this paper, a hybrid multilevel inverter (MLI) topology is proposed for active-power-filter applications. The proposed MLI is a combination of two standard topologies: the cascaded H-bridge and the three-phase cascaded voltage source inverter. This configuration enhances the voltage levels of the proposed MLI while using fewer switches than typical MLI topologies. The proposed MLI was developed in the MATLAB/Simulink environment, and a closed-loop control technique was used to achieve a unity power factor connection of the PV modules to the grid, as well as to compensate for harmonics caused by nonlinear loads. To demonstrate that the configuration was working correctly and that the control was precise, the proposed MLI was constructed in a laboratory. A MicroLabBox real-time controller handled data acquisition and switch gating. The proposed topology was experimentally connected to the grid and the MLI was experimentally used as an active power filter to compensate for the harmonics generated due to nonlinear loads. This control technique was able to generating a sinusoidal grid current that was in phase with the grid voltage, and the grid current's total harmonic distortion was within acceptable limits. To validate the practicability of the proposed MLI, both simulation and experimental results are presented.

Keywords: multilevel inverter; hybrid MLI; cascaded two-level VSI; active power filter

Citation: Noman, A.M.; Alkuhayli, A.; Al-Shamma'a, A.A.; Addoweesh, K.E. Hybrid MLI Topology Using Open-End Windings for Active Power Filter Applications. *Energies* **2022**, *15*, 6434. https://doi.org/10.3390/en15176434

Academic Editors: Najib El Ouanjli, Saad Motahhir, Mustapha Errouha and Marco Pau

Received: 21 May 2022
Accepted: 29 July 2022
Published: 2 September 2022

1. Introduction

Multilevel inverter (MLI) topologies have attracted interest in a variety of applications, including grid-connected PV applications, static VAR compensators, and active-power-filter (APF) applications. MLIs have many advantages over conventional inverters, such as reducing the total harmonic distortion (THD) of the produced voltage and current, reducing the size of filters, reducing the switching frequency, and improving the inverter efficiency [1]. Many cascaded inverter topologies have been proposed in the literature [2]. Of these, cascaded H-bridge (CHB) topologies have advantages over other MLI topologies (e.g., layout simplicity, extreme modularity, and construction and control simplicity) because they are free of voltage-balance issues. Moreover, compared to other MLI topologies, CHB topologies use the fewest components at the same voltage levels [3,4]. Asymmetrical CHB topologies have been proposed in which the DC voltages are not symmetrical. The combination of these asymmetrical DC voltages results in the generation of increased voltage levels [5,6]. On the other hand, hybrid MLI topologies presented in the literature have been based on a three-phase voltage source inverter (VSI) unit with a CHB. In these topologies, H-bridge units are coupled to floating capacitors whose voltages must be carefully regulated. This complicates the control algorithm [7,8].

Recently, research into the development of APF applications has increased [9,10]. APFs are electronic devices that can be accurately used to eliminate the harmonic contents in a current. The harmonics in power systems can be created by nonlinear loads connected to the public grid. Some of these loads are computers with a switched-mode power supply, motor drives, and electronic light ballasts (fax machines, medical equipment, etc.). These harmonics may cause unwanted effects, such as heating of sensitive electrical equipment or of the generators and transformers, which leads to increased core loss and may cause transformer and equipment failure, random circuit breaker tripping, flickering lights, high neutral currents due to zero sequence harmonics, and conductor losses [11]. As a result, these harmonics must be eliminated. There are three different types of harmonic filters: passive, active, and hybrid. The advantages of APFs over passive filters are numerous. They can, for example, suppress supply-current harmonics as well as reactive currents. On the other hand, certain harmonics will involve their own passive filter, hence there will be more passive filters needed if there are more harmonics to remove. Because only large harmonics with lower frequencies are usually removed, the filters will be bulky due to the geometric size of the inductor [12]. In contrast to passive power filters, APFs can eliminate any harmonics by using a suitable control and the performance of APFs is unaffected by the characteristics of the power distribution system [9,11,13,14].

Several APF topologies using the conventional three-phase VSI have been proposed [15–17]. In addition, some of the proposed APFs were based on the classical three-level H-bridge inverter [18]. Recently, the use of MLI topologies in APF applications has become a hot topic [19,20]. An APF based on a seven-level neutral-point clamped (NPC) MLI was proposed in [21]. The authors used LS–PWM and fuzzy control approaches to eliminate current harmonics generated by nonlinear loads. Others [19] used a three-level NPC for active-power-filter applications. The authors used a fuzzy-logic controller and a fractional-order proportional-integral (PI) controller to control the proposed system. From another point of view, an APF based on a three level NPC-T type was proposed in [20,22]. The authors in [23] proposed a five-level HB-NPC MLI for APF applications with an experimental validation. Other authors used a CHB for APF applications due it the advantages of the CHB such as simplicity and possession of a high modularity [24,25]. Both CHN topologies were proposed for APFs either by directly connecting the H-bridges in series or by cascading the H-bridges using transformers [26–28]. Moreover, cascaded MLI topologies are used for APF applications [29,30]. The authors in [30] proposed a shunt APF based on cascaded MLI topology using a single power source with three-phase transformers. The proposed scheme could adjust for polluted loads with high harmonics and a poor power factor. To compute compensating currents, the dq theory was applied. A prototype was built to validate the simulation results by the experimental results.

In this paper, a hybrid MLI topology is proposed for APF applications. The proposed MLI topology conjoins the cascaded H-bridge MLI and the cascaded three-phase VSI topology using open-end windings and was first proposed and tested under different load parameters, as presented in [31]. In addition, this new topology has been patented, as seen in [32]. However, the previous paper focused only on the topology itself. The topology was not used in any applications. This paper is the first to detail the utilization of the new topology for the active power filter and for grid-connected PVs. The proposed MLI is used to compensate the harmonic currents of the nonlinear load, while the grid supplies the fundamental positive-sequence currents of the load. A closed-loop control is proposed for both applications—the APF and the PV–grid connection. Finally, the proposed MLI was experimentally implemented in the lab, and it was tested for the two applications.

The rest of the paper is structured as follows: the suggested APF topology and corresponding control approaches are discussed in Section 2, the simulation results are described in Section 3, and experimental validation of the proposed topology for the two applications is described in the Section 4.

2. The Proposed Hybrid Active-Power-Filter Topology

A three-phase multilevel high-voltage/high-power converter is used in this topology. It is a hybrid arrangement that combines two common multilevel configurations: a cascaded H-bridge MLI and a three-phase cascaded VSI. These two configurations are joined together to create the hybrid configuration. Figure 1 depicts the suggested cascaded

MLI architecture. This novel topology has been given a patent with the number US 10, 141, 865 B1. It is suitable for a wide variety of grid-connected applications, including power-factor correction, static-VAR compensation, and grid-connected photovoltaic (PV) systems. The proposed architecture is made up of two components linked by an open-end windings transformer.

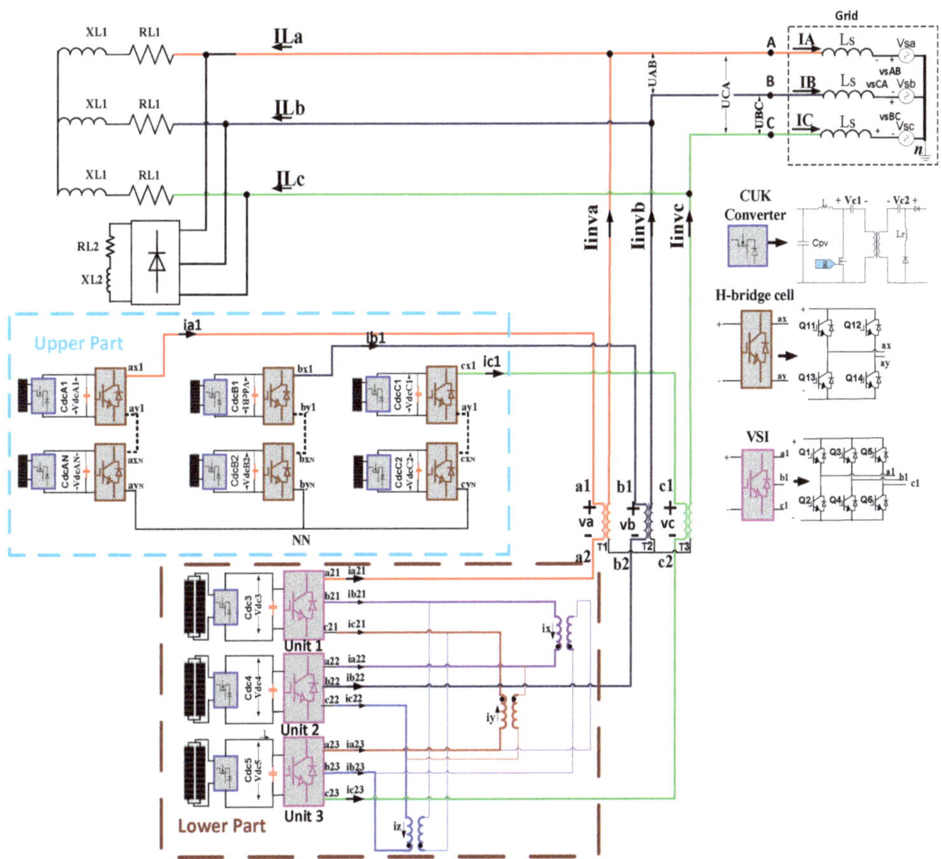

Figure 1. Proposed APF topology. Upper part: This part is the conventional cascaded H-bridge MLI which consists of three-phase systems. Each phase consists of N H-bridge cells that are cascaded as shown in the stage 1 of Figure 1 such that the a_{y1} terminal of H-bridge cell 1 in phase a, for example, is connected to the a_{x2} terminal of the next H-bridge cell. In addition, the $a_{y(N-1)}$ terminal is connected to the a_{xN} terminal of the last H-bridge cell. The same idea is applied to phases b and c. Lower part: This part is a three-phase triple-voltage source inverter, which is made up of three VSIs. Each unit is a three-leg, two-level inverter, and the three units are linked together in a chain, as shown in the lower part of Figure 1. The three VSIs are cascaded with one another by using open-end winding transformers. Finally: The upper part and the lower part are connected to each other via open-end windings. The terminals a_{yN}, b_{yN}, and c_{yN} of stage 1 are connected together to a common point NN. The terminals a_{x1}, b_{x1}, and c_{x1} of the upper part are connected to the points a_1, b_1, and c_1 of the open-end windings transformers, respectively. The three points a_{21}, a_{22}, and a_{23} of the lower part are connected to the points a_1, b_1, and c_1 of the open-end windings transformers, respectively. The secondary sides of the transformer are connected in Y connection and the terminals are connected to the grid and to the nonlinear loads.

As shown in this figure, the proposed MLI is capable of supplying the extracted PV power to the nonlinear loads and of injecting power into the grid. The idea is to compensate the harmonics generated by the nonlinear loads while supplying balanced three-phase currents to the grid. The control scheme proposed in [33] has been extended to be used for the proposed APF in this paper. The proposed control scheme is used to inject balanced three-phase low harmonic currents into the grid. If the three-phase loads are nonlinear, then the load currents will contain harmonics, which may reduce the current quality of the grid. Therefore, the harmonic contents of the load currents must be extracted to be compensated by the MLI. Consequently, the grid current will be balanced and contain low harmonic contents. The current of the nonlinear loads may contain positive-, negative-, or zero-sequence harmonics.

Therefore, the following steps are considered:

1. Extract the fundamental positive-sequence component of the load current using the following equation:

$$\begin{bmatrix} i_{Lzero} \\ i_{Lpos} \\ i_{Lneg} \end{bmatrix} = \frac{1}{3} \begin{bmatrix} 1 & 1 & 1 \\ 1 & a & a^2 \\ 1 & a^2 & a \end{bmatrix} \begin{bmatrix} i_{La} \\ i_{Lb} \\ i_{Lc} \end{bmatrix} \tag{1}$$

where $a = e^{j2\pi/3}$, i_{Lzero}, i_{Lpos}, and i_{Lneg} are the zero, positive, and negative components of the load currents, respectively.

2. The maximum value of the fundamental component of the load current is extracted, and the harmonic contents of the load currents can be extracted by the following equation:

$$\begin{bmatrix} i_{LA(harm)} \\ i_{LB(harm)} \\ i_{LC(harm)} \end{bmatrix} = i_{Lpos\_max} \begin{bmatrix} sin(\theta) \\ sin(\theta - 120^\circ) \\ sin(\theta + 120^\circ) \end{bmatrix} - \begin{bmatrix} i_{La} \\ i_{Lb} \\ i_{Lc} \end{bmatrix} \tag{2}$$

where $i_{LA(harm)}$, $i_{LB(harm)}$, and $i_{LC(harm)}$ are the extracted harmonic components of the load currents of phases a, b, and c, respectively.

3. The extracted harmonic components of the load currents are then converted into a dq frame as shown in Figure 2.

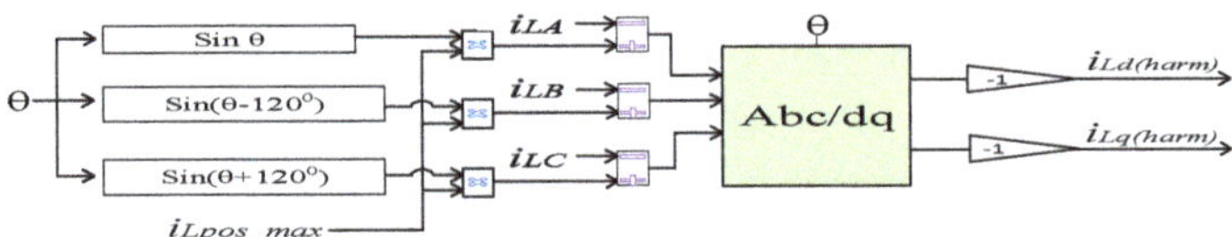

Figure 2. Extraction of the load harmonic currents.

From another perspective, the main goal of the control scheme is to produce reference currents that deliver only available active power to the grid while maintaining unity power factor. The DC links share the same active phase grid currents. These DC-link voltages are compared to their respective reference voltages.

If the DC-link voltages are accurately regulated by the control system, then:

$$V_{dcA} = V_{dcB} = V_{dcC} = V_{dctot} \tag{3}$$

where V_{dcA}, V_{dcB}, and V_{dcC} are the corresponding DC-link voltages of the phases a, b, and c, respectively.

The d-q components of the terminal voltages of the proposed MLI are given by:

$$V_d = L_s \frac{di_d}{dt} - (\omega L_s)i_q + d_{nd}V_{dctot}$$
$$V_q = L_s \frac{di_q}{dt} + (\omega L_s)i_d + d_{nq}V_{dctot} \tag{4}$$

The sequential function d_{nX} is given by:

$$\begin{bmatrix} d_{nA} \\ d_{nB} \\ d_{nC} \end{bmatrix} = \begin{bmatrix} C_A \\ C_B \\ C_C \end{bmatrix} - \frac{1}{3}(C_A + C_B + C_C) \tag{5}$$

The DC link current is defined as follows:

$$C_{dc}\frac{dV_{dctot}}{dt} = d_{nd}i_d + d_{nq}i_q \tag{6}$$

Analyzing nonlinearity problems requires presention of a new model. These inputs could be written as:

$$\begin{aligned} u_d &= (\omega L_s)i_q - d_{nd}V_{dctot} + V_d \\ u_q &= -(\omega L_s)i_d - d_{nq}V_{dctot} + V_q \end{aligned} \tag{7}$$

To ensure the unity power factor, the reactive current i_q in Equation (6) should be set to zero. Therefore,

$$u_{dc} = C_{dc}\frac{dV_{dctot}}{dt} = d_{nd}\frac{u_{dc}}{d_{nd}} \tag{8}$$

In normal operation the following properties apply:

$$V_d \approx d_{nd}V_{dctot} = d_{nd}V_{dctot} = \sqrt{\frac{3}{2}}V_{max} \tag{9}$$

where V_{max} is the maximum grid voltage. Substituting Equation (8) in (9) yields:

$$i_d = \sqrt{\frac{2}{3}}\frac{u_{dc}}{V_{max}}V_{dctot} \tag{10}$$

The active current, i_d, is used to regulate the DC-link capacitors. As shown in Equation (10), the active reference current can be given as:

$$i_{dA} = \sqrt{\frac{2}{3}}\frac{u_{dcA}}{V_{max}}V_{dcA} \tag{11}$$

The reference active current of the grid is the sum of the three i_d active currents:

$$i_{dref} = i_{dA} + i_{dB} + i_{dC} = \sqrt{\frac{2}{3}}\frac{u_{dcA}}{V_{max}}V_{dcA} + \sqrt{\frac{2}{3}}\frac{u_{dcB}}{V_{max}}V_{dcB} + \sqrt{\frac{2}{3}}\frac{u_{dcC}}{V_{max}}V_{dcC} \tag{12}$$

The DC-link voltage controllers of the proposed topology based on the analysis above can be seen in Figure 3. As seen in the analysis, the resulting signal from the DC-link controllers is the reference active current of the inverter, i_{dref}. The inner current controller can be seen in Figure 4.

Where the proposed MLI is used in a PV application, the reference active grid current i_{dref} is compared with the actual grid current, and the reference grid reactive current $\left(i_{qref} = 0\right)$ is compared with the actual reactive grid current i_q to ensure unity power factor. However, where the proposed MLI is used in an APF application, the harmonic load current in the d axes $i_{Ld(harm)}$ given from Figure 2 is then added to the reference current in the d axes i_{dref} generated form the DC-link controllers. The total reference inverter currents are given as:

$$\begin{aligned} i_{drefTOT} &= i_{dref} + i_{Ld(harm)} \\ i_{qrefTOT} &= i_{Lq(harm)} \end{aligned} \tag{13}$$

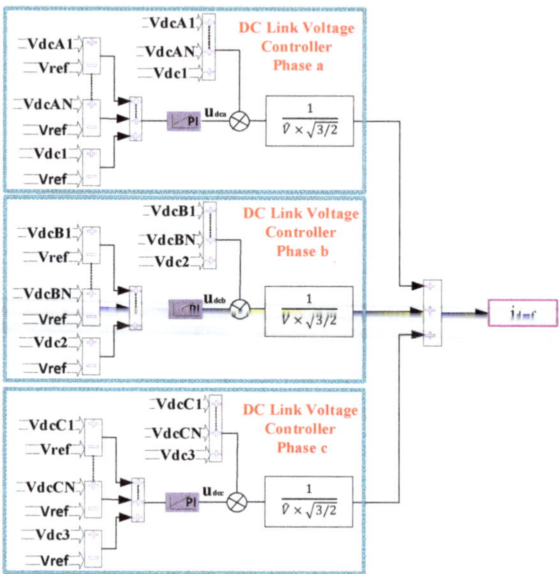

Figure 3. Proposed voltage control scheme.

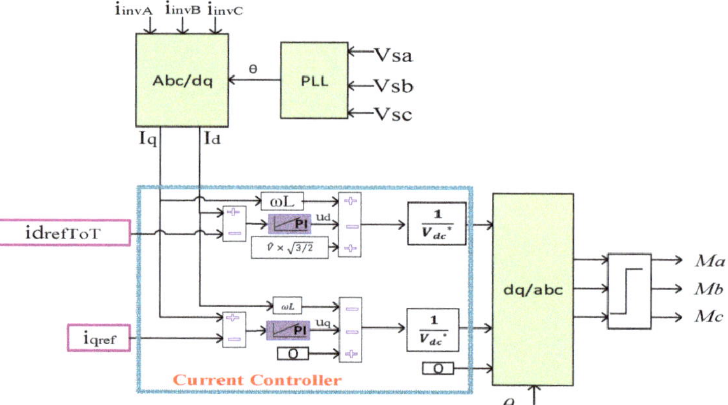

Figure 4. The current controllers. $V_{dc}{}^*$ is the reference DC voltage.

Equation (13) states that the reference active current of the inverter equals the active current produced by the DC-links and the active harmonic load current. The complete proposed control scheme for APF application can be seen in Figure 5.

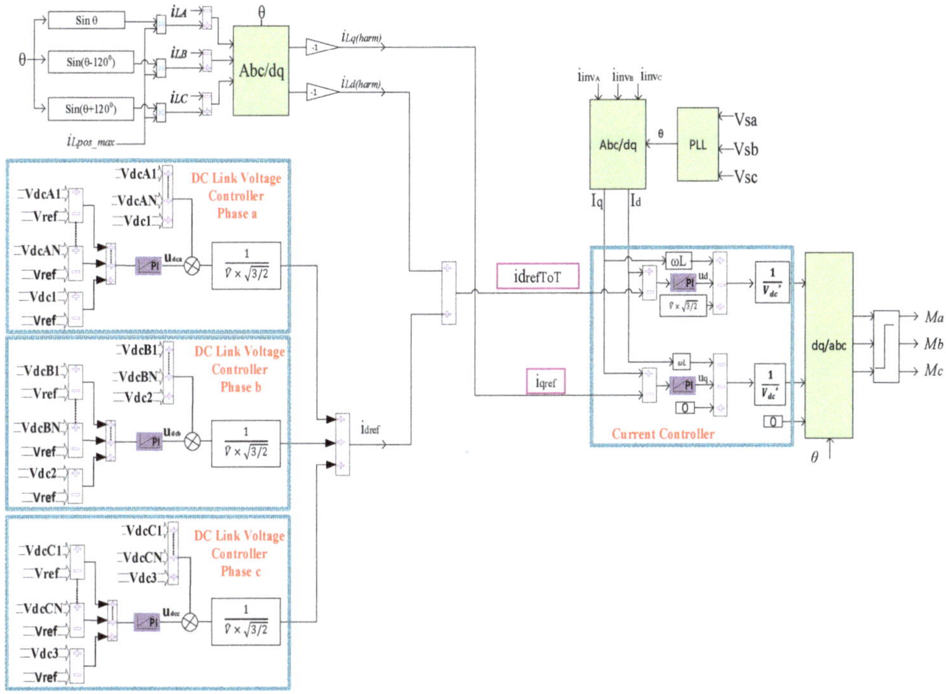

Figure 5. The proposed control scheme for APF application.

According to the proposed control scheme, two cases can be applied:

1. **The proposed inverter acts as an active filter only.**

   The reference currents of the inverter are the harmonic load currents. The inverter is responsible for compensating the harmonic load currents while the grid supplies the fundamental positive-sequence currents of the load.

2. **The proposed inverter is responsible for delivering the PV power to the grid and the load.**

   If there is enough power at the DC side, e.g., from the PV modules, the control system drives the inverter to supply the balanced positive-sequence current to the grid, as well as supplies the active load current and compensates harmonic load currents.

   In both cases, Equation (13) is applied to perform the current control stage. The load current harmonics are then transferred into the dq frame (Figure 2). These currents are then used as reference MLI currents, which are then used in Equation (13). The resulting signals from Figure 4, d_{nd} and d_{nq}, obtained from Equation (7), are then transformed into abc reference frame (M_a, M_b, and M_c). These values are then compared to the triangle carrier waveforms in Figure 6 to create switching pulses to operate the IGBTs of the proposed hybrid MLI. To produce the required pulses for the IGBTs of the upper part of the proposed MLI, the carrier waveform of the H_{2a} is shifted from the H_{1a} by 180°/N. Only the carrier waveforms of phase a in the upper part are displayed in Figure 5. Moreover, the modulation waveforms M_a, M_b, and M_c are also used to fire the IGBTs of the lower part in the proposed configuration by comparing them with the phase-shifted carrier signals, as seen in Figure 6. The carrier signal, which is used to generate the pulses of unit 2, is shifted by (T/3) from that of unit 1, and the carrier signal of unit 3 is shifted by (T/3) from that of unit 2.

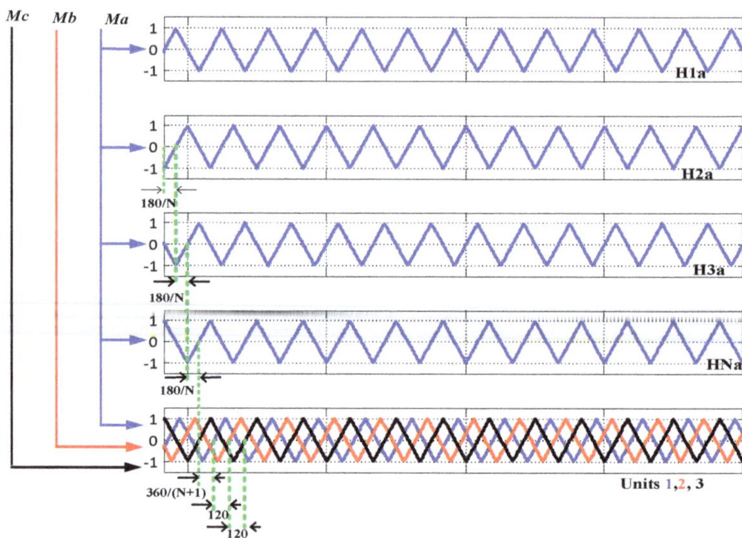

Figure 6. Phase-shifted PWM technique.

3. Simulation Results

The proposed topology was built in the SIMULINK environment. For simplicity, two H-bridge units were used per phase in the proposed topology. The system consisted of 12 HIT-N220A01 PV modules; six of them were connected to upper part of the proposed topology such that one PV module was linked to each H-bridge unit. The other six PV modules were connected to the lower part of the proposed MLI such that two parallel PV modules were linked to each VSI unit. Two cases were used in the simulation:

A. Grid-Connected PV Application Case

The system parameters used for simulation are shown in Table 1, while Table 2 shows the PV module parameters. The nonlinear load is disconnected from Figure 1. Therefore, no harmonic current term is applied in Equation (2). The reference reactive current of the inverter was set to zero to guarantee unity power factor. The proposed topology was used only to connect the PV modules to the grid. The perturb and observe (P & O) algorithm was used to extract the maximum power from the PV modules. The closed-loop control scheme shown in Figures 3 and 4 was used for grid-connection purposes.

Table 1. System parameter for simulation and experiment.

DC-Link Capacitor	4 mF
Interface inductor	4.2 mH
Ćuk switching frequency	20 kHz
Grid-rated RMS voltage	120 V
Reference voltage of each VSI unit (upper part)	55 V
Reference voltage of each H-bridge cell (lower part)	40 V
Inverter switching frequency	1.5 kHz

Table 2. The PV module parameters.

Maximum Power (P_{max})	245 W
Maximum voltage (V_{max})	28.8 V
Maximum current (I_{max})	8.5 A
Open circuit voltage (V_{oc})	31.5 V
Short circuit current (I_{sc})	9.5 A

The generated three-phase voltages were measured across the primary windings of the transformers T_1, T_2, and T_3. The number of voltage steps produced from the proposed MLI was 22, as shown in Figure 7. The three-phase currents injected into the grid can be seen in Figure 8. In addition, the harmonic spectrum of the line current, i_a, is displayed in Figure 9. The total harmonic distortion (THD) of the grid current, i_a, was 2.22%, which was less than the IEEE-519 standard limit of 5%.

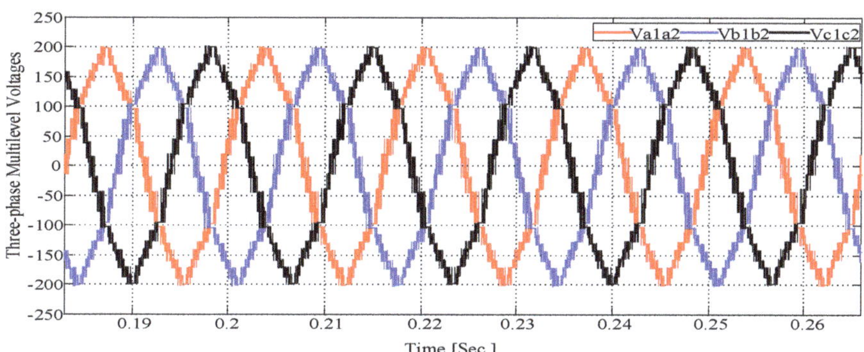

Figure 7. The simulated three-phase generating voltage.

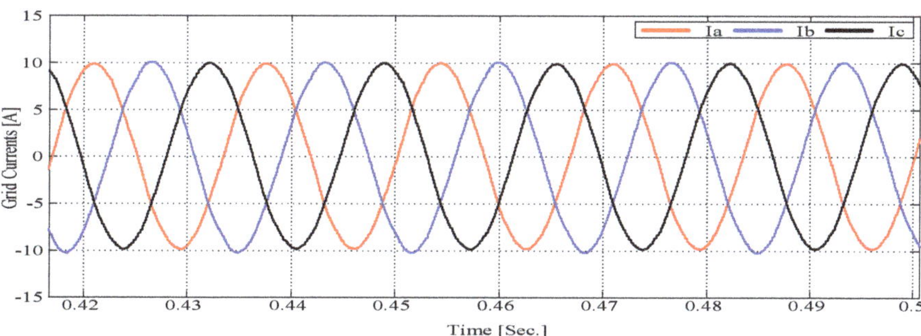

Figure 8. The simulated three-phase grid currents.

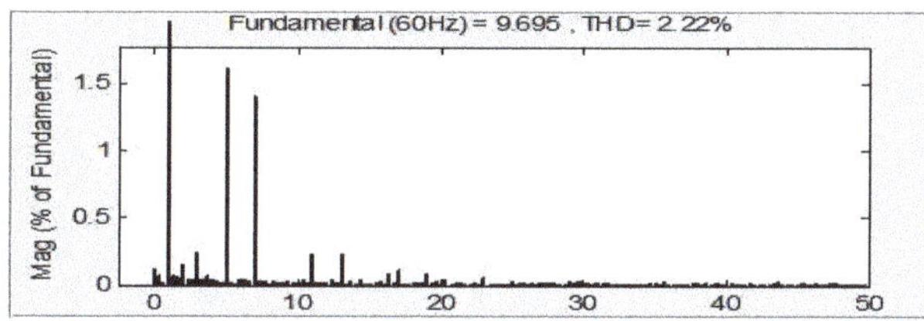

Figure 9. Harmonic spectrum of the phase a grid current.

The THD can be calculated according to the following equation:

$$THD = \frac{\sqrt{\sum_{h=2}^{\infty}(I_h)^2}}{I_1} \tag{14}$$

where h is the harmonic order, I_h is the RMS value of the current at h order, and I_1 is the RMS value of the fundamental current.

As seen in Figure 9, the first harmonic bands appear with an amplitude of 0.2% at two times the fundamental frequency. Then the third, fifth, and seventh harmonic orders appear with amplitudes of 0.3%, 1.6%, and 1.5%, respectively.

B. Active-Power-Filter Application Case

The closed-loop control scheme shown in Figure 5 is used for APF applications. In this case, the harmonic current term in Equation (2) will be included to be compensated by the proposed MLI. The PV module parameters are shown in Table 2, while the system parameters are seen in Table 3. As discussed in the previous section, the proposed control scheme can be used for APFs only or it is able to work as an APF and integrate the PVs into the grid at the same time. These two scenarios are investigated.

Table 3. System parameter for simulation and experiment (APF application).

DC-Link Capacitor	4 mF
Connection inductor	4.2 mH
Ćuk switching frequency	20 KHz
Grid-rated RMS voltage	80 V
Reference voltage of each VSI unit (upper part)	40 V
Reference voltage of each H-bridge cell (lower part)	40 V
Inverter switching frequency	1.5 KHz

<u>**Scenario 1:**</u> **The control scheme is able to work as an APF only:**

In this scenario, the grid is responsible for supplying the balanced three-phase currents to the nonlinear load, and the proposed MLI is controlled to only compensate for the harmonic currents of the load. The values of the nonlinear load parameters were ($R_{L1} = R_{L2} = 48\ \Omega$, $X_{L1} = X_{L2} = 154$ mH). To allow the grid to supply the active power to the load, the duty cycle of the DC–DC converter was kept constant at 0.3 (no MPPT was used). The DC-link voltages were regulated to 40 V. Table 3 presents the system parameters that were considered during the simulation. The nonlinear load currents can be seen in Figure 10a, while the three-phase low harmonized grid currents can be seen in Figure 10b. It should be noted that the grid current flowed from the grid to the nonlinear load. As revealed in Figure 10b, the grid currents were balanced, and the nonlinear load did not affect the grid currents. On the other hand, the harmonic currents extracted from the nonlinear load currents can be seen in Figure 11a. These currents were the reference currents of the proposed MLI. As revealed in Figure 11b, the MLI currents followed the extracted harmonic currents of the nonlinear load. Figure 12 shows that the control scheme was successful in maintaining the power-factor unity. The grid currents are shown out of phase with the grid voltages because the grid current direction was from the grid to the load.

<u>**Scenario 2:**</u> **The control scheme is able to work as an APF and integrate the PVs into the grid at the same time.**

In this scenario, 24 PV modules were used to guarantee that the inverter supplies the nonlinear load currents and at the same time inject currents into the grid. Twelve PV modules were connected to the upper part of the proposed MLI such that two parallel PV modules were connected to each H-bridge cell. On the other hand, the other 12 PV modules were connected to the lower part of the MLI such that four parallel PV modules were connected to each VSI unit. The proposed control scheme will generate the reference inverter currents that equal the grid current and the nonlinear load currents. The MLI will supply the distorted nonlinear load currents and at the same time will inject currents into the grid with low distortion.

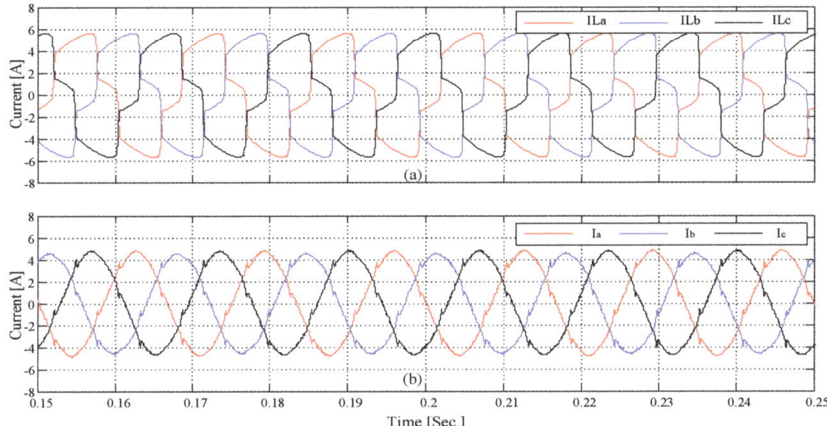

Figure 10. The simulated grid currents and the load currents.

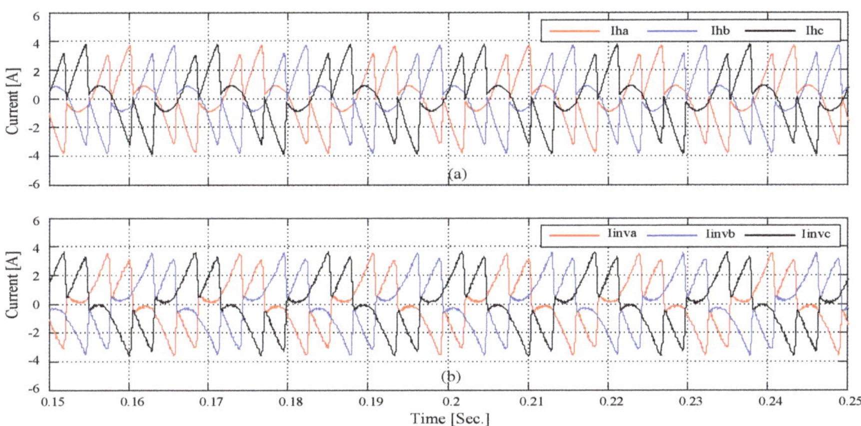

Figure 11. The simulated harmonic load currents and the active power filter currents.

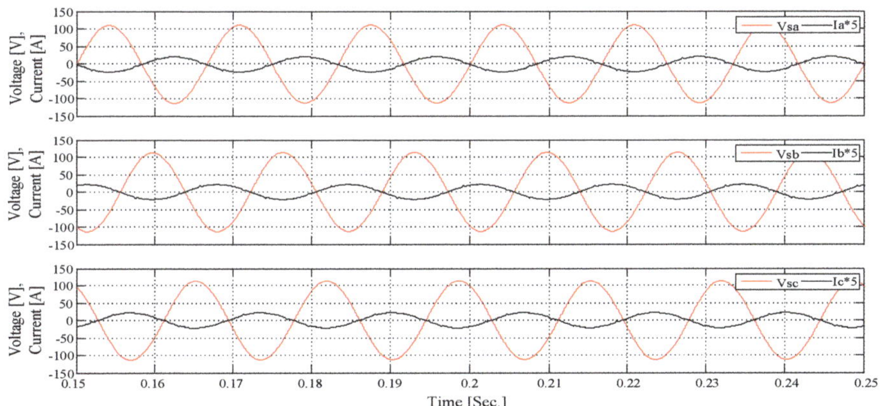

Figure 12. The simulated grid currents with the grid voltages.

Figure 13a shows the extracted harmonic currents of the nonlinear load. In addition, the actual MLI currents are seen in Figure 13b. As can be seen, the inverter currents were not a mirror to the load harmonic load current. The inverter currents contained the nonlinear load currents and the injected currents to the grid. Moreover, Figure 14a shows the nonlinear load currents while Figure 14b shows the grid currents.

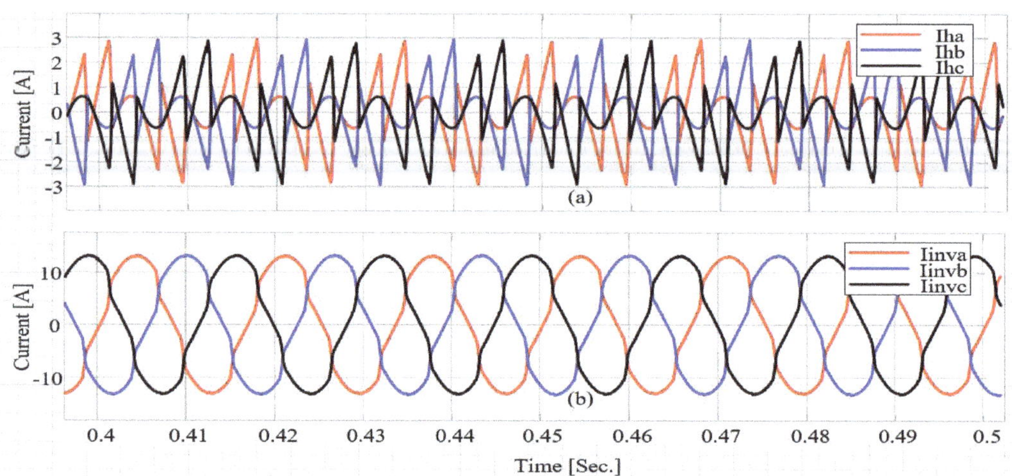

Figure 13. (a) The extracted harmonic currents of the nonlinear load and (b) the MLI currents.

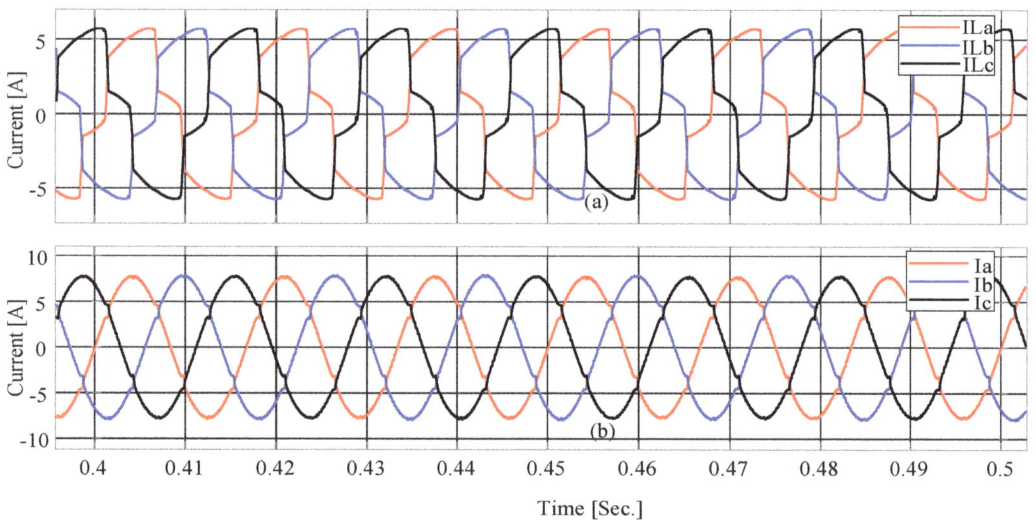

Figure 14. (a) The nonlinear load currents and (b) the grid currents.

As can be seen, the proposed control scheme succeeded in keeping the grid current in phase with the grid voltage, so that the power factor was kept unity, as seen in Figure 15.

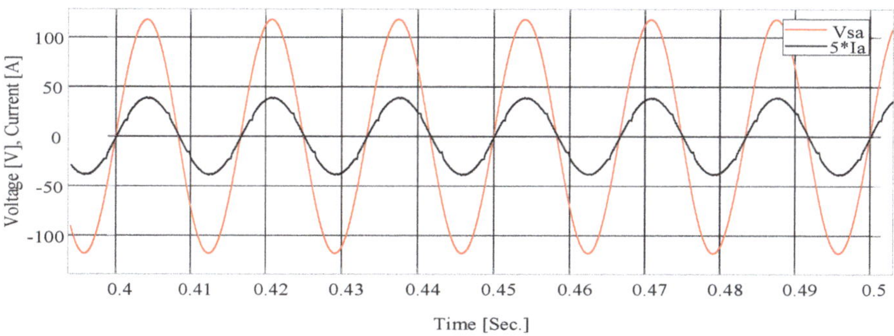

Figure 15. The grid current and the grid voltage.

4. Experimental Results

The proposed MLI was built in the lab and tested for PV–grid connections as well as for APF applications. The hardware setup of the proposed MLI is seen in Figure 16. The PV module parameters are seen in Table 2. The closed-loop control scheme explained in Section 3 was used to perform the closed-loop control of the proposed system. The PWM pulses that were generated were transferred to the IGBTs through the use of a DS1202 board. Two cases were performed, PV-grid connections and APF applications, as seen in the simulation section. The same two cases were experimentally implemented.

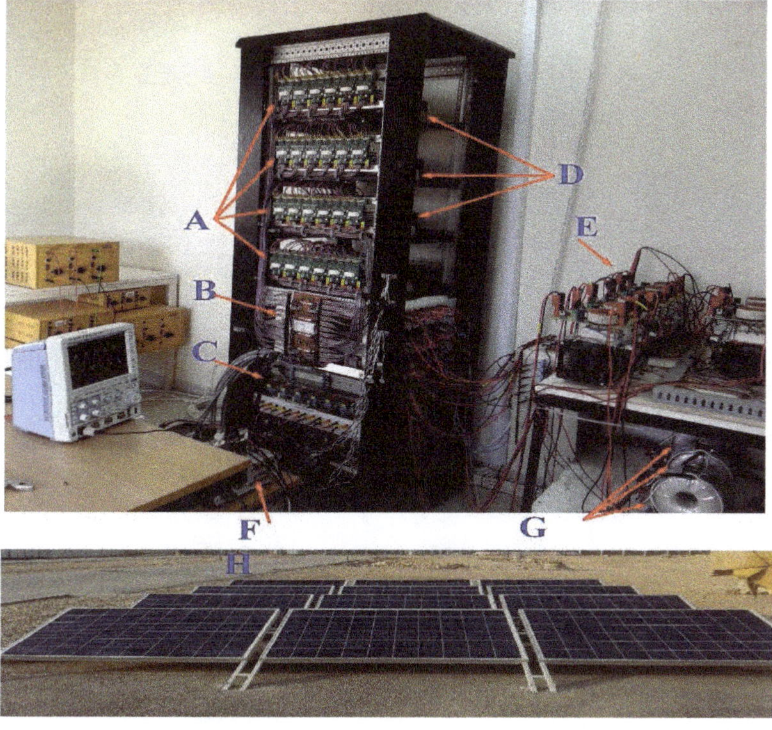

Figure 16. Hardware setup of the proposed MLI. A: IGBT drives, B: level shifter, C: voltage/current sensors, D: IGBTs and DC-link capacitors, E: CUK converters, F: dSPACE 1102, G: open-end winding transformers, and H: PV modules.

A. Grid-Connected PV Application Case

A 4.2 mH interface inductor was used to establish the connection between the proposed topology and the grid during experimental testing. The parameters used for the experiment can be seen in Table 1. The entire closed-loop control scheme depicted in Figures 3 and 4 was used to execute the constructed inverter. In this case, the nonlinear load was disconnected from the constructed inverter; therefore, the harmonic current extraction term from (2) was zero. In addition, the reactive current reference of the constructed inverter was set to zero to guarantee unity-power factor. The proposed topology was used only to connect the PV modules to the grid. Twelve PV modules were used in the experimental setup; six of them are connected to the upper part such that one PV module was assigned for each H-bridge cell and the other six PV modules were connected to the lower part such that two parallel PV modules were assigned to each VSI unit. The P&O MPPT algorithm was used to track the maximum power of the PV modules by using a DC–DC Ćuk converter.

Figure 17 shows the experimental DC-link voltages of the proposed topology. This figure demonstrates that the control scheme was successful in maintaining the DC-link voltages at the reference voltages. On the other hand, the three-phase grid voltages were measured using voltage sensor model LV 25-P to be used for the phase-locked loop. A snapshot of the generated voltages (v_{a1a2}, v_{b1b2}, v_{c1c2} across the primary windings of the T_1, T_2, and T_3 transformers) measured via an oscilloscope is shown in Figure 18. Figure 19 reveals the harmonic spectrum of the generated voltage v_{a1a2}. As displayed in this figure, the THD was 8.57%. Figure 20 shows the grid currents measured by the Hall-effect current sensor model LTS 25-NP.

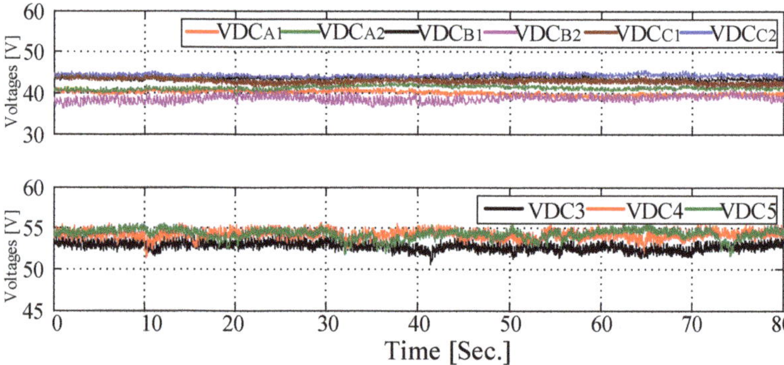

Figure 17. Experimental DC-link voltages.

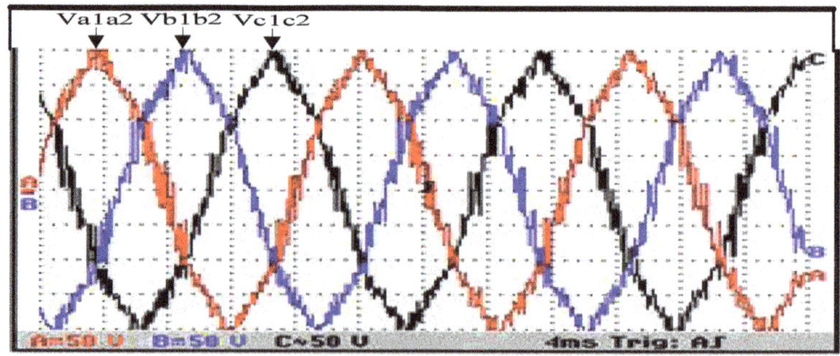

Figure 18. The snapshot of the three-phase generated voltages.

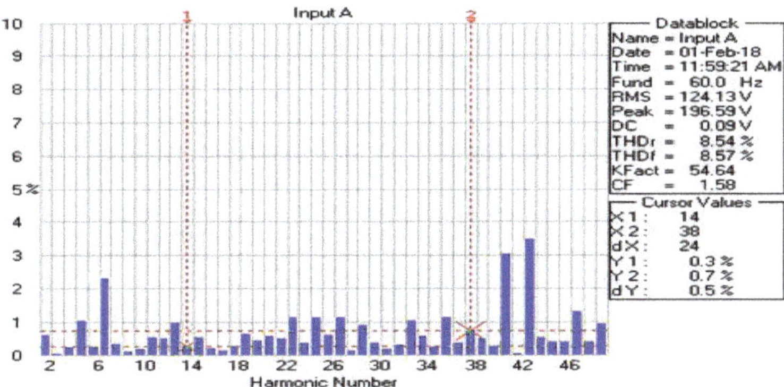

Figure 19. Experimental harmonic spectrum of the generated voltage v_{a1a2}.

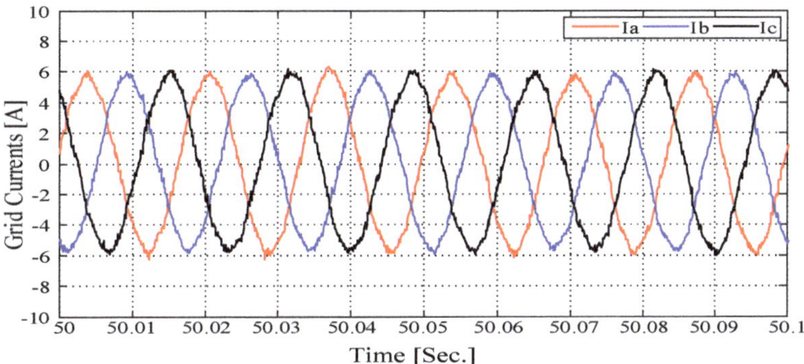

Figure 20. Experimental three-phase grid currents.

The proposed control scheme succeeded in keeping the grid currents in phase with the grid voltages so that the power factor was unified, as demonstrated in Figure 21. From another point of view, the THD of the grid current, i_a, was 4.41%, which is within the IEEE-519 standard limit. The harmonic spectrum of the grid current in the proposed MLI can be seen in Figure 22.

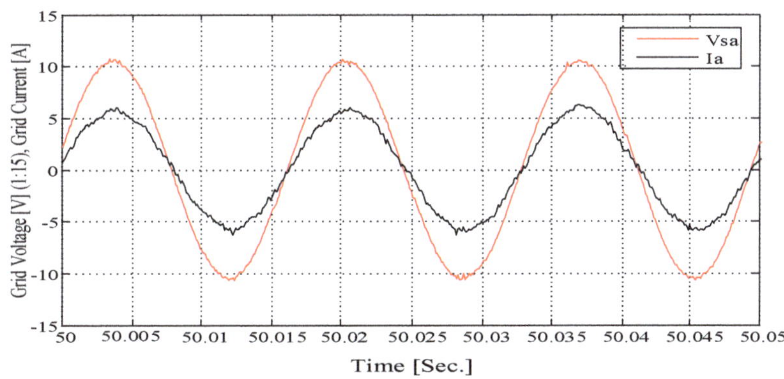

Figure 21. Experimental grid voltage and grid currents.

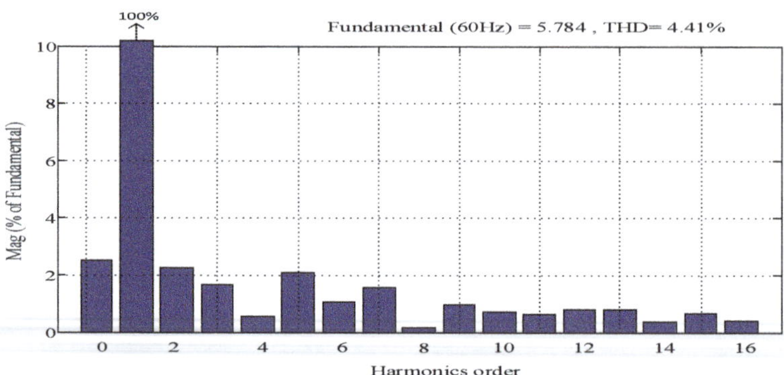

Figure 22. Experimental harmonic spectrum of the grid current i_a.

B.   APF-Application Case

In this case, the constructed inverter was experimentally used for APF application. That is, the control scheme was able to work as an APF only. The parameters used for the experiment can be seen in Table 3. In addition, the harmonic current term in Equation (2) was be included so as to be compensated by the MLI. The resulting PWM pulses were then sent to the IGBTs via the DS1202 board. The constructed MLI was controlled to only compensate the harmonic contents of the load currents, while the grid supplied the three-phase positive-sequence currents of the nonlinear load.

The values of the nonlinear load parameters were $R_{L1} = R_{L2} = 48\ \Omega$, and $X_{L1} = X_{L2} = 154$ mH. To allow the grid to feed power to the nonlinear load, the duty cycle of the DC-DC converter was kept constant at 0.3 (no MPPT is used). The DC-link controllers were successful in maintaining the DC-link voltages at reference levels, as seen in Figure 23. These voltages were measured via analog-to-digital converters by using voltage sensors and plotted in the control desk of the MicroLabBox.

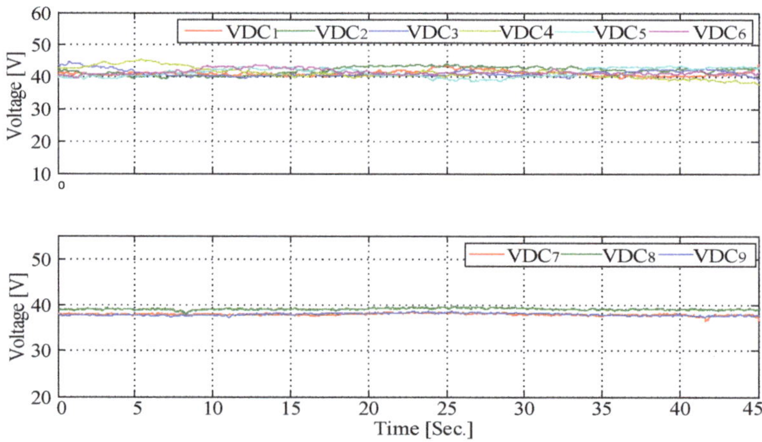

Figure 23. The experimental DC-link voltages.

On the other hand, Figure 24 shows the experimental load currents measured by using Hall-effect current sensors. As seen in Figure 24, the load currents are not sine waves since they are highly distorted due to the nonlinear loads. As explained in the control scheme, the harmonic currents of the load were extracted from the load current so as to be compensated by the inverter. The experimental extracted harmonic load currents (the experimental currents of APF) are shown in Figure 25.

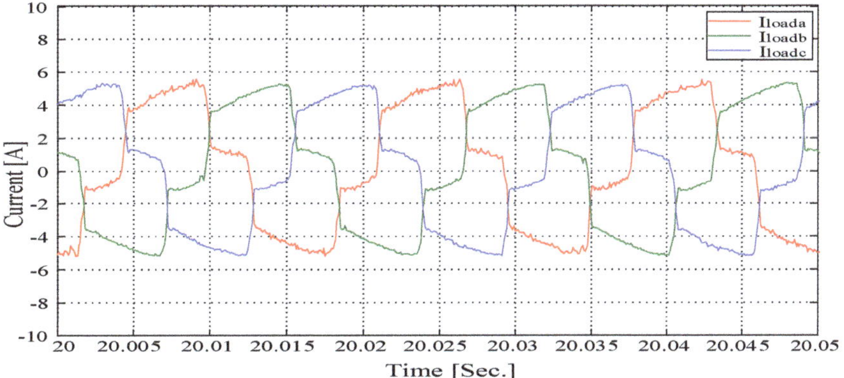

Figure 24. The experimental nonlinear load currents.

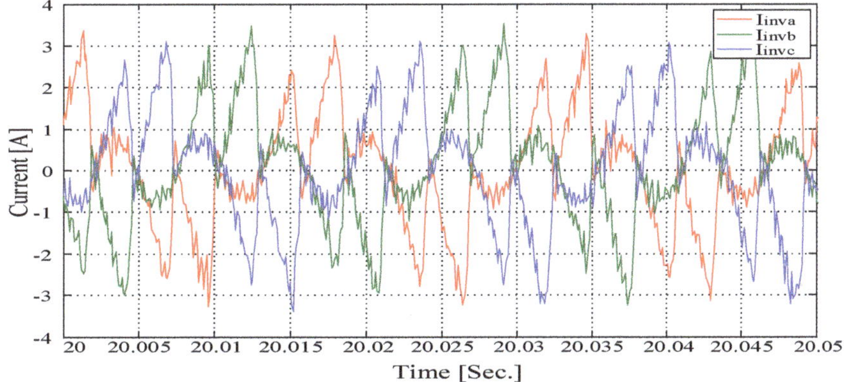

Figure 25. The experimental APF currents.

These harmonic currents were then used as a reference for the inverter currents so that the inverter compensated only these load current harmonics, and the grid supplied the nonharmonic currents. The experimental three-phase grid currents are shown in Figure 26.

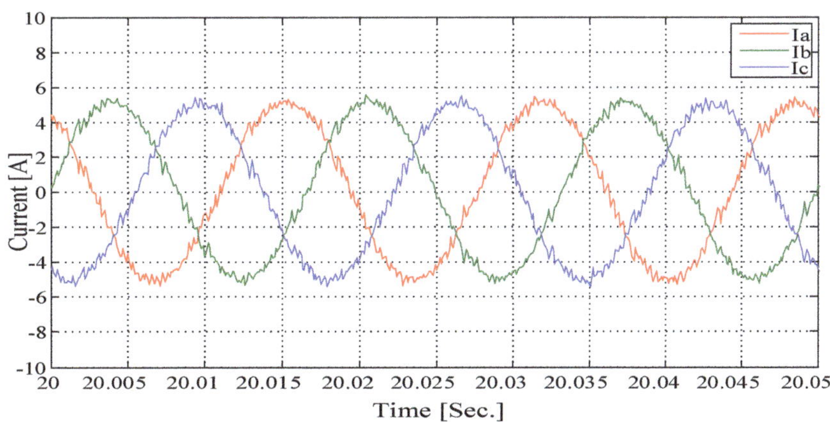

Figure 26. The experimental three-phase grid currents.

The line currents and the grid voltages were out of phase, as revealed in Figure 27, so that the power factor was kept unified. The out-of-phase grid currents were due to the current sensors measuring the current from the grid to the load. On the other hand, the THD of the line currents was 4.8%, which had acceptable harmonic content of less than 5%, as defined by the IEEE-519 standard. The harmonic spectrum of the grid current, i_a, is shown in Figure 28.

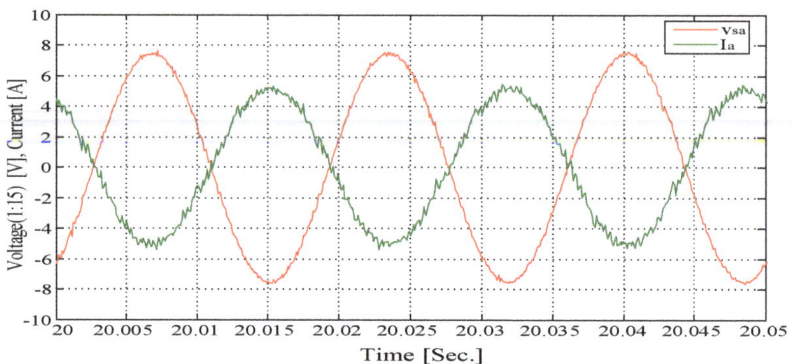

Figure 27. The experimental grid current and grid voltage.

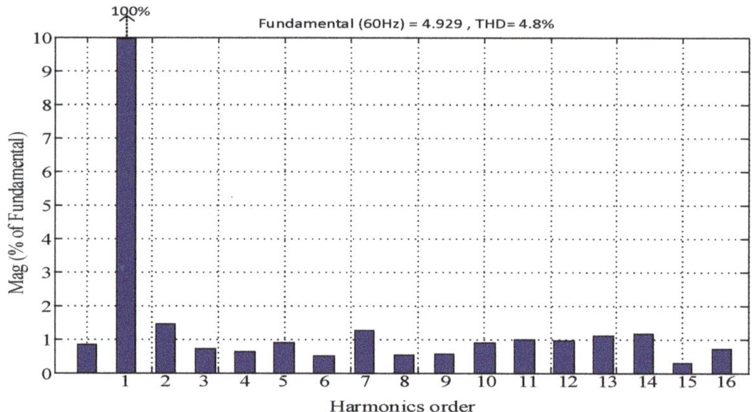

Figure 28. Harmonic spectrum of the grid current i_a.

5. Conclusions

A three-phase hybrid cascaded MLI configuration for APF applications has been proposed in this paper. The proposed configuration conjoins the cascaded H-bridge MLI configuration and the three-phase cascaded VSI configuration via OEWs. This topology uses fewer switches than typical MLI topologies while improving the voltage levels of the proposed MLI. Twenty-two voltage levels are generated per phase from this topology by using only 42 switches. Table 4 shows a comparison of the proposed topologies with the conventional MLI topologies to produce 21 voltage levels (one level lower than the number of levels generated by the proposed topology). The proposed topologies have fewer switches than the prior art MLI topologies. The modularity of the proposed topologies is higher and the proposed topologies have fewer DC-link capacitors than other MLI topologies. For a comparison of conduction, the voltage and current stresses of the other topologies were then compared using the same number of switches as in the proposed topology. A control scheme was proposed to execute the proposed topology for APF application. The proposed MLI-based APF was used to compensate the harmonic load currents while the grid supplied the fundamental positive-sequence currents of the load. The proposed APF topology was simulated by using the SIMULINK environment. The nonlinear load was implemented by using a three-phase diode bridge circuit connected to the resistive inductive loads. To validate the good performance of the proposed MLI and the accuracy of the control, the MLI was built in a lab, and the generated switching pulses were implemented on the MicroLabBox data acquisition system and fired into the IGBTs through gate drives. The proposed topology and its related proposed control scheme were experimentally tested for the use for a PV–grid connection and for APF applications. The simulation and the experimental results show that the proposed APF is functions well in eliminating the unwanted harmonics generated by the nonlinear load. The THD of the line currents was within the acceptable limits as defined by IEEE-519 standard.

Table 4. Comparison results.

	NPC	FC	CHB	MMC	Proposed Topology
No. of switches	120	120	120	240	42
DC-link capacitors	20	20	30	60	9
No. of inductors	0	0	0	6	3
No. of diodes	1140	0	0	0	0
No. of flying capacitors	0	570	0	0	0
Modularity	No	No	Yes	No	Yes
Voltage-balancing problem	Yes	Yes	No	Yes	No
Separated DC sources	No	No	Yes	No	Yes
Voltage stress	$\frac{\sqrt{2}V}{7\sqrt{3}}$	$\frac{\sqrt{2}V}{7\sqrt{3}}$	$\frac{\sqrt{2}V}{3.5\sqrt{3}}$	$\frac{\sqrt{2}V}{4\sqrt{3}}$	Upper part: $\frac{\sqrt{2}V}{4\sqrt{3}}$ Lower part: $\frac{\sqrt{2}V}{6\sqrt{3}}$
Current stress	$\sqrt{2}I$	$\sqrt{2}I$	$\sqrt{2}I$	$\sqrt{2}I$	$\sqrt{2}I$

Author Contributions: Conceptualization, A.M.N. and A.A.A.-S.; Data curation, A.M.N. and A.A.; Formal analysis, A.M.N. and A.A.A.-S.; Funding acquisition, A.M.N. and A.A.; Investigation, A.A.A.-S. and K.E.A.; Methodology, A.M.N. and A.A.A.-S.; Project administration, A.A. and K.E.A.; Software, A.M.N., A.A., A.A.A.-S. and K.E.A.; Supervision, A.A. and K.E.A.; Writing—original draft, A.M.N. and A.A.A.-S.; Writing—review & editing, A.M.N. and A.A. All authors have read and agreed to the published version of the manuscript.

Funding: This research received no external funding.

Informed Consent Statement: Not applicable.

Acknowledgments: This work was supported by the National Plan for Science, Technology, and Innovation, King Abdulaziz City for Science and Technology, Kingdom of Saudi Arabia, Award Number (13-ENE1157-02).

Conflicts of Interest: The authors declare no conflict of interest.

References

1. Choudhury, S.; Bajaj, M.; Dash, T.; Kamel, S.; Jurado, F.J.E. Multilevel Inverter: A Survey on Classical and Advanced Topologies, Control Schemes, Applications to Power System and Future Prospects. *Energies* **2021**, *14*, 5773. [CrossRef]
2. Al-Shammáa, A.A.; Noman, A.M.; Addoweesh, K.E.; Alabduljabbar, A.A.; Alolah, A.I. Multilevel Converter by Cascading Two-Level Three-Phase Voltage Source Converter. *Energies* **2018**, *11*, 843. [CrossRef]
3. Galván, L.; Gómez, P.J.; Galván, E.; Carrasco, J.M.J.E. Optimization-Based Capacitor Balancing Method with Customizable Switching Reduction for CHB Converters. *Energies* **2022**, *15*, 1976. [CrossRef]
4. Marzoughi, A.; Burgos, R.; Boroyevich, D.; Xue, Y. Investigation and comparison of cascaded H-bridge and modular multilevel converter topologies for medium-voltage drive application. In Proceedings of the 40th Annual Conference of the IEEE Industrial Electronics Society-IECON, Dallas, TX, USA, 29 October–1 November 2014; pp. 1562–1568.

5.	Espinosa, E.; Melín, P.; Baier, C.; Espinoza, J.; Garcés, H.J.E. An Efficiency Analysis of 27 Level Single-Phase Asymmetric Inverter without Regeneration. *Energies* **2021**, *14*, 1459. [CrossRef]
6.	Chattopadhyay, S.K.; Chakraborty, C. Performance of three-phase asymmetric cascaded bridge (16: 4: 1) multilevel inverter. *IEEE Trans. Ind. Electron.* **2015**, *62*, 5983–5992. [CrossRef]
7.	Sivakumar, K.; Das, A.; Ramchand, R.; Patel, C.; Gopakumar, K. A hybrid multilevel inverter topology for an open-end winding induction-motor drive using two-level inverters in series with a capacitor-fed H-bridge cell. *IEEE Trans. Ind. Electron.* **2010**, *57*, 3707–3714. [CrossRef]
8.	Rajeevan, P.; Sivakumar, K.; Patel, C.; Ramchand, R.; Gopakumar, K. A seven-level inverter topology for induction motor drive using two-level inverters and floating capacitor fed H-bridges. *IEEE Trans. Power Electron.* **2011**, *26*, 1733–1740. [CrossRef]
9.	Das, S.R.; Ray, P.K.; Sahoo, A.K.; Ramasubbareddy, S.; Babu, T.S.; Kumar, N.M.; Elavarasan, R.M.; Mihet-Popa, L. A Comprehensive Survey on Different Control Strategies and Applications of Active Power Filters for Power Quality Improvement. *Energies* **2021**, *14*, 4589. [CrossRef]
10.	Lumbreras, D.; Gálvez, E.; Collado, A.; Zaragoza, J.J.E. Trends in power quality, harmonic mitigation and standards for light and heavy industries: A review. *Energies* **2020**, *13*, 5792. [CrossRef]
11.	Urquizo, J.; Singh, P.; Kondrath, N.; Hidalgo-León, R.; Soriano, G. Using D-FACTS in microgrids for power quality improvement: A review. In Proceedings of the IEEE Second Ecuador Technical Chapters Meeting (ETCM), Salinas, Ecuador, 16–20 October 2017; pp. 1–6.
12.	Rachmildha, T.D. Optimized Combined System of Shunt Active Power Filters and Capacitor Banks. *Int. J. Electr. Eng. Inform.* **2011**, *3*, 326–335. [CrossRef]
13.	Bag, A.; Subudhi, B.; Ray, P.K. Comparative analysis of sliding mode controller and hysteresis controller for active power filtering in a grid connected PV system. *Int. J. Emerg. Electr. Power Syst.* **2018**, *19*, 1–13. [CrossRef]
14.	Pou, J.; Boroyevich, D.; Pindado, R. Effects of imbalances and nonlinear loads on the voltage balance of a neutral-point-clamped inverter. *IEEE Trans. Power Electron.* **2005**, *20*, 123–131. [CrossRef]
15.	Liu, Q.; Li, Y.; Hu, S.; Luo, L. A transformer integrated filtering system for power quality improvement of industrial DC supply system. *IEEE Trans. Ind. Electron.* **2019**, *67*, 3329–3339. [CrossRef]
16.	Liu, Q.; Li, Y.; Hu, S.; Luo, L.; Systems, E. A controllable inductive power filtering system: Modeling, analysis and control design. *Int. J. Electr. Power Energy Syst.* **2019**, *105*, 717–728. [CrossRef]
17.	Ali, H.H.; Kamal, N.A.; Elbasuony, G.S. Two-Level Grid-Side Converter-Based STATCOM and Shunt Active Power Filter of Variable-Speed DFIG Wind Turbine-Based WECS Using SVM for Terminal Voltage. *Int. J. Serv. Sci. Manag. Eng. Technol. (IJSSMET)* **2021**, *12*, 169–202.
18.	Jignesh, P.; Jadeja, R.; Trivedi, T. Cascaded Three Level Inverter Based Shunt Active Power Filter with Modified Three Level Hysteresis Current Control. *Int. Rev. Model. Simul.* **2018**, *11*, 125–197.
19.	Sihem, G.; Azar, A.T.; Dib, D. Three-level (NPC) shunt active power filter based on fuzzy logic and fractional-order PI controller. *Int. J. Autom. Control.* **2021**, *15*, 149–169.
20.	Dawid, B.; Jarek, G.; Michalak, J.; Zygmanowski, M. Control method of four wire active power filter based on three-phase neutral point clamped T-type converter. *Energies* **2021**, *14*, 8427.
21.	Salim, C.H.E.N.N.A.I. Shunt Active Power Filter Performances based on Seven-level NPC Inverter using Fuzzy and LS-PWM Control Scheme. *Electroteh. Electron. Autom.* **2022**, *1*, 23–30.
22.	Adrikowski, T.; Bula, D.; Pasko, M. Three-phase active power filter with T-NPC type inverter. In Proceedings of the 2018 Progress in Applied Electrical Engineering (PAEE), Koscielisko, Poland, 18–22 June 2018; pp. 1–5.
23.	Escobar, G.; Martinez-Rodriguez, P.R.; Iturriaga-Medina, S.; Vazquez-Guzman, G.; Sosa-Zuñiga, J.M.; Langarica-Cordoba, D. Control Design and Experimental Validation of a HB-NPC as a Shunt Active Power Filter. *Energies* **2020**, *13*, 1691. [CrossRef]
24.	Soumyadeep, R.; Gupta, N.; Gupta, R.A. A novel non-linear control for three-phase five-level cascaded H-bridge inverter based shunt active power filter. *Int. J. Emerg. Electr. Power Syst.* **2018**, *19*, 1–11. [CrossRef]
25.	Wang, H.; Liu, S. An optimal operation strategy for an active power filter using cascaded H-bridges in delta-connection. *Electr. Power Syst. Res.* **2019**, *175*, 105918. [CrossRef]
26.	Sochor, P.; Hirofumi, A. Theoretical and experimental comparison between phase-shifted PWM and level-shifted PWM in a modular multilevel SDBC inverter for utility-scale photovoltaic applications. *IEEE Trans. Ind. Appl.* **2017**, *53*, 4695–4707. [CrossRef]
27.	Nair, V.; Gopakumar, K.; Franquelo, L.G. A very high resolution stacked multilevel inverter topology for adjustable speed drives. *IEEE Trans. Ind. Electron.* **2017**, *65*, 2049–2056. [CrossRef]
28.	Behrouzian, E.; Bongiorno, M. DC-link voltage modulation for individual capacitor voltage balancing in cascaded H-bridge STATCOM at zero current mode. In Proceedings of the 20th European Conference on Power Electronics and Applications, Riga, Latvia, 17–21 September 2018; pp. 1–10.
29.	Mariem, T.; Guermazi, A.; Ghariani, M. Cascaded Multilevel Inverter for PV-Active Power Filter Combination into the Grid-Tied Solar System. *Int. J. Renew. Energy Res. (IJRER)* **2020**, *10*, 1810–1819.
30.	Nageswar, R.B.; Suresh, Y.; Panda, A.K.; Naik, B.S.; Jammala, V. Development of cascaded multilevel inverter based active power filter with reduced transformers. *CPSS Trans. Power Electron. Appl.* **2020**, *5*, 147–157.

31. Noman, A.M.; Al-Shammáa, A.A.; Addoweesh, K.E.; Alabduljabbar, A.A.; Alolah, A.I.J.E. Cascaded multilevel inverter topology based on cascaded H-bridge multilevel inverter. *Energies* **2018**, *11*, 895. [CrossRef]
32. Noman, A.M.A.; Alshammaa, A.A.; Addoweesh, K.E.; Alabduljabbar, A.A.A.; Alolah, A.I. Hybrid CHB-TVSI Multilevel Voltage Source Inverter. U.S. Patent US10141865B1, 2018.
33. Hamadi, A.; Rahmani, S.; Al-Haddad, K.; Al-Turki, Y.A. A three-phase three wire grid-connect photovoltaic energy source with sepic converter to track the maximum power point. In Proceedings of the 37th Annual Conference of the IEEE Industrial Electronics Society-IECON, Melbourne, VIC, Australia, 7–10 November 2011; pp. 3087–3092.

*energies*

MDPI

Article

Optimization of a Solar Water Pumping System in Varying Weather Conditions by a New Hybrid Method Based on Fuzzy Logic and Incremental Conductance

Abdelilah Hilali [1], Najib El Ouanjli [2], Said Mahfoud [3], Ameena Saad Al-Sumaiti [4,*] and Mahmoud A. Mossa [5,*]

[1] Electronics, Communication Systems and Energy Optimization Team, Faculty of Sciences, Moulay Ismail University, Meknes 11201, Morocco
[2] Laboratory of Mechanical, Computer, Electronics and Telecommunications, Faculty of Sciences and Technology, Hassan First University, Settat 26000, Morocco
[3] Industrial Technologies and Services Laboratory, Higher School of Technology, Sidi Mohamed Ben Abdellah University, Fez 30000, Morocco
[4] Advanced Power and Energy Center, Department of Electrical and Computer Engineering, Khalifa University, Abu Dhabi 127788, United Arab Emirates
[5] Electrical Engineering Department, Faculty of Engineering, Minia University, Minia 61111, Egypt
* Correspondence: ameena.alsumaiti@ku.ac.ae (A.S.A.-S.); mahmoud_a_mossa@mu.edu.eg (M.A.M.)

Abstract: The present work consists of developing a new hybrid FL-INC optimization algorithm for the solar water pumping system (SWPS) through a SEPIC converter whose objective is to improve these performances. This technique is based on the combination of the fuzzy logic of artificial intelligence and the incremental conductance (INC) technique. Indeed, the introduction of fuzzy logic to the INC algorithm allows the extraction of a maximum amount of power and an improvement in the efficiency of the SWPS. The performance of the system through the SEPIC converter is compared with those of the direct coupling to show the interest of the indirect coupling, which requires an adaptation stage driven by an optimal control algorithm. In addition, a comparative analysis between the proposed hybrid algorithm and the conventional optimization techniques, namely, P&O and INC Modified (M-INC), was carried out to confirm improvements related to the SWPS in terms of efficiency, tracking speed, power quality, tracking of the maximum power point under different weather changes, and pumped water flow.

Keywords: solar water pumping system; FL-INC; SEPIC converter; MPPT; optimization algorithm; centrifugal water pump

Citation: Hilali, A.; El Ouanjli, N.; Mahfoud, S.; Al-Sumaiti, A.S.; Mossa, M.A. Optimization of a Solar Water Pumping System in Varying Weather Conditions by a New Hybrid Method Based on Fuzzy Logic and Incremental Conductance. *Energies* **2022**, *15*, 8518. https://doi.org/10.3390/en15228518

Academic Editors: Sérgio Cruz and Oscar Barambones

Received: 5 October 2022
Accepted: 1 November 2022
Published: 14 November 2022

Publisher's Note: MDPI stays neutral with regard to jurisdictional claims in published maps and institutional affiliations.

1. Introduction

The right to a healthy and enabling environment is a main principle for all living beings. This is one of the reasons why the world is moving towards an effective policy for managing this sector. The environment is constantly suffering from greenhouse gas emissions, which are mainly generated by fossil fuels. The world must therefore begin a process to manage energy issues as part of a comprehensive and successful strategy. Despite the major energy carriers being fossil fuels and their contribution to environmental pollution, the energy demand is increasing. Renewable energy is a competitive alternative, and their economics has been assessed in comparison with various renewable energy sources as in [1] aside from a regulation strategy suggested in [2,3]. Various studies have accounted for their planning and integration into the distribution grid, as discussed in [4,5]. This is why the world is turning to new energy resources that are promising, socially compatible, renewable, and sustainable. These sources include solar power [6], wind power [7], biomass [8], geothermal power [9], and ocean power [10].

Countries located between altitudes of −40 and +40 have a very high amount of sunshine throughout the year [11]. In addition, observed decreases in rainfall have resulted in insufficient resources available to meet agricultural demands. Therefore, pumped irrigation has become an inevitable necessity. Two approaches are considered to achieve the pumping activity, either by connecting to an electricity source or by using fuels, such as diesel or butane, for pumping. However, the latter alternative is not pollution-free aside from requiring financial resources, raising the production cost. An alternative approach is the use of photovoltaic energy, which is a competitive alternative in terms of being environmentally friendly and is a competing alternative from the economic aspect [12].

The use of solar energy to partially or totally solve the problem of pumping is an option that has existed for a long time. Photovoltaic panels are components of a solar system designed to collect solar energy and convert it into electricity. However, PV systems have low efficiency due to their dependence on weather conditions, including temperature, irradiance, and other system equipment [13]. The generator photovoltaic (GPV) has a nonlinear static current–voltage characteristic of a maximum power point (MPP). This characteristic depends mainly on the level of irradiation, the temperature of the cells, and the aging process [14]. The operating point of a photovoltaic system is very complex to determine. In addition, the load connected to the GPV determines the operating point, which leads to a significant deviation from the maximum power point. The occurrence of multiple irradiance variations can lead to irrecoverable yield decreases, variable power, outages, and even insufficient solar pumping [15].

To ensure optimal operation of the GPV, it is necessary to adjust a solar water pumping system (SWPS) in such a way that the generator is forced to deliver its maximum power. Thanks to a static DC–DC converter, this function is ensured by providing an MPPT control system that acts on the converter's duty cycle, thus ensuring optimal adaptation to the variation of the load characteristic. DC–DC converters are commonly used in renewable energy installations, including photovoltaic panels [16].They are employed to adapt the input voltage to the desired output voltage. In some cases, it is sufficient to lower or raise the output voltage; in other cases, it is better to use a controller that can lower or raise the voltage to achieve optimal energy extraction based on MPPT optimization algorithms [17].

There are numerous MPPT algorithms proposed in the literature to track the maximum power point, which are divided into two categories: (i) typical optimization techniques, including perturbation and observation (P&O) [18], hill climbing (HC) [19], and incremental conductance (INC) [20]. These algorithms are characterized by their simple design and implementation. However, they have low performance in terms of tracking speed, efficiency, and power quality. (ii) Modern optimization techniques are based on artificial intelligence and metaheuristic approaches, including neural networks, ABC-PO [21], genetic algorithms [22], and so on [23]. These algorithms offer better performance and efficiency for PV systems. However, they have drawbacks that lie in the complexity of design, implementation, and very high computational time.

In this paper, a hybrid FL-INC algorithm based on fuzzy logic and incremental technique is developed and applied to the solar pumping system to improve its functioning in terms of efficiency, tracking speed, power quality, steady-state oscillations, complexity, and tracking of the maximum power point under different weather changes. In addition, the benefits of indirect coupling through an adaptation stage composed of a SEPIC converter compared with direct coupling of the system are analyzed to demonstrate the efficiency of such a system. As another contribution, the present work shows a comparative study between the proposed FL-INC algorithm and the P&O and M-INC techniques to confirm improvements related to the pumped water flow for a solar pumping system application.

The remainder of this research article is structured as follows: Section 2 is devoted to the description of the solar pumping system. Section 3 describes the constitution of the solar pumping system. Section 4 discusses the MPPT optimization techniques used. Section 5 presents the analysis of the obtained research results. Section 6 is devoted to a comparative study of the MPPT optimization techniques. Finally, Section 7 presents a conclusion of this work and future perspectives.

2. Description of the Solar Water Pumping System

Solar pumping system installations typically require the following components: a GPV, a static power converter to improve system efficiency and reliability to optimize the amount of water pumped, and a motor pump assembly. These elements mentioned above are coupled in two ways. As shown in Figure 1a, direct coupling is the basic topology that can be made, in which a motor pump unit is directly connected to the GPV. The main reason for this option is the lack of other electronic devices to control the system, which results in a low-cost solution [23]. However, the direct influence of the coupling on the power delivered by the GPV is reflected by the intersection between the GPV and motor pump characteristics [13]. The main disadvantage of this configuration mode is that it does not allow any regulation of the voltage at the motor terminals, while the transfer of the maximum available power to the GPV is not ensured. In fact, to overcome the latter, an indirect coupling is used, in which a matching stage is inserted between the GPV and the motor pump assembly, as shown in Figure 1b. In fact, this stage acts as an interface between the two equipment and ensures the permanent transfer of maximum power supplied by the generator.

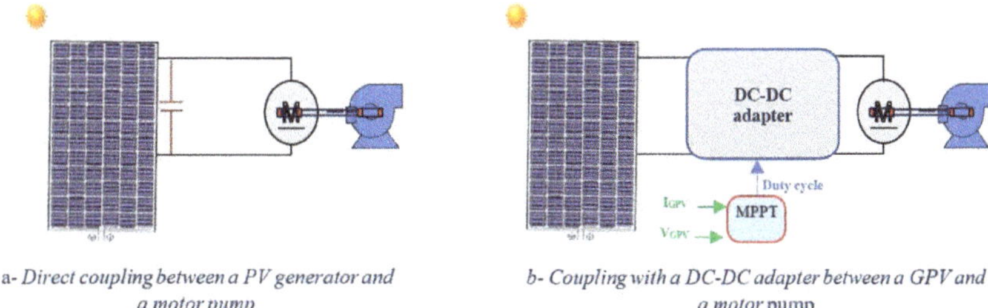

a- *Direct coupling between a PV generator and a motor pump*

b- *Coupling with a DC-DC adapter between a GPV and a motor pump*

Figure 1. Coupling types between the motor pump and the GPV.

3. Solar Pumping System Constitutions

3.1. PV Generator

PV cells are made of a semiconductor material that forms the basic element of a photovoltaic module. A generator typically includes several modules connected in series and in parallel to produce the required current and voltage. Figure 2 shows the most commonly used equivalent schematic of a solar cell, which includes a current source, I_{ph}; a diode; and two parasitic resistors, R_s and R_{sh}.

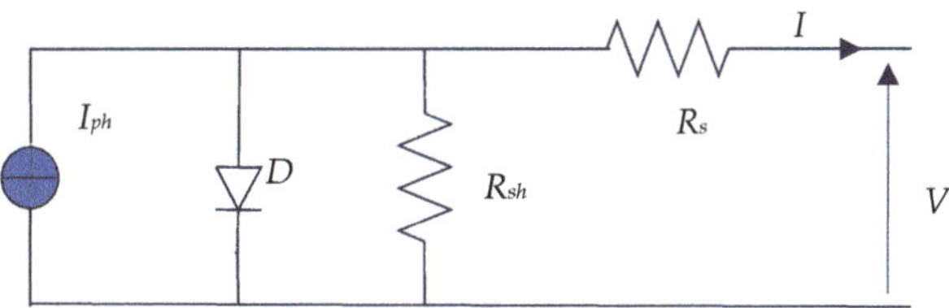

Figure 2. Electrical model of a photovoltaic cell [24].

The actual current–voltage characteristic can be represented by the following equation:

$$I = I_{ph} - I_{obs} = I_{ph} - I_s \left[exp^{\frac{V+I.R_s}{nV_{th}}} - 1 \right] - \frac{V + I.R_s}{R_{sh}} \tag{1}$$

where I_{ph} (A) is the saturation current, V_{th} (V) is the thermodynamic potential, K ($J.K^{-1}$) is Boltzmann's constant, T (K) is the effective cell temperature, e (c) is the electron's charge, n

is the nonideality factor of the junction, I (A) is the current provided by the cell, and V (V) is the voltage across the cell.

Table 1 represents the electrical characteristics of the GPV proposed in this study. Indeed, it is composed of 5 PV panels (REC_330NP), and each panel can generate a peak power of 330 W peak in standard conditions. These panels are connected in series, which allows for generating a voltage of (V_{MPP}) = 173 V and delivering a total peak power of 1650 W peak in standard conditions.

Table 1. GPV under investigation specifications.

V_{oc} (V)	I_{sc} (A)	P_{mpp} (W)	R_{sh} (kΩ)	R_s (mΩ)	V_{mpp} (V)	I_{mpp} (A)
41	10.3	330	1	73	34.6	9.55

Figure 3 shows the current and power versus voltage evolution for different irradiance patterns and for a cell temperature held constant at 25 °C. From the analysis of power and current, it is interesting to note that the variation of irradiance has a significant effect on these two variables. Indeed, the increase in this characteristic leads to an increase in the power and the current. Table 2 summarizes the progression of the peak power as a function of irradiance.

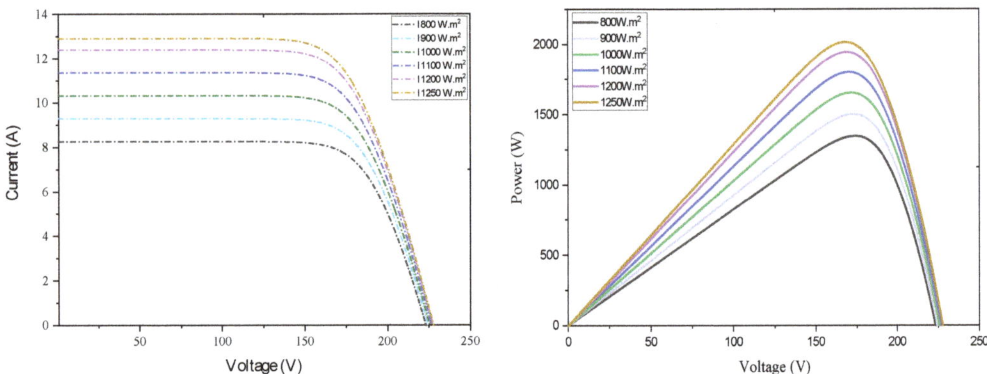

Figure 3. Current-Voltage and Power-Voltage characteristics of GPV under variable irradiance.

Table 2. Peak power evolution as a function of irradiance.

Irradiance (W·m^{-2})	GPV Temperature (°C)	Peak Power (W·m^{-2})
800	25	1347
900	25	1508
1000	25	1657
1100	25	1803
1200	25	1950
1250	25	2018

3.2. Pump Centrifuge

A necessary requirement of the SPS is the motor pump. The latter is made up of two parts: an electric motor coupled to a pump. In this study, the motor used to turn the pump (SQF 0-6-2 DC) is characterized by a nominal current I_n = 8.4 A, an input voltage ranging from 30 to 300 VDC. In the field of technology, two types of pump technology are generally used in pumping applications: the positive displacement pump and the centrifugal pump [25,26]. The first technology is distinguished by a constant load, whereas centrifugal pumps are distinguished by an aerodynamic characteristic. The latter are characterized by a load of torque that develops with the quadratic form of the drive speed. This characteristic clearly indicates that the drive torque of the pump is almost zero at the start-up. Therefore, the pump can be operated with very low solar irradiation, and as the solar irradiation increases, the drive motor can have a high speed. The proportionality of the different parameters as a function of the speed is expressed by the following equations [27]:

- Torque (C_r)—speed (ω):

$$C_r = A.\omega^2 \tag{2}$$

- Flow (Q)—speed (ω):

$$\frac{Q_2}{Q_1} = \frac{\omega_2}{\omega_1} \tag{3}$$

- Height (H)—speed (ω):

$$\frac{H_2}{H_1} = \left(\frac{\omega_2}{\omega_1}\right)^2 \tag{4}$$

3.3. Matching Stage: SEPIC Converter

The SEPIC is a direct-voltage-to-direct-voltage converter that will allow the output to be greater than, less than, or equal to the input voltage [28]. This is because the voltage of the converter is controlled by the duty cycle. In addition, the SEPIC is analogous to a buck-boost converter with the benefit of having a noninverting output. In effect, it ensures low current ripples. In terms of construction, the SEPIC converter consists of a power switch (IGBT; MOSFET, Thyristor, etc.), a diode, inductors (L_1, L_2), and capacitors (C_1, C_2). Figure 4 shows the electrical diagram of this converter. The input inductance L_1 of the circuit makes the whole SEPIC converter look like a boost converter. This converter also has the advantage of isolating the input and output by means of the coupling capacitor C_1, which protects against a short circuit or an overload at the output. Moreover, the SEPIC has the advantage of being able to cut its output voltage down to zero voltage, unlike a boost converter where the lowest output voltage is equal to the input voltage.

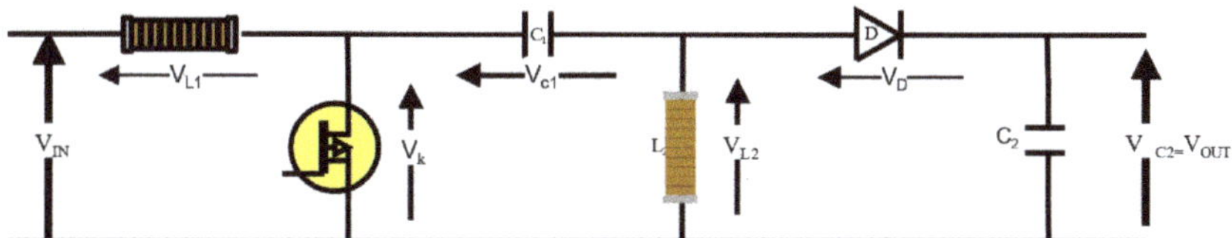

Figure 4. Electrical diagram of a SEPIC converter [28].

The relation between the load voltage and the converter input voltage is given by [28,29]:

$$V_{OUT} = \frac{1-\alpha}{\alpha} V_{IN} \tag{5}$$

where V_{OUT} is the output voltage, V_{IN} is the input voltage, and α is the duty cycle.

Table 3 summarizes the sizing results for the SEPIC converter applied to the studied solar pumping system.

Table 3. SEPIC converter features.

Component	Designate	Value
Coupling capacitor	C_1	250 µF
Inductor	$L_1 = L_2$	1.5 mH
Filtering capacitor	C_2	500 µF
Hashing frequency	f	10 kHz

4. Solar Water Pumping System Optimization Techniques

The optimization of the solar pumping system using an MPPT tracker is shown in Figure 5. In addition, the system structure consists of a GPV that powers the electric motor. The latter drives a centrifugal pump through an MPPT controller. Then a series of studies are carried out to give an overview of the influence of MPPT techniques on the performance of the pumping system.

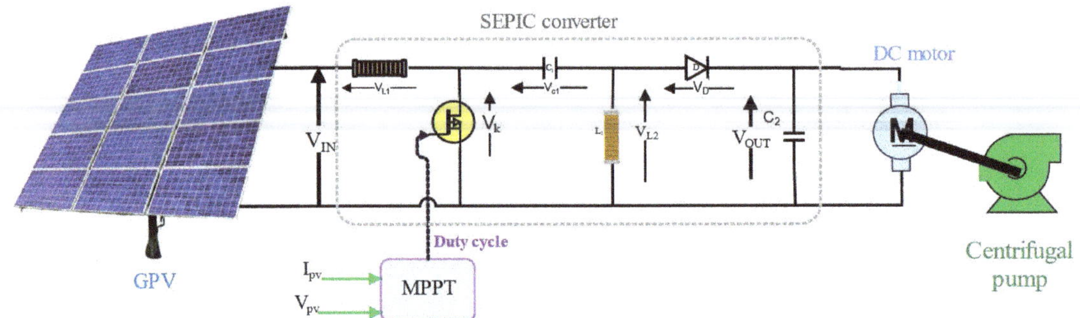

Figure 5. Solar water pumping system optimization structure.

4.1. P&O Optimization Technique

The perturb and observe (P&O) optimization technique is probably the most natural method to find the maximum power point MPP, and its principle is given by the flowchart in Figure 6. Moreover, MPP tracking is independent of temperature and/or irradiance variations. The basic principle of this optimization technique can therefore be extracted from the flowchart in Figure 6:

- At a fixed voltage V(k), the corresponding power P(k) delivered by the generator is measured;
- After a certain delay, the algorithm imposes a voltage V(k + 1) = V(k) + ΔV and also measures the corresponding power P(k + 1);
- If P(k + 1) is greater than P(k), the algorithm seeks to apply for a higher-voltage V(k + 1) = V(k) + ΔV;
- Otherwise, the algorithm instead looks to decrease the voltage V(k) = V(k + 1) − ΔV.

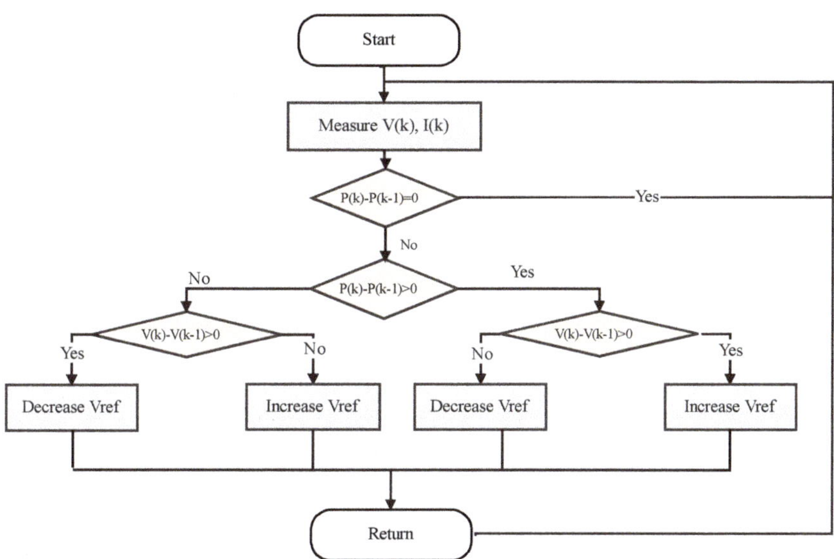

Figure 6. Flowchart of the P&O algorithm [27].

4.2. Incremental Conductance Modified (M-INC) Optimization Technique

Under unstable irradiance and temperature conditions, a variation in voltage leads to a variation in the current. On the other hand, a voltage perturbation leads to a current perturbation when the photovoltaic field is subjected to a sudden change in atmospheric conditions [30]. Contrary to the classical incremental method [31], the M-INC algorithm distinguishes between these operating conditions and, thus, avoids the divergence caused by atmospheric disturbances. Indeed, the modified incremental conductance (M-INC) algorithm is supposed to act in the opposite way of the INC method when the system operates in the partial shading situation. This method is based on the power derivative being zero at MPP. It can be expressed as follows [32]:

$$\frac{dP}{dV} = \frac{d(IV)}{dV} = V\frac{dI}{dV} + I = 0 \tag{6}$$

where dP is the power derivative, dV is the voltage derivative, and dI is the current derivative.

Equation (6) can be expressed as follows:

$$\frac{\Delta I}{\Delta V} = \frac{dI}{dV} = -\frac{I}{V} \tag{7}$$

where ΔV and ΔI are the increments of voltage and current. The characteristics of the method can be obtained from the P–V curve and can be written as follows [32]:

$$\frac{dI}{dV} = -\frac{I}{V} \text{ at maximum power point} \tag{8}$$

$$-\frac{I}{V} \prec \frac{dI}{dV} \text{ left to maximum power point} \tag{9}$$

$$-\frac{I}{V} \succ \frac{dI}{dV} \text{ right to maximum power point} \tag{10}$$

The duty cycle of this technique is calculated by:

$$\alpha_i(n) = \alpha_i(n-1) \pm k_i \left| \frac{P(n) - P(n-1)}{V(n) - V(n-1)} \right| = \alpha_i(n-1) \pm \delta\alpha_i \tag{11}$$

Figure 7 depicts a flowchart of the modified INC algorithm.

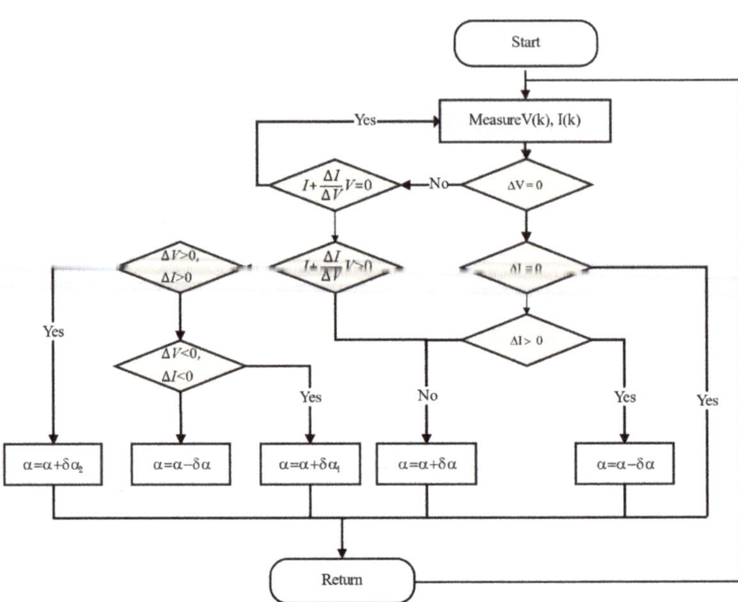

Figure 7. Flowchart of the M-INC algorithm.

4.3. FL-INC Hybrid Optimization Technique

The fuzzy logic of artificial intelligence is introduced in the incremental conductance technique to develop a new hybrid optimization method named FL-INC. The latter is applied to ensure a better optimization of the SWPS in case of very unstable weather conditions. Fuzzy logic allows the study and representation of imprecise knowledge and approximate reasoning [33]. Figure 8 shows the basic design of a fuzzy logic controller, which is composed of three stages. The first step is the fuzzification step, which consists in converting the input variables (physical variables) into linguistic variables (fuzzy variables), by establishing membership functions for the different input variables. The second step is the inference engine, which includes the inference block and the rule base to determine the output of the fuzzy controller from the inputs resulting from the fuzzification. Finally, defuzzification is the last step of the fuzzy controller; it allows for converting the fuzzy data provided by the inference mechanism into a physical or numerical quantity to define the decision process [34].

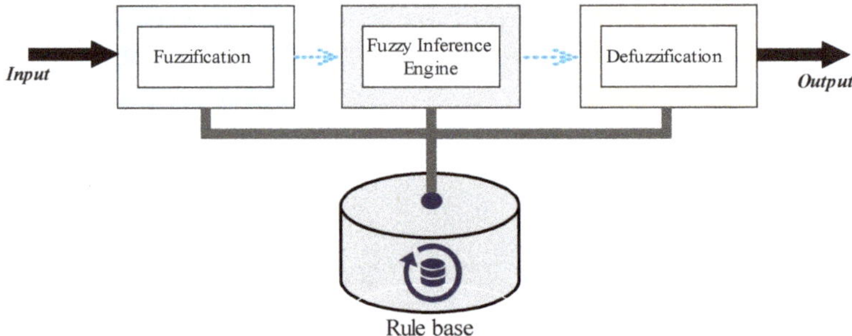

Figure 8. Fuzzy logic controller structure.

The optimization using the fuzzy logic technique can significantly reach the global power point. However, the performance of this control technique depends mainly on the human expertise. Indeed, the rules developed from the human operator's expertise are expressed in linguistic form. Moreover, these rules determine the dynamic performance of the fuzzy controller [35]. Therefore, applying this reasoning to the incremental algorithm improves the performance of the extracted power. Figure 9 shows a flowchart of the developed hybrid FL-INC technique.

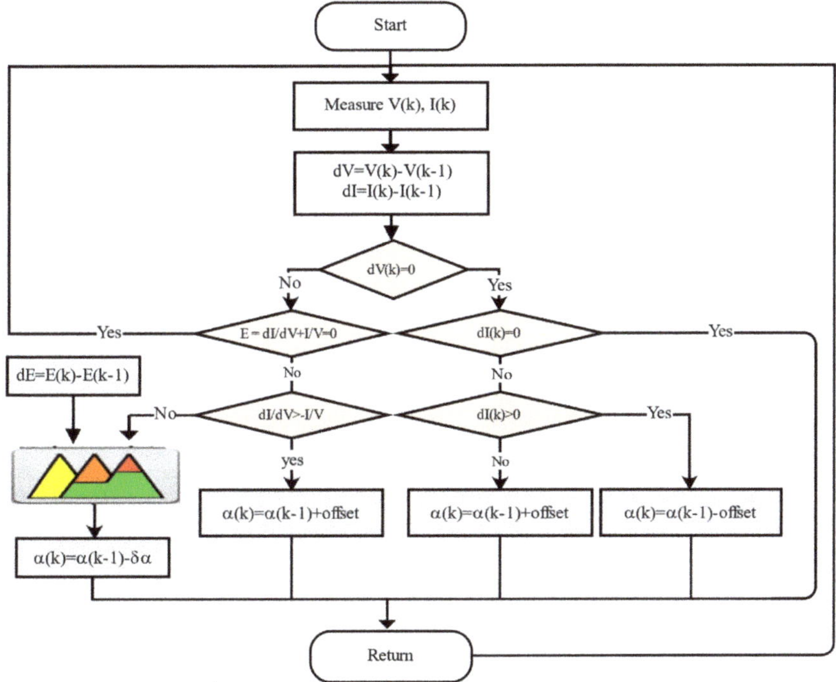

Figure 9. Flowchart of the hybrid technique FL-INC.

The fuzzy controller proposed in this technique consists of two inputs, $E(k)$ and $\Delta E(k)$, which are defined by the following equations:

$$E(k) = \frac{I(k) - I(k-1)}{V(k) - V(k-1)} + \frac{I(K)}{V(K)} \tag{12}$$

$$\Delta E = E(k) - E(k-1) \tag{13}$$

where $E(k)$ is the derivative of the conductance calculated from the measured voltage and current. It is nullified when the operating point attains the MPP. Additionally, $\Delta E(k)$ is the error of the input $E(k)$.

Additionally, the output of this fuzzy controller represents the change in the duty cycle ($\delta\alpha$). Figure 10 shows the selected membership functions for E, ΔE, and $\delta\alpha$.

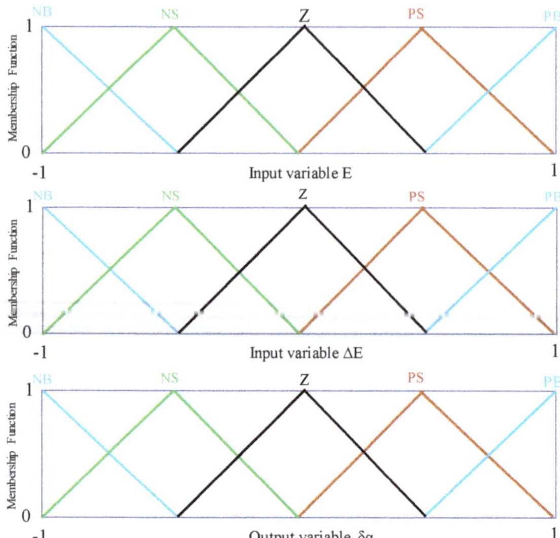

Figure 10. Membership functions of the input (E, ΔE) and output (δα) variables.

The triangular membership function is chosen for all fuzzy sets because of its simplicity. The boundaries of the fuzzy variable range are usually normalized between −1 and +1. The control rules are used to make the decision and decide the action at the output of the fuzzy controller, which consists of 25 fuzzy rules, and are grouped in Table 4.

Table 4. Base of the fuzzy controller rules used for FL-INC.

E \ ΔE	NB	NS	Z	PS	PB
NB	Z	Z	PB	PB	PB
NS	Z	Z	PS	PS	PS
Z	PS	Z	Z	Z	NS
PS	NS	NS	NS	Z	Z
PB	NB	NB	NB	Z	Z

Linguistic variables are expressed as: PB: positive big; PS: positive small; Z: zero; NS: negative small; and NB: negative big.

The control rules can be expressed as follows: rule: if (E is X) and (ΔE is Y), then (δα is W), where: X, Y, and W are the fuzzy sets of input and output variables.

The implementation of the different control rules is based on the instructions below:

- If the conductance value is very big, the variation of the duty cycle (δα) must be big so as to quickly bring this conductance to zero;
- If the conductance value is close to zero, slight variations in the duty cycle should be applied;
- If the conductance value is close to zero and approaches it quickly, the duty cycle must be constant to avoid strong overshoot;
- If the conductance reaches zero and the output voltage is not stable, the duty cycle should be varied a little to reduce fluctuations;
- If the conductance reaches zero and the output voltage of the converter is stable, the duty cycle should be kept constant;
- If the value of the conductance variation is greater than zero, the variation of the duty cycle is negative and vice versa.

5. Simulation Results of the Solar Water Pumping System

5.1. Simulation Conditions

In this section, the MPPT simulation results are obtained through the PSIM simulation software. Indeed, the efficiency of MPPT by optimization techniques is evaluated under irradiation conditions similar to the real conditions. Moreover, a variable irradiance profile (from 200 to 1250 $W \cdot m^{-2}$) is chosen to test the system performance. This profile includes fast, slow, strong, and stable fluctuations, as shown in Figure 11.

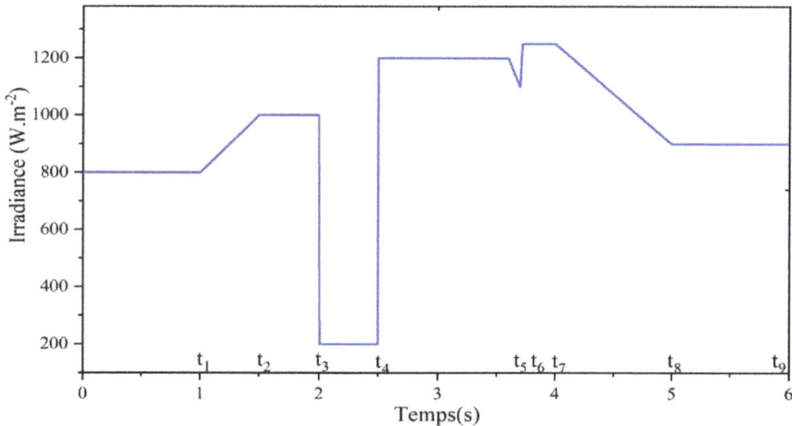

Figure 11. Irradiance profile used for simulation.

The graph in Figure 11 shows the different levels of irradiance fluctuations as a function of time t_i, which are modeled by:

- Constant irradiance in the intervals: $(0, t_1)$, (t_2, t_3), (t_3, t_4), (t_4, t_5), (t_6, t_7), and (t_8, t_9);
- Variable irradiance in the intervals: (t_1, t_2) and (t_7, t_8);
- Abrupt changes in irradiance at times t_3 and t_4;

This irradiation profile is very unstable in terms of meteorology, which is a good test to evaluate the behavior of the solar pumping system.

5.2. Simulation Results of the Direct Coupling

Figure 12 shows the simulation results of the SWPS in the direct coupled case. It should be noted that the irradiance variation leads to a significant power fluctuation. In addition, the power transmitted to the pump is considerably less than that generated by the GPV, as shown in Table 2. Under standard conditions where the irradiance has a value of 1000 $W \cdot m^{-2}$, the motor pump receives a power of 961 W, which presents 58% of the peak power (1650 W) produced by the GPV. Moreover, a significant variation of the motor pump voltage is undesirable for the safety of this type of system. Moreover, this configuration has as a problem the dependence between the power supplied by the GPV and the load characteristic. Indeed, the optimization of the energy requires another configuration in which an adaptation unit is introduced downstream of the GPV.

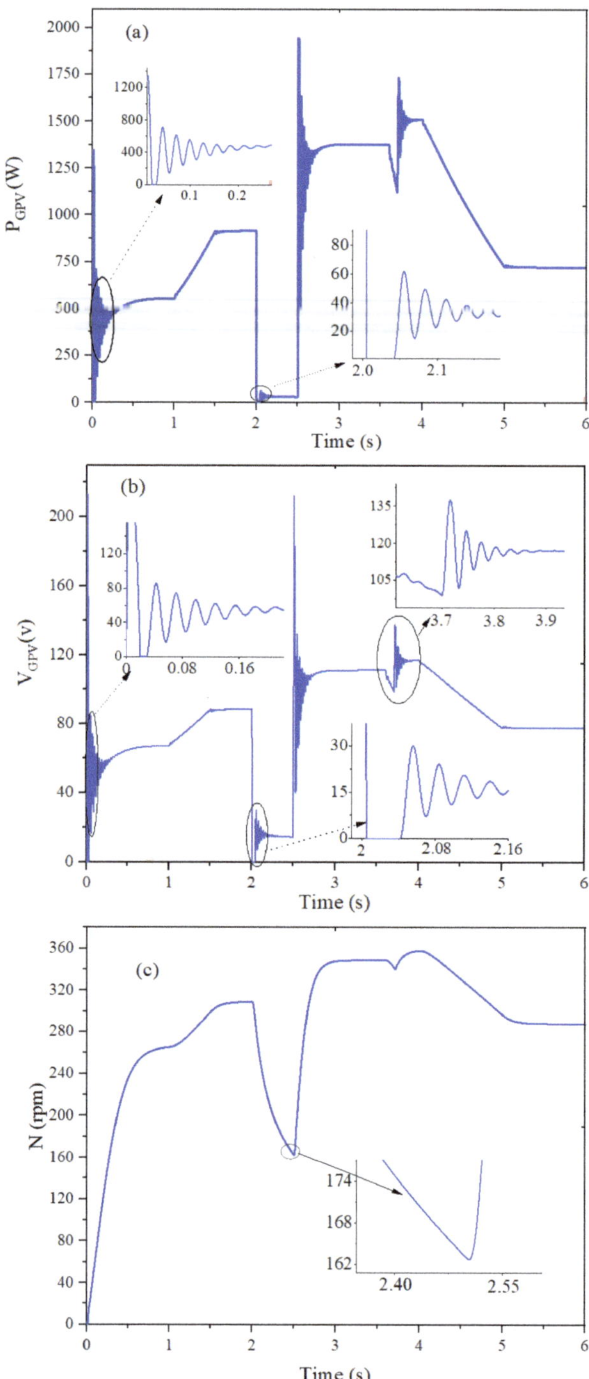

Figure 12. Simulation results for direct coupling. (**a**) Instantaneous power extracted from the GPV, (**b**) terminal voltage at the pump motor, (**c**) rotation speed of the pump motor.

5.3. Simulation Results Using the P&O Optimization Algorithm

Figure 13 shows the evolution of the variables in the indirect coupling between the GPV and the motor pump using the SEPIC converter in the case of optimization by the P&O technique. Figure 13a–c shows, respectively, the variation of the instantaneous electrical power, voltage, and rotational speed of the motor pump as a function of time. Under the standard irradiation conditions applied to the pumping system, the extracted power remains much higher and presents an increase of 47% compared with the direct coupling. In this regard, it is important to note that a clear improvement of the system from the power point of view with the developed controller is observed. Indeed, the power is perfectly improved compared with the direct coupling. However, this technique presents oscillations of about ±2 W, while with the decrease in the duty cycle, it is possible to decrease these oscillations, but the tracking will be much slower. To test the behavior of the MPPT–P&O technique under a fast variation of the irradiance, the instant t_3 presents an instantaneous drop of this one. Under these climatic conditions, the extracted power undergoes a decrease proportional to the irradiance. Compared with the direct coupling, the extracted power remains much higher. Finally, the P&O technique allows a good optimization of the pumping system. However, the decrease in the increment step of the duty cycle leads to a decrease in the power fluctuation, while the tracking speed does not adapt quickly to the irradiance changes. From Figure 13b, compared with direct coupling, the terminal voltage of the pump motor is significantly improved, especially during abrupt changes in irradiance (t_3, t_4, and t_5). Moreover, under steady-state conditions, the voltage has fluctuations of the order of 0.05 V, which is quite acceptable for pump systems. From Figure 13c, the speed of the motor pump is significantly improved and is almost generated for the entire range of weather variations. Compared with the direct coupling, this speed is significantly increased. In fact, it shows a significant improvement in that it has become 667 rpm instead of 310 rpm, a gain of 46%.

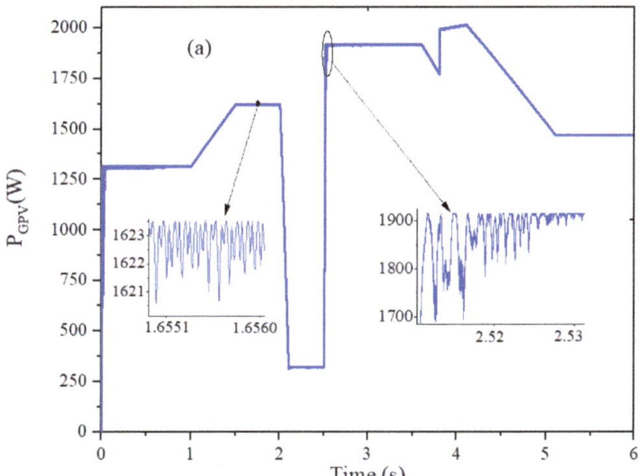

Figure 13. *Cont.*

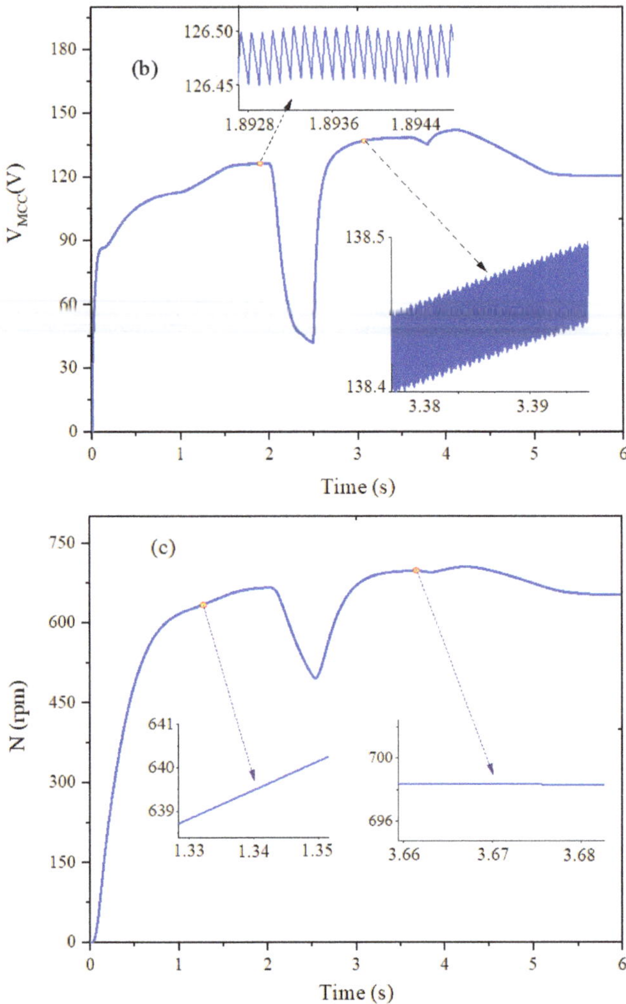

Figure 13. Simulation results associated with the P&O technique. (**a**) Instantaneous power extracted from the GPV, (**b**) terminal voltage at the pump motor, (**c**) rotation speed of the pump motor.

5.4. Simulation Results Using the M-INC Optimization Algorithm

Figure 14 shows the simulation results of the indirect coupling between the GPV and the motor pump using the SEPIC converter endowed with the M-INC technique. Figure 14a–c shows, respectively, the variation of the instantaneous electrical power, voltage, and rotational speed of the motor pump as a function of time. From Figure 14, it is important to note that the tracking speed is significantly higher than that obtained by the P&O technique. Indeed, under standard conditions, a value of 1654 W is noted, which is a gain of almost 3% compared with the P&O technique. In addition, the behavior of the M-INC control is quite satisfactory. Indeed, the tracking speed is significantly improved due to the fact that the system adapts quickly to changes. However, in the case of sudden changes in irradiance (t_3), the system suffers from overshoot, which affects the quality of the energy transmitted to the motor pump. The speed of the motor pump with this technique is significantly improved compared with the P&O technique, as shown in Figure 14c. The same behavior is observed for the voltage characteristic curve. Finally, the results obtained highlight the drawbacks of coupling by the M-INC technique for an application where the irradiance variation is very unstable. The power is significantly improved and does not deviate much from the MPP under unstable weather conditions, except that it suffers from power overshoot, which will affect the power quality.

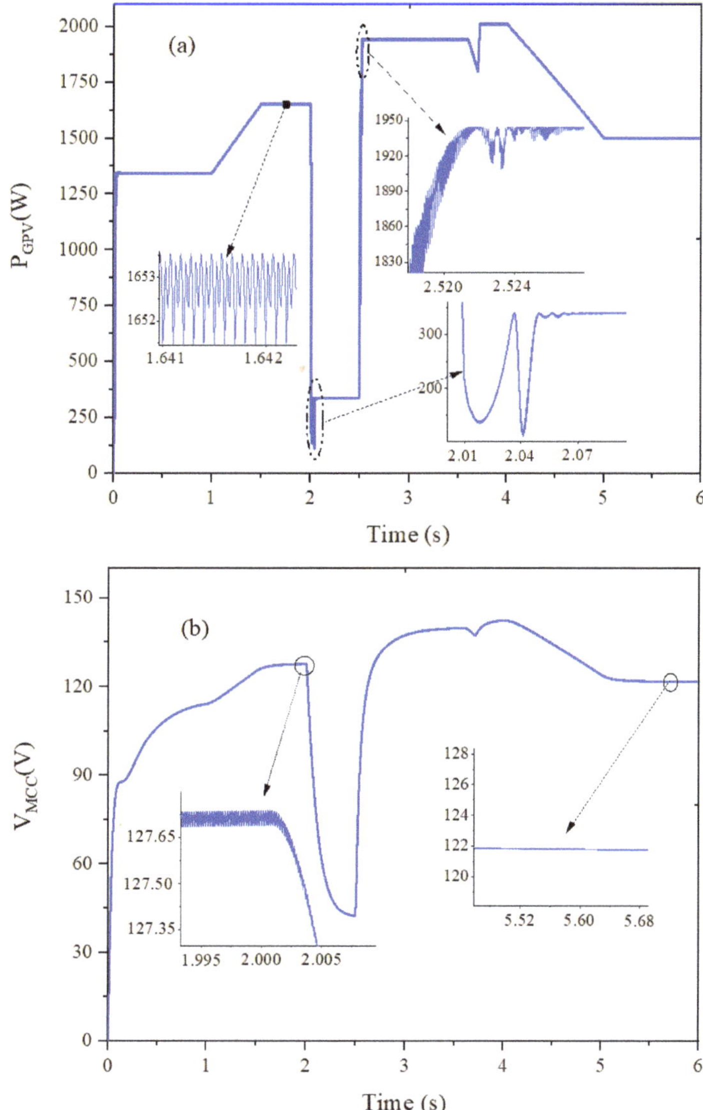

Figure 14. *Cont.*

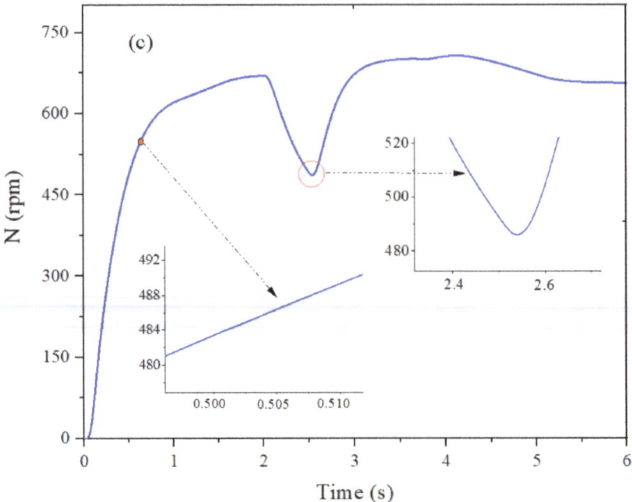

Figure 14. Simulation results associated with the M-INC technique. (**a**) Instantaneous power extracted from the GPV, (**b**) terminal voltage at the pump motor, (**c**) rotation speed of the pump motor.

5.5. Simulation Results Using the FL-INC Optimization Algorithm

Simulation results of the hybrid FL-INC technique of indirect coupling by a SEPIC converter with this algorithm are shown in Figure 15. It is interesting to note that this technique has a significant effect on overshoot, which is greatly attenuated compared with the M-INC technique. In addition, this technique improves the quality of the power transmitted to the SWPS while ensuring the required performance for such a system. Finally, the extracted power has a more improved character compared with the other techniques, as the tracking rate quickly reaches the maximum power point under different weather conditions. In addition, this power reflects the available solar energy potential while contributing to the preservation of the environment and the reduction of greenhouse gas emissions and climate change mitigation. This is due to the fact that this power presents a sufficient amount of solar water pumping applications.

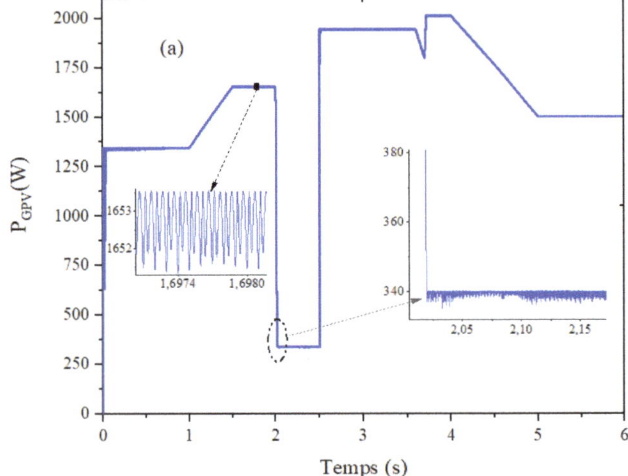

Figure 15. *Cont.*

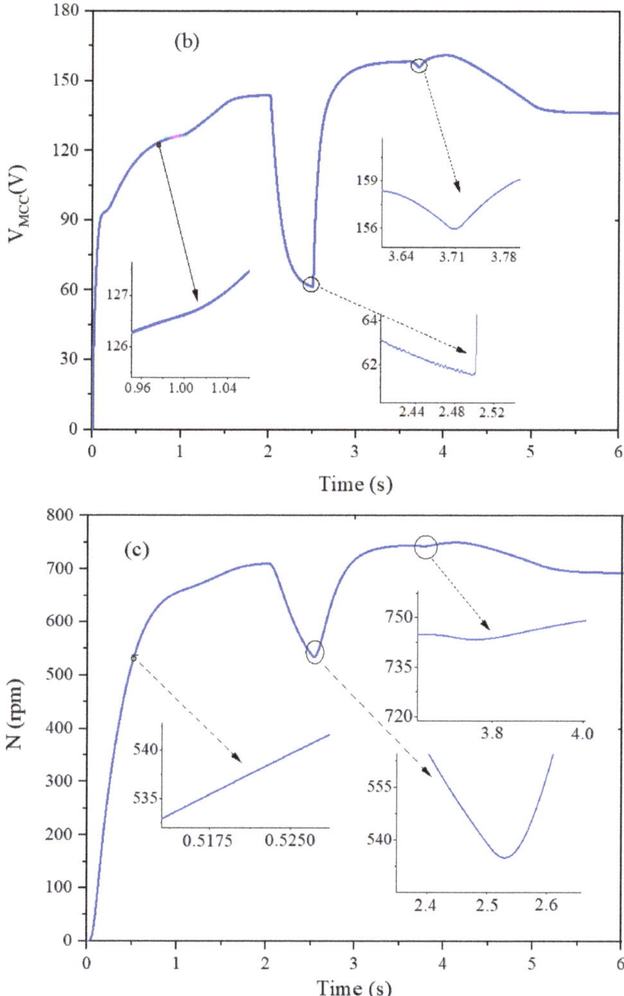

Figure 15. Simulation results associated with the FL-INC technique. (**a**) Instantaneous power extracted from the GPV, (**b**) terminal voltage at the pump motor, (**c**) rotation speed of the pump motor.

6. Comparative Study between the Studied Optimization Techniques

Figure 16 shows the simulation results of the instantaneous power extracted from the GPV across the four types of coupling, that is, direct coupling and coupling by an MPPT tracker. The latter is controlled by three types of algorithms: P&O, modified incremental, and hybrid FL-INC. The efficiency of the SWPS is very influenced by the weather conditions and the operating point of the system. The results in Figure 16 show that with direct coupling, the system efficiency is poor and the system operates farther from the maximum power point. The SEPIC converter with MPPT control has variable efficiency and tracking dynamics. With the P&O technique, the tracking of the power point is significantly improved over direct coupling, but the system is not able to have a better tracking speed during sudden changes in irradiance. The M-INC technique can overcome the tracking problem, but suffers from power overshoot during sudden irradiance changes compared with the hybrid technique.

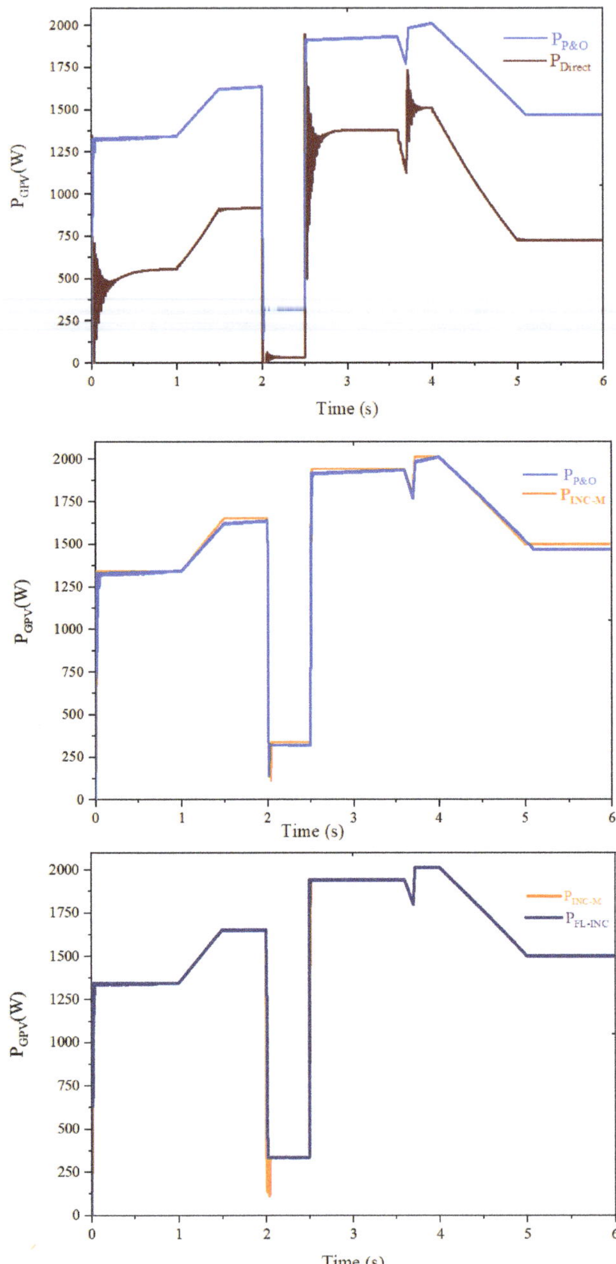

Figure 16. Instantaneous power of different solar water pumping techniques studied.

To show the influence of the power point tracking techniques on the solar water pumping flow rate, the histogram in Figure 17 presents a comparative study of the flow rate for each technique used. In fact, the flow rate in this simulation is calculated from a rotational speed that varies over the entire range of the simulation. According to Equation (3), the calculated flow rate is directly in proportion to the speed motor in the simulation. From the results, the hybrid technique has a better pumping rate than the others, that is, 55% rate compared with direct coupling, 8.5% compared with the P&O technique, and 5% rate compared with M-INC. Finally, the hybrid technique based on fuzzy logic for the artificial intelligence and the incremental technique has effectively improved the performance of SWPS.

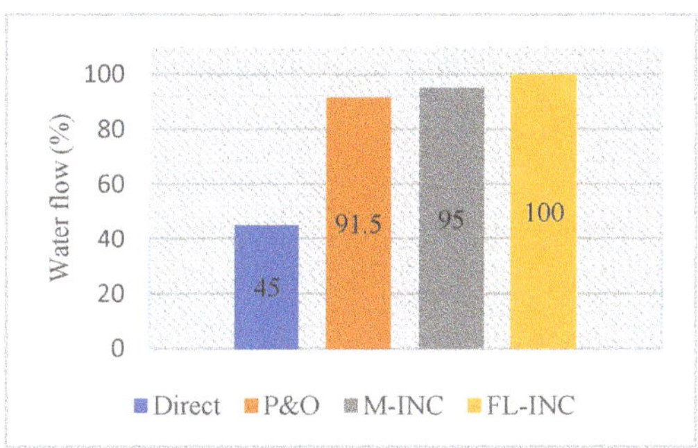

Figure 17. Water flow pumped from different techniques.

7. Conclusions

In this paper, the behavior and functioning of a solar water pumping system are studied. In a direct coupling, the performance of the system is related to the operating point. The latter can be more or less distant from the MPP as a function of climatic conditions. In this case, the efficiency of the MPPT depends strongly on the meteorological data and the characteristics of the motor pump. In this perspective, the improvement of the efficiency and power transfer quality of the system is provided by the SEPIC converter proposed in this study. Then, different control techniques of the SEPIC converter have been developed to optimize the power transfer. Finally, the analysis of the simulation results reveals that the combined hybrid method of fuzzy logic and the INC technique provides excellent static and dynamic performance compared with other techniques in terms of robustness, efficiency, stability, and tracking speed. According to the simulation results, the hybrid FL-INC technique provides a pumping rate gain of 5% over the M-INC technique, 8.5% over the P&O technique, and 55% over the direct coupling. The perspectives of this research work would be to test the developed optimization techniques in a real operating context.

Author Contributions: Conceptualization, A.H., M.A.M. and N.E.O.; methodology, M.A.M., S.M. and A.H.; software, A.H., S.M. and M.A.M.; validation, A.H., M.A.M., A.S.A.-S. and N.E.O.; formal analysis, A.H., M.A.M., S.M. and N.E.O.; investigation, S.M., M.A.M., N.E.O. and A.S.A.-S.; resources, S.M., M.A.M., N.E.O. and A.H.; data curation, A.H., S.M. and N.E.O.; writing—original draft preparation, A.H., N.E.O. and M.A.M.; writing—review and editing, A.H., S.M., M.A.M. and A.S.A.-S.; visualization, A.H., M.A.M. and S.M.; supervision, M.A.M. and A.S.A.-S.; project administration, M.A.M. and A.S.A.-S.; funding acquisition, A.S.A.-S. and M.A.M. All authors have read and agreed to the published version of the manuscript.

Funding: This research is supported by Khalifa University as part of research excellence award 8474000427 and ASPIRE under the ASPIRE Virtual Research Institute (VRI) Program, Award Number VRI20-07.

Data Availability Statement: The data are available upon request from the corresponding author (M.A.M.).

Conflicts of Interest: The authors declare no conflict of interest.

Abbreviations

SWPS	solar water pumping system
FL-INC	fuzzy logic and incremental conductance
SEPIC	single-ended primary inductance converter
INC	incremental conductance
M-INC	modified incremental conductance
P&O	perturb and observe
HC	hill climbing
ABC	artificial bee colony
PV	photovoltaic
IGBT	insulated gate bipolar transistor
MOSFET	metal oxide semiconductor field effect transistor
GPV	generator photovoltaic
MPP	maximum power point
MPPT	maximum power point tracking

References

1. Al-Sumaiti, A.S.; Ahmed, M.H.; Salama, M. Residential Load Management Under Stochastic Weather Condition in Developing Countries. *Electr. Power Compon. Syst.* **2014**, *42*, 1452–1473. [CrossRef]
2. Mossa, M.A.; Gam, O.; Bianchi, N.; Quynh, N.V. Enhanced Control and Power Management for a Renewable Energy-Based Water Pumping System. *IEEE Access* **2022**, *10*, 36028–36056. [CrossRef]
3. Mossa, M.A.; Gam, O.; Bianchi, N. Dynamic Performance Enhancement of a Renewable Energy System for Grid Connection and Stand-Alone Operation with Battery Storage. *Energies* **2022**, *15*, 1002. [CrossRef]
4. Mossa, M.A.; Echeikh, H.; Quynh, N.V.; Bianchi, N. Performance dynamics improvement of a hybrid wind/fuel cell/battery system for standalone operation. *IET Renew. Power Gener.* **2022**, 1–27. [CrossRef]
5. Banhidarah, A.K.; Al-Sumaiti, A.S. Heuristic search algorithms for optimal locations and sizing of distributed generators in the grid: A brief recent review. In Proceedings of the 2018 Advances in Science and Engineering Technology International Conferences (ASET), Dubai, Sharjah, Abu Dhabi, United Arab Emirates, 6 February–5 April 2018; pp. 1–5. [CrossRef]
6. Sharma, G.D.; Shah, M.I.; Shahzad, U.; Jain, M.; Chopra, R. Exploring the nexus between agriculture and greenhouse gas emissions in BIMSTEC region: The role of renewable energy and human capital as moderators. *J. Environ. Manag.* **2021**, *297*, 113316. [CrossRef]
7. Mossa, M.A.; Echeikh, H.; Iqbal, A. Enhanced control technique for a sensor-less wind driven doubly fed induction generator for energy conversion purpose. *Energy Rep.* **2021**, *7*, 5815–5833. [CrossRef]
8. Liu, X.; Liu, Z. Exploring the impact of media discourse on social perceptions towards biomass energy utilization in China. *Energy Strategy Rev.* **2022**, *42*, 100896. [CrossRef]
9. Li, J.; Yang, Z.; Yu, Z.; Shen, J.; Duan, Y. Influences of climatic environment on the geothermal power generation potential. *Energy Convers. Manag.* **2022**, *268*, 115980. [CrossRef]
10. Fujiwara, R.; Fukuhara, R.; Ebiko, T.; Miyatake, M. Forecasting design values of tidal/ocean power generator in the strait with unidirectional flow by deep learning. *Intell. Syst. Appl.* **2022**, *14*, 200067. [CrossRef]
11. World Health Organization. *Progress on Sanitation and Drinking Water: 2015 Update and MDG Assessment*; World Health Organization: Geneva, Switzerland, 2015.
12. Al-Sumaiti, A.S.; Kavousi-Fard, A.; Salama, M.; Pourbehzadi, M.; Reddy, S.; Rasheed, M.B. Economic Assessment of Distributed Generation Technologies: A Feasibility Study and Comparison with the Literature. *Energies* **2020**, *13*, 2764. [CrossRef]
13. Diab, A.A.Z.; Sultan, H.M.; Do, T.D.; Kamel, O.M.; Mossa, M.A. Coyote Optimization Algorithm for Parameters Estimation of Various Models of Solar Cells and PV Modules. *IEEE Access* **2020**, *8*, 111102–111140. [CrossRef]
14. Belhaouas, N.; Mehareb, F.; Assem, H.; Kouadri-Boudjelthia, E.; Bensalem, S.; Hadjrioua, F.; Aissaoui, A.; Bakria, K. A new approach of PV system structure to enhance performance of PV generator under partial shading effect. *J. Clean. Prod.* **2021**, *317*, 128349. [CrossRef]
15. Hilali, A.; Mardoude, Y.; Mahfoud, M.E.; Essahlaoui, A.; Houssam, M.; Rahali, A. Photovoltaic Water Pumping System: Modeling and Simulation of Characteristics for Direct Coupling. In *Digital Technologies and Applications*; Springer: Cham, Switzerland, 2022; pp. 651–660. [CrossRef]
16. Spier, D.W.; Oggier, G.G.; da Silva, S.A.O. Dynamic modeling and analysis of the bidirectional DC-DC boost-buck converter for renewable energy applications. *Sustain. Energy Technol. Assess.* **2019**, *34*, 133–145. [CrossRef]
17. Sakthivel, K.; Krishnasamy, R.; Balasubramanian, K.; Krishnakumar, V.; Ganesan, M. Averaged state space modeling and the applicability of the series Compensated Buck-Boost converter for harvesting solar Photo Voltaic energy. *Sustain. Energy Technol. Assess.* **2022**, *53*, 102611. [CrossRef]
18. Hilali, A.; Mardoude, Y.; Akka, Y.B.; Alami, H.E.; Rahali, A. Design, modeling and simulation of perturb and observe maximum power point tracking for a photovoltaic water pumping system. *Int. J. Electr. Comput. Eng. IJECE* **2022**, *12*, 4. [CrossRef]

19. Jately, V.; Azzopardi, B.; Joshi, J.; Venkateswaran, V.B.; Sharma, A.; Arora, S. Experimental Analysis of hill-climbing MPPT algorithms under low irradiance levels. *Renew. Sustain. Energy Rev.* **2021**, *150*, 111467. [CrossRef]
20. Kumar, R.; Khandelwal, S.; Upadhyay, P.; Pulipaka, S. Global maximum power point tracking using variable sampling time and p-v curve region shifting technique along with incremental conductance for partially shaded photovoltaic systems. *Sol. Energy* **2019**, *189*, 151–178. [CrossRef]
21. Pilakkat, D.; Kanthalakshmi, S. Single phase PV system operating under Partially Shaded Conditions with ABC PO as MPPT algorithm for grid connected applications. *Energy Rep.* **2020**, *6*, 1910–1921. [CrossRef]
22. Hilali, A.; Mardoude, Y.; Essahlaoui, A.; Rahali, A.; Ouanjli, N.E. Migration to solar water pump system: Environmental and economic benefits and their optimization using genetic algorithm Based MPPT. *Energy Rep.* **2022**, *8*, 10144–10153. [CrossRef]
23. Diab, A.A.Z.; Mohamed, M.A.; Al-Sumaiti, A.; Sultan, H.; Mossa, M. A Novel Hybrid Optimization Algorithm for Maximum Power Point Tracking of Partially Shaded Photovoltaic Systems. In *Advanced Technologies for Solar Photovoltaics Energy Systems*; Motahhir, S., Eltamaly, A.M., Eds.; Springer International Publishing: Berlin/Heidelberg, Germany, 2021; pp. 201–230. [CrossRef]
24. Eltamaly, A.M. Musical chairs algorithm for parameters estimation of PV cells. *Sol. Energy* **2022**, *241*, 601–620. [CrossRef]
25. Wang, C.; Shi, W.; Wang, X.; Jiang, X.; Yang, Y.; Li, W.; Zhou, L. Optimal design of multistage centrifugal pump based on the combined energy loss model and computational fluid dynamics. *Appl. Energy* **2017**, *187*, 10–26. [CrossRef]
26. Fatigati, F.; di Battista, D.; Cipollone, R. Design improvement of volumetric pump for engine cooling in the transportation sector. *Energy* **2021**, *231*, 120936. [CrossRef]
27. Yao, F.; Sun, H. The stability criterion of pumps. In Proceedings of the 2011 International Conference on Electronic Mechanical Engineering and Information Technology, Harbin, China, 12–14 August 2011; pp. 342–344. [CrossRef]
28. Rais, Y.; Aroudi, A.E.; Zakriti, A.; Khamlichi, A. Modeling and design of voltage-mode controlled PV-fed SEPIC converter. *IFAC-Pap.* **2022**, *55*, 514–519. [CrossRef]
29. Chellakhi, A.; el Beid, S.; Abouelmahjou, Y. An improved adaptable step-size P&O MPPT approach for standalone photovoltaic systems with battery station. *Simul. Model. Pract. Theory* **2022**, *121*, 102655. [CrossRef]
30. Mossa, M.A.; Gam, O.; Bianchi, N. Performance Enhancement of a Hybrid Renewable Energy System Accompanied with Energy Storage Unit Using Effective Control System. *Int. J. Robot. Control Syst.* **2022**, *2*, 140–171. [CrossRef]
31. Tey, K.S.; Mekhilef, S. Modified incremental conductance MPPT algorithm to mitigate inaccurate responses under fast-changing solar irradiation level. *Sol. Energy* **2014**, *101*, 333–342. [CrossRef]
32. Motahhir, S.; Chalh, A.; el Ghzizal, A.; Derouich, A. Development of a low-cost PV system using an improved INC algorithm and a PV panel Proteus model. *J. Clean. Prod.* **2018**, *204*, 355–365. [CrossRef]
33. Ganjeh-Alamdari, M.; Alikhani, R.; Perfilieva, I. Fuzzy logic approach in salt and pepper noise. *Comput. Electr. Eng.* **2022**, *102*, 108264. [CrossRef]
34. Reyes-García, C.A.; Torres-García, A.A. Chapter 8—Fuzzy logic and fuzzy systems. In *Biosignal Processing and Classification Using Computational Learning and Intelligence*; Torres-García, A.A., Reyes-García, C.A., Villaseñor-Pineda, L., Mendoza-Montoya, O., Eds.; Academic Press: Cambridge, MA, USA, 2022; pp. 153–176. [CrossRef]
35. Bose, B.K. Chapter 11—Fuzzy logic and applications. In *Power Electronics and Motor Drives*, 2nd ed.; Bose, B.K., Ed.; Academic Press: Cambridge, MA, USA, 2021; pp. 789–874. [CrossRef]

 *energies*

Article

Feedforward Artificial Neural Network (FFANN) Application in Solid Insulation Evaluation Methods for the Prediction of Loss of Life in Oil-Submerged Transformers

Bonginkosi A. Thango

Department of Electrical and Electronic Engineering Technology, University of Johannesburg, Johannesburg 2028, South Africa; bonginkosit@uj.ac.za; Tel.: +27-65-564-7287

Abstract: In this work, the application of a feed-forward artificial neural network (FFANN) in predicting the degree of polymerization (DP) and loss of life (LOL) in oil-submerged transformers by using the solid insulation evaluation method is presented. The solid insulation evaluation method is a reliable technique to assess and predict the DP and LOL as it furnishes bountiful information in examining the transformer condition. Herein, two FFANN models are proposed. The first model is based on predicting the DP when only the 2-Furaldehyde (2FAL) concentration measured from oil samples is available for new and existing transformers. The second FFANN model proposed is based on predicting the transformer LOL when the 2FAL and DP are available to the utility owner, typically for the transformer operating at a site where un-tanking the unit is a daunting and unfeasible task. The development encompasses constructing numerous FFANN designs and picking networks with superlative performance. The training and testing procedures databank is based on the dataset of the 2FAL and DP from a fleet of transformers and measured from laboratory analysis. The correlation coefficient of 0.964 was ascertained when the DP was predicted using the 2FAL measured in oil. In the FFANN model, a correlation coefficient of 0.999 against the practical data where one can make a reliable prediction of transformer LOL concerning 2FAL was generated and the amount of DP present produced. This model can be used to predict the DP and LOL of new and existing transformers at the manufacturer's premises and operating in the field, respectively. To the knowledge of the authors, no research work has been published addressing the methods proposed in this work.

Keywords: 2-furaldehyde (2FAL); feedforward artificial neural network (FFANN); degree of polymerization (DP); loss of life (LOL); transformers

Citation: Thango, B.A. Feedforward Artificial Neural Network (FFANN) Application in Solid Insulation Evaluation Methods for the Prediction of Loss of Life in Oil-Submerged Transformers. *Energies* **2022**, *15*, 8548. https://doi.org/10.3390/en15228548

Academic Editors: Najib El Ouanjli, Saad Motahhir and Mustapha Errouha

Received: 14 October 2022
Accepted: 13 November 2022
Published: 15 November 2022

Publisher's Note: MDPI stays neutral with regard to jurisdictional claims in published maps and institutional affiliations.

1. Introduction

Oil-submerged cellulose paper is extensively used as a solid dielectric material in electrical power transformers [1]. During the operational lifetime of the power transformer, cellulose paper progressively decomposes on account of the multi-pronged stresses it is susceptible to [2]. Considering that the insulation condition is a crucial factor to ascertain the reliable operation of power transformers, assessment of the insulation condition has obtained considerable attention from various researchers in the power industry [3]. The effective residual duration of the operation of a power transformer is determined in keeping with the condition of its paper insulation by measuring the degree of polymerization (DP) [4]. According to [5], the earliest DP value of newly manufactured cellulose paper is in the range between 1000 and 1200, which plunges to about 200 to 300 once the cellulose paper attains the end of its lifecycle. Whereas the DP is the most artful indication to examine the decomposition of a cellulose paper insulation, it is seldomly employed by the manufacturing industry due to the intricacy of harvesting the physical paper samples, particularly from hotspot regions of the unit in-service. Intrinsically, considerable research attempts have been performed to associate the DP value with various measurable dissolved-in oil indicators [6–10]. Cellulose is a straight polymer of glucose molecules inextricably linked to glycosidic bonds [5,6]. The cause of towering thermal stresses, hydrogen bonds have the propensity to disintegrate, which is caused by the reduced cellulose molecular chain and the formation of chemical derivatives that are then suspended in the dielectric oil. Furanic compounds and moisture are the principal derivatives of cellulose decomposition [7,8].

In recent times, several DP and LOL models for power transformers have been presented, however, no model has been universally accepted. It is critical to identify the major impediments. Previous reviews in this context have concentrated on a single solution based on specific case studies and alternative routes that can be taken by the manufacturers based on data availability have not been addressed. To corroborate the contribution to the current research, a summary of recent research works is tabulated in Table 1.

Table 1. A summary of recent research works.

Ref. No	Journal Ranking	Year	Applied Method	Summary
[11]	Q2	2021	Statistical tool (multiple linear regression)	A study of the relationship between the DP and various transformer parameters (DGA, breakdown voltage, furans, oil interfacial tension, moisture content, etc.) is carried out. A strong correlation between the DP and furan was discovered. There was no strong correlation between the DP and other parameters except for furans
[12]	Q2	2021	ANN	The study attempted to predict the DP using methanol, carbon oxides, and hydrogen released in the insulating oil. The classification of the study was, however, tailor-made to the individual maintenance and scheduling strategy of a power utility
[13]	Q1	2021	ANN	The study predicted the furans using the temperature, carbon dioxide, carbon monoxide, and moisture to determine the DP
[14]	Q1	2022	Empirical modeling	The study predicted DP using methanol. The relative error yielded 7%
[15]	Q1	2021	Regression modeling	Prediction of the DP using the furfural indicator at various oil over pressboard ratios and oil change status
[16]	Q1	2021	Frequency domain spectroscopy	The DP was predicted by employing the frequency conditional dielectric modulus technique
[17]	Q2	2020	Fuzzy controller	The DP was predicted using a fuzzy logic controller and considering the furans
[18]	Q1	2019	Adaptive neuro-fuzzy inference system (ANFIS)	DP was predicted using ANFIS and considering furans, carbon dioxide, and carbon monoxide
[19]	Q2	2018	Using Monte Carlo algorithm and ANN	The transformer LOL was predicted considering the loading data, carbon dioxide, breakdown voltage, and acidity
[20]	Q1	2022	Genetic-algorithm optimized support vector machine	The PD was predicted using methanol and ethanol

It can be observed from the most recent works that the methodologies proposed herein have not been reported. It is therefore vital to disseminate the proposed methods to support transformer manufacturers and dielectric laboratories with reliable alternatives to these methods. In this work, the remnant loss of life of the cellulose paper insulation was predicted by proposing two FFANN approaches. The first model is based on predicting the DP when only the 2-furaldehyde (2FAL) concentration measured from the oil samples is available for new and existing transformers. The second FFANN model proposed is based on predicting the transformer LOL when the 2FAL and DP are available to the utility owner.

1.1. Manuscript Contribution

This work presents comprehensive research on developing novel prediction models for loss of life in oil-submerged transformers. The research underpins two models based on FFANN. The contributions of the current research study are as follows:

- A model capable of predicting the DP when only the 2-furaldehyde (2FAL) concentration measured from oil samples is available for new and existing transformers using FFANN was developed
- An FFANN model was developed to predict the transformer LOL when the 2FAL and DP are available to the utility owner, typically for the transformer operating at a site where un-tanking the unit will be an impractical task.

1.2. The Novelty of Current Research

The fundamental goal of this research work was to obtain the loss of life of solid insulation in transformers using different available data alternatively. Even though numerous investigators have worked on the transformer loss of life prediction, as shown in Table 1, no existing research has reported on the application of FFANN to suggest each of the methods proposed in the current study. On the two novel approaches proposed in the current work, the first model was based on predicting the DP when only the 2-furaldehyde (2FAL) concentration measured from the oil samples is available for new and existing transformers. The second FFANN model proposed was based on predicting the transformer LOL when the 2FAL and DP are available to the utility owner, typically for the transformer operating at a site where un-tanking the unit will be a daunting and unfeasible task. These approaches are crucial in the development of prediction techniques for the DP and LOL of new and existing transformers at the manufacturer's premises and operating in the field, respectively.

1.3. The Manuscript Organization

The rest of this manuscript is organized as follows. Section 2 introduces the fundamental principle of artificial neural networks and proposed feedforward artificial neural network. Section 3 presents the results of thee developed models for predicting DP when only 2FAL is available and predicting LOL using the predicted DP and measured 2FAL. Finally, Section 4 presents a detailed conclusion.

2. Materials and Methods

2.1. The Fundamental Principle of Artificial Neural Network

Artificial neural networks (ANNs) are computing techniques motivated by the genetic neural networks that are composed in animal brains [18–20]. An ANN is constructed on an assemblage of coupled nodes so-called artificial neurons, which are essentially archetypal of the neurons in a genetic animal brain. Respective connections such as the synapses in the animal brain can diffuse information to coupled neurons. An artificial neuron obtains information and subsequently processes them and can inform coupled neurons. The "information" at a link is a real number, plus the output of the respective neuron is calculated by a certain non-linear function of the summation of its inputs [18–20]. The links are so-called edges. Generally, neurons and edges are characteristically composed of a weight that adapts as learning progresses. The weight rises or drops the power of the information at a connection. Neurons can have a permissible range in a manner where information is driven exclusively when the total signal intersects that permissible range. Classically, neurons are amassed into layers [18–20]. Distinctive layers can carry out various conversions on the respective inputs. Information travels from the input layer to the output layer, conceivably following crisscrossing the layers numerous times.

2.2. Proposed Feedforward Artificial Neural Network

Two computational schemes based on FFANN were designed to estimate the residual lifespan of oil-submerged transformers. The first model was based on predicting the DP when only the 2-furaldehyde (2FAL) concentration measured from the oil samples is available for new and existing transformers. The second FFANN model proposed was based on predicting the transformer LOL when the 2FAL and DP are available to the utility owner, typically for the transformer operating at a site where un-tanking the unit will be a daunting and unfeasible task.

The choice of inputs, outputs, and network structure in the FFANN prototype is dependent on the efficiency of the FFANN models. The measurements for the gas concentrations concerning the emerging transformer faults were gathered from the real data of a power transformer. In this work, the development of an FFANN was split into four phases: data gathering and processing, FFANN modeling, training, and testing.

2.2.1. Data Gathering and Processing

In the data gathering and processing stage, various transformers from 315 kVA to about 40 MVA 132 kV were considered in the study. These units had been removed from service and were processed in a workshop to diagnose their conditions. The oil sample of individual units was analyzed in the laboratory to attain the 2FAL concentration existing in the oil sample. Concurrently, the sample of the solid insulation material was extracted and analyzed in the laboratory to attain the degree of polymerization. For the respective unit, the lifetime of the unit in-service was collected from the original equipment manufacturer. It follows that the data were divided into two datasets: one contained the measured DP and the measured 2FAL, and in the other dataset, a dataset with column vectors comprising the DP, 2FAL and LOL of the unit was assembled. It should be noted that the LOL is essentially the remnant life of the unit (i.e., the designed services life of the unit minus the lifetime in operation). In processing the datasets, the inputs as well as target data are determined and fed into the ANN matrix for training and validation. In the first model, the 2FAL was specified to be the input, whereas the corresponding DP was specified as the target. In the second proposed model, the transformer remnant life was defined as the target response whereas the available DP and 2FAL dataset were utilized as the inputs. These datasets were classified into three categories: learning, verification, and evaluation. The learning sample comprised 70% of the entire dataset, with the remaining 30% used for verification and evaluation.

2.2.2. ANN Models

In this work, the MATLAB/SIMULINK tool was utilized to execute two ANN models. To identify the optimal ANN model, the learning or training rate (LR) and momentum cost (MC) was changed between 0 to 0.9. However, since all variables were altered iteratively, underfitting as well as overfitting systems remained possible. Overfitting happens whenever a system is proficient in memorizing the system but is unable to extrapolate new input for the system. The early halting approach has been used to achieve an optimized performance to address the overfitting challenge. The termination criteria were obtained by evaluating the mean square error of the learning data during training using data that are limited in size.

2.2.3. Training Phase

A backpropagation approach is a simplified delta function for a feed-forward system with numerous layers that are used during the training phase. This is due to its ability to calculate the slope of each layer, continuously utilizing the chain principle. In practice, quadratic activation functions are utilized to improve performance due to their non-linearity and suitability with a feed-forward backpropagation training (FFBPT) approach. The LM was adopted as the learning classifier in this work because it is a rapid, simple, and stable approach. As a result, the FFBPT approach was chosen as the system structure for the ANN design. The optimal ANN settings with the maximum precision, which is equivalent to R, were obtained by modifying the number of hidden layers, the number of neurons as well as the transfer function. In this work, a three-layer system with 10-hidden layers and a 1-output layer was adopted for both ANN models. While one hidden layer is sufficient for nonlinear modeling, a system with 2-hidden layers outperforms systems with 1- as well as 3-hidden layers in terms of the number of iterations, precision, and sophistication. Furthermore, the 3-layer system helps solve the challenge of slow learning rates.

2.2.4. Testing Phase

An additional batch of data was used to evaluate the trained system. The trained ANN was used to mimic the response output of the additional data batch. The optimal-trained system demonstrates that the modeled output matched the desired output accurately. The efficiency of the trained system was determined utilizing the R correlation value.

Table 2 illustrates a comparative analysis of the ANN types when using the first proposed model dataset. These results justify the selection for the FFANN, which yielded an MSE of 0.324 and an R2 of 0.96431.

Table 2. Comparative analysis of the ANN types.

Parameter	FFANN	RBNN	MPM	RNN
MSE	0.324	2.567	5.876	8.26
R2	0.96431	0.845	0.765	0.56

RBNN—Radial basis neural network, MPM—Multilayer perception model, RNN—Recurrent Neural network, MSE—Mean Square error, R2—correlation coefficient.

3. Results

3.1. Predicting DP When 2FAL Only Is Available

Figure 1 depicts the modeling configuration that predicts DP when only 2FAL is available. The coaching efficiency graph in Figure 2 clearly shows that the learning ability of the ANN is satisfactory. The average mean square error of the trained ANN was less than the predetermined minimum of 0.0001. The mean square error was 692.943 after the coaching of the ANN. As a result, this model was selected as the definitive option for the specified input and output. The trial set error as well as the verification set error had comparable features, and zero notable overfittings transpired after iteration 14. This was the region where the optimum validation result had been achieved.

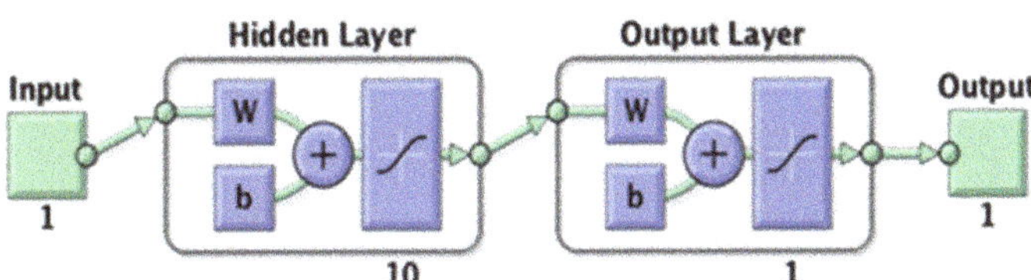

Figure 1. The ANN classifier for the LOL estimation for model one.

Mineral-based oil units were considered. The condition of the oil was taken as is for analysis as it represents the accurate amounts of furans present. The dataset was utilized to train the ANN. Following ANN training, its efficiency was verified by showing the linear regression graph that correlated the targets to the outputs, as illustrated in Figure 3. The correlation coefficient indicates how effectively the ANN's targets may detect changes in outputs. A correlation of 0 means that there is no correlation at all, whereas a correlation of 1 means that there is a complete correlation. A regression result suggests a close relationship between the outputs and targets. The correlation coefficient in this scenario was observed to be 0.964, indicating a high correlation.

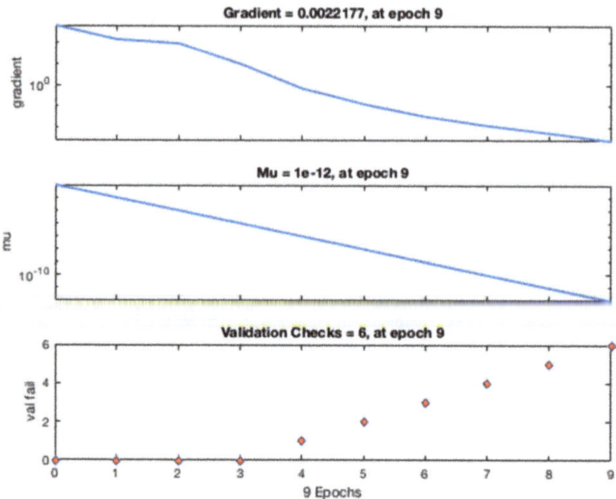

Figure 2. Performance of the training process of model one.

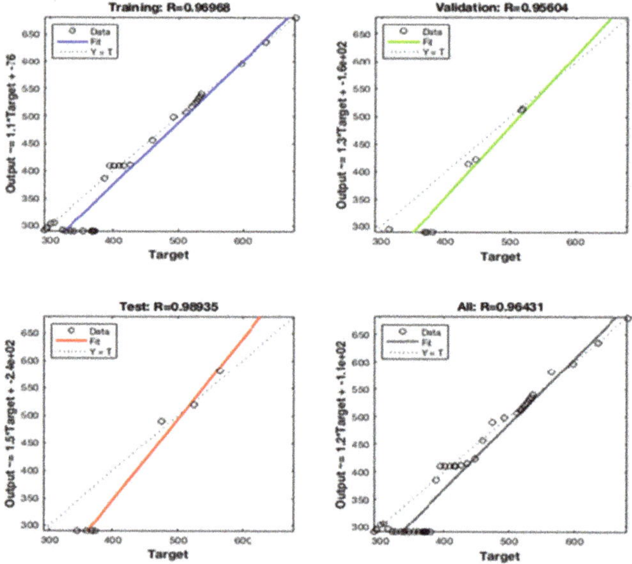

Figure 3. Model linear fitting in training and testing for model one.

Figure 4 depicts the accuracy of the validation checks using the validation set. When the validation error begins to rise, the ANN terminates the learning session, regardless of whether the target has still not been achieved. Therefore, the ANN has a high degree of generalizability. This will halt the learning session when the abstraction performance has reached its maximum. Figure 5 demonstrate the performance parameters of the proposed FFANN.

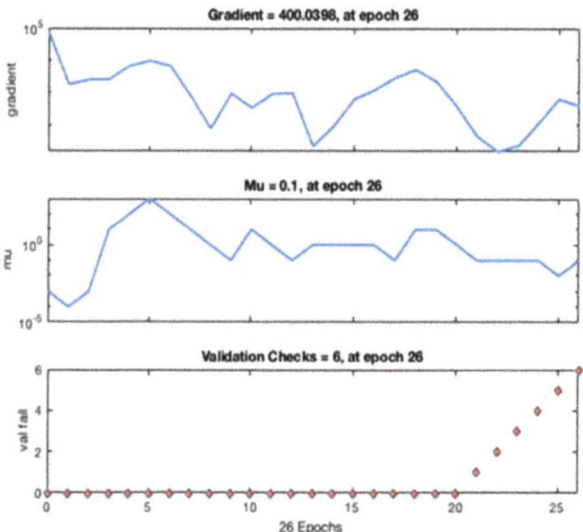

Figure 4. Validation in the training phase of model one.

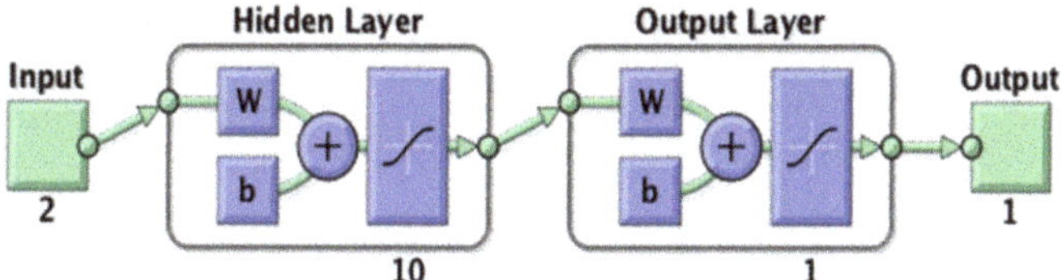

Figure 5. ANN model training of model one.

3.2. Predicting LOL Using Predicted DP and Measured 2FAL

Figure 6 illustrates the model setup for predicting LOL based on the predicted DP and measured 2FAL.

Figure 6. The ANN classifier for LOL estimation for model two.

The cumulative mean square deviation of the generated ANN was 0.0006684, and Figure 7 illustrates that the testing and verification curves had comparable properties, indicating efficient training. The trial set error and the verification set error had similar characteristics, and no significant overfitting occurred after iteration 13. The correlation coefficient represents how successfully the ANN's goals can make corrections in outputs, with 0 representing no correlation at all, while 1 represents perfect correlation. The function of the trained ANN was evaluated in two methods. Initially, the linear regression that ties the goals to the outputs is illustrated in Figure 8. The correlation coefficient in this scenario was observed to be 0.999, indicating a strong correlation.

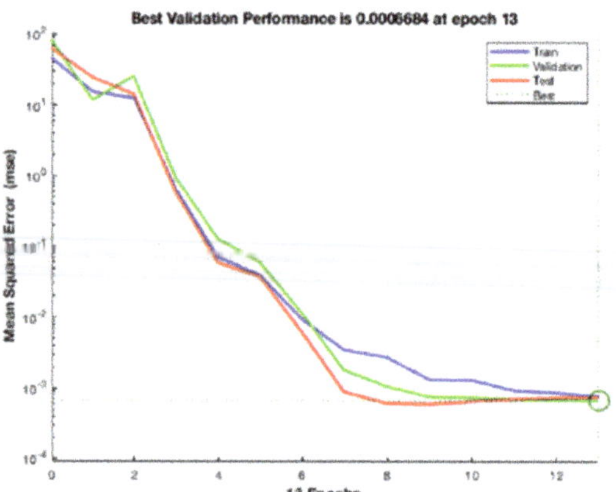

Figure 7. Performance of the training process for model two.

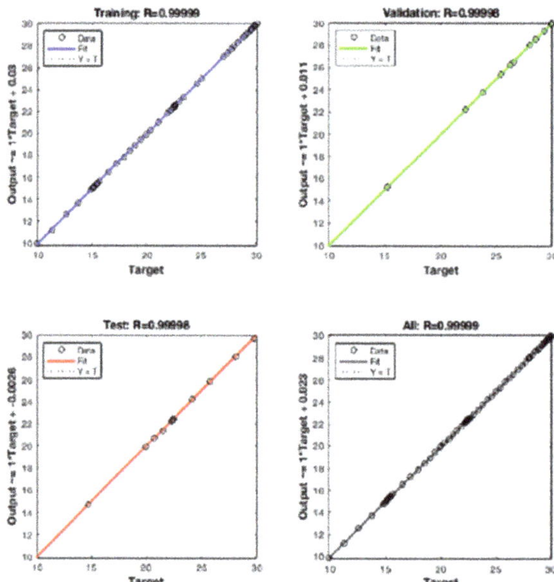

Figure 8. Model linear fitting in training and testing model two.

It was observed that the model could be used to predict the DP and LOL of new and existing transformers at the manufacturer's premises and operating in the field, respectively.

In Figure 9, the Gradient, mu (i.e., control parameter for the algorithm used to train the neural network) and validation failure (val fail) results are presented.

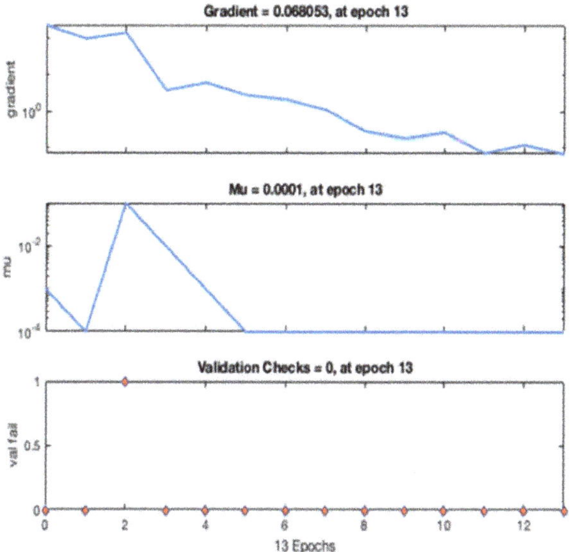

Figure 9. Validation in the training phase of model two.

The output parameters of the training model for model two are illustrated in Figure 10.

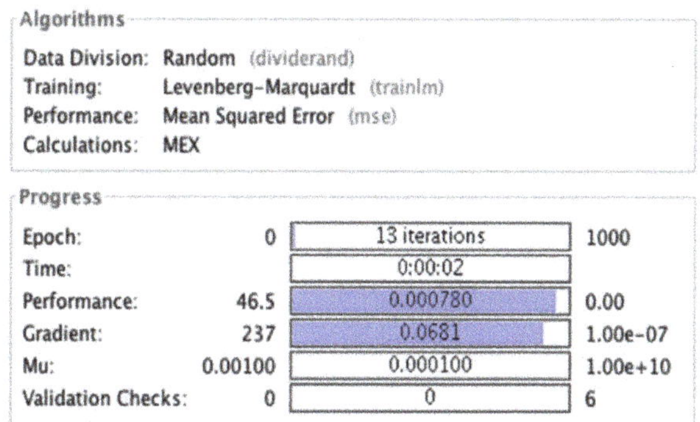

Figure 10. ANN model training of model two.

4. Conclusions

Several publications have suggested that power transformers erected in the 1980s are still in service and that some of these transformers are in an acceptable state based on the data analyses such as DP, furan, CO_2, and others. The service durability of these transformers exceeds 21 years, although the conventional technique predicts 21 years. This sustained efficiency is dependent on operational situations. These transformers are most likely to be used within the manufacturers' specified thresholds. Furthermore, the residual lifespan of the insulation is dependent on the transformer overloading cases and may be evaluated utilizing software iterations, considering the variance of overloading and LOL in the past years.

In this work, we proposed an approach for estimating the lifespan reduction and expansion of transformers. To acquire better consistent results, it is preferable to adopt a knowledge-based technique that accommodates all sets of data to accurately estimate the LOL of transformers. The ANN was applied to predict the DP and LOL in oil-submerged transformers by using the solid insulation evaluation. The proposed approach makes it simple to ascertain the extent of lifetime reduction and expansion for transformers, providing for improved accurate prediction of residual serviceability.

In this work, two ANN models were proposed. The first model was based on predicting the DP when only the 2FAL concentration measured from oil samples is available for new and existing transformers. The second ANN model proposed was based on predicting the transformer LOL when the 2FAL and DP are available to the utility owner, typically for the transformer operating at the site where un-tanking the unit will be a daunting and unfeasible task. The training and testing procedures databank was based on the dataset of the 2FAL and DP from a fleet of transformers and measured from laboratory analysis. The correlation coefficient of 0.964 was ascertained when the DP was predicted using the 2FAL measured in oil. On the ANN model, a correlation coefficient of 0.999 was obtained, against the practical data where one can make a reliable prediction of transformer LOL concerning the 2FAL generated and the amount of DP present produced. It was found that this model can be used to predict the DP and LOL of new and existing transformers at the manufacturer's premises and operating in the field, respectively.

This work provides critical knowledge for the electrical energy industry as well as beneficial attributes for future preparation. Operational planning for electrical generation, transmission, and distribution networks can perhaps be designed with greater reliability.

Funding: This research received no external funding.

Institutional Review Board Statement: Not applicable.

Informed Consent Statement: Not applicable.

Data Availability Statement: Not applicable.

Conflicts of Interest: The author declares no conflict of interest.

References

1. Ghoneim, S.S.; Taha, I.B.; Elkalashy, N.I. Integrated ANN-based proactive fault diagnostic scheme for power transformers using dissolved gas analysis. *IEEE Trans. Dielectr. Electr. Insul.* **2016**, *23*, 1838–1845. [CrossRef]
2. Singh, A.; Verma, P. A review of intelligent diagnostic methods for condition assessment of insulation system in power transformers. In Proceedings of the 2008 International Conference on Condition Monitoring and Diagnosis, Beijing, China, 21–24 April 2008; pp. 1354–1357.
3. Husain, E.; Mohsin, M. Transformer insulation condition monitoring using artificial neural network. In Proceedings of the 2001 IEEE 7th International Conference on Solid Dielectrics (Cat. No. 01CH37117), Eindhoven, The Netherlands, 25–29 June 2001; pp. 295–298.
4. Nezami, M.; Equbal, D.; Khan, S.; Sohail, S.; Ghoneim, S. Classification of Cellulosic Insulation State Based on Smart Life Prediction Approach (SLPA). *Processes* **2021**, *9*, 981. [CrossRef]
5. Sameh, W.; Gad, A.H.; Eldebeikey, S.M. An intelligent classifier of electrical discharges in oil immersed power transformers. In Proceedings of the 2019 21st International Middle East Power Systems Conference (MEPCON), Cairo, Egypt, 17–19 December 2019; pp. 866–871.
6. Ahmad, A.; Othman, M.L.; Zainab, K.K.B.; Hizam, H. Adaptive ANN based differential protective relay for reliable power transformer protection operation during energisation. *IAES Int. J. Artif. Intell.* **2019**, *8*, 307–316. [CrossRef]
7. Thango, B.A.; Akumu, A.O.; Sikhosana, L.S.; Nnachi, A.F.; Jordaan, J.A. Study of the Impact of Degree of Polymerization on the Remnant Life of Transformer Insulation. In Proceedings of the 2021 IEEE PES/IAS PowerAfrica, Nairobi, Kenya, 23–27 August 2021; pp. 1–5. [CrossRef]
8. Žarković, M.; Stojković, Z. Analysis of artificial intelligence expert systems for power transformer condition monitoring and diagnostics. *Electr. Power Syst. Res.* **2017**, *149*, 125–136. [CrossRef]
9. Thango, B.A.; Akumu, A.O.; Sikhosana, L.S.; Nnachi, A.F.; Jordaan, J.A. Empirical Based Approaches to Evaluating the Residual Life for Oil-Immersed Transformers—A Case Study. In Proceedings of the 2021 IEEE AFRICON, Arusha, Tanzania, 13–15 September 2021; pp. 1–4. [CrossRef]
10. Thango, B.A.; Jordaan, J.A.; Nnachi, A.F. A Novel Approach for Estimating the Service Lifetime of Transformers within Distributed Solar Photovoltaic (DSPV) Systems. *Adv. Sci. Technol. Eng. Syst. J.* **2021**, *6*, 126–130. [CrossRef]
11. Ghoneim, S. The Degree of Polymerization in a Prediction Model of Insulating Paper and the Remaining Life of Power Transformers. *Energies* **2021**, *14*, 670. [CrossRef]
12. Behjat, V.; Emadifar, R.; Pourhossein, M.; Rao, U.; Fofana, I.; Najjar, R. Improved Monitoring and Diagnosis of Transformer Solid Insulation Using Pertinent Chemical Indicators. *Energies* **2021**, *14*, 3977. [CrossRef]

13. Oria, C.; Méndez, C.; Carrascal, I.; Ortiz, A.; Ferreño, D. Impact of the Use of Vegetable Oil on the Mechanical Failure of the Cellulosic Insulation of Continuously Transposed Conductors in Power Transformers. In *IEEE Transactions on Dielectrics and Electrical Insulation*; IEEE: Piscataway, NJ, USA, 2022; Volume 29, pp. 607–613. [CrossRef]
14. Chen, Q.; Li, C.; Cheng, S.; Sun, W.; Chi, M.; Zhang, H. Study on Aging Assessment Model of Transformer Cellulose Insulation Paper Based on Methanol in Oil. *IEEE Trans. Dielectr. Electr. Insul.* **2022**, *29*, 591–598. [CrossRef]
15. Liu, J.; Zhang, H.; Geng, C.; Fan, X.; Zhang, Y. Aging Assessment Model of Transformer Insulation Based on Furfural Indicator under Different Oil/Pressboard Ratios and Oil Change. *IEEE Trans. Dielectr. Electr. Insul.* **2021**, *28*, 1061–1069. [CrossRef]
16. Liu, J.; Zhang, H.; Fan, X.; Zhang, Y.; Zhang, C. Aging evaluation for transformer oil-immersed cellulose insulation by using frequency dependent dielectric modulus technique. *Cellulose* **2021**, *28*, 2387–2401. [CrossRef]
17. Elmashtoly, A.M.; Chang, C.-K. Prognostics Health Management System for Power Transformer with IEC61850 and Internet of Things. *J. Electr. Eng. Technol.* **2020**, *15*, 673–683. [CrossRef]
18. Mharakurwa, E.T.; Nyakoe, G.N.; Akumu, A.O. Transformer Insulation Degree of Polymerization Estimation through Adaptive Neuro Fuzzy Inference System Approach. In Proceedings of the 2019 IEEE Electrical Insulation Conference (EIC), Calgary, AB, Canada, 16–19 June 2019; pp. 160–162. [CrossRef]
19. Jahromi, M.Z.; MehrabanJahromi, M.; Tajdinian, M.; Allahbakhshi, M. A Novel method to estimate Economic Replacing Time of Transformer Using Monte Carlo Algorithm and ANN. *IIUM Eng. J.* **2018**, *19*, 54–67. [CrossRef]
20. Wu, S.; Zhang, H.; Wang, Y.; Luo, Y.; He, J.; Yu, X.; Zhang, Y.; Liu, J.; Shuang, F. Concentration Prediction of Polymer Insulation Aging Indicator-Alcohols in Oil Based on Genetic Algorithm-Optimized Support Vector Machines. *Polymers* **2022**, *14*, 1449. [CrossRef] [PubMed]

Article

Dissolved Gas Analysis and Application of Artificial Intelligence Technique for Fault Diagnosis in Power Transformers: A South African Case Study

Bonginkosi A. Thango

Citation: Thango, B.A. Dissolved Gas Analysis and Application of Artificial Intelligence Technique for Fault Diagnosis in Power Transformers: A South African Case Study. *Energies* **2022**, *15*, 9030. https://doi.org/10.3390/en15239030

Academic Editors: José Matas, Saad Motahhir, Najib El Ouanjli and Mustapha Errouha

Received: 14 October 2022
Accepted: 24 November 2022
Published: 29 November 2022

Publisher's Note: MDPI stays neutral with regard to jurisdictional claims in published maps and institutional affiliations.

Department of Electrical and Electronic Engineering Technology, University of Johannesburg, Johannesburg 2000, South Africa; bonginkosit@uj.ac.za; Tel.: +27-65-564-7287

Abstract: In South Africa, the growing power demand, challenges of having idle infrastructure, and power delivery issues have become crucial problems. Reliability enhancement necessitates a life-cycle performance analysis of the electrical power transformers. To attain reliable operation and continuous electric power supply, methodical condition monitoring of the electrical power transformer is compulsory. Abrupt breakdown of the power transformer instigates grievous economic detriment in the context of the cost of the transformer and disturbance in the electrical energy supply. On the condition that the state of the transformer is appraised in advance, it can be superseded to reduced loading conditions as an alternative to unexpected failure. Dissolved gas analysis (DGA) nowadays has become a customary method for diagnosing transformer faults. DGA provides the concentration level of various gases dissolved, and consequently, the nature of faults can be predicted subject to the concentration level of the gases. The prediction of fault class from DGA output has so far proven to be not holistically reliable when using conventional methods on account of the volatility of the DGA data in line with the rating and working conditions of the transformer. Several faults are unpredictable using the IEC gas ratio (IECGR) method, and an artificial neural network (ANN) has the hindrance of overfitting. Nonetheless, considering that transformer fault prediction is a classification problem, in this work, a unique classification algorithm is proposed. This applies a binary classification support vector machine (BCSVM). The classification precision is not reliant on the number of features of the input gases dataset. The results indicate that the proposed BCSVM furnishes improved results concerning IECGR and ANN methods traceable to its enhanced generalization capability and constructional risk-abatement principle.

Keywords: dissolved gas analysis (DGA); IEC gas ratio; transformer; faults; binary classification support vector machine (BCSVM)

1. Introduction

It is well documented that the power supply system is a composite network comprising several customers, i.e., residential, industrial, etc., operating at distinct voltage levels [1,2]. This distinct voltage level is facilitated using power and distribution transformers [2]. Consequently, the high degree of operational reliability and lucrative operation of the electrical transformers and, thereby, the power system are of considerable engineering significance. It is recognized that the performance and planned life span of a transformer is assignable to the decomposition of the dielectric system [3]. Moreover, this chain of events requires efficacious condition-monitoring procedures and diagnostic tools. For this reason, condition monitoring of dielectric oil and cellulosic paper insulation is an absolute priority to achieve a planned operational lifetime and to evade destructive failures of electrical transformers. Dissolved gas analysis (DGA) is a robust and generally acknowledged diagnostic tool in the transformer manufacturing industry. This is because DGA has the potential to divulge the electrical and thermal stresses predominating with transformer dielectric oil and cellulosic paper insulation. Prospective examination approaches for DGA fault diagnosis are reported in [4]. By and large, DGA fault diagnostic methods demonstrate high equivocation in examining the fault gases, given the nonlinear comportment of the fault gases and various problems with DGA fault diagnosis methods. The production of dissolved gases has a nonsequential connection with transformer duration of operation and insulation aging indicators, i.e., interfacial tension, furan, acidity, etc. [5]. This nonsequential comportment gives rise to intricacy in fault recognition when artificial intelligence algorithms are employed.

A handful of investigators have employed diverse intelligent methods comprising artificial neural networks (ANNs), fuzzy logic (FL), decision trees, etc. for diagnosing DGA faults [6,7]. As a result of the dubious accuracy and high equivocation in the classification of DGA faults, modern computational methods have been reported in recent years. These methods comprise machine learning algorithms and optimization algorithms. It is recognized that multidimensional and astronomical data samples are necessitated for training these algorithms to be efficacious. A handful of researchers have adopted these algorithms for diagnosing transformer DGA faults [8–10].

The current research contribution—This work presents a detailed investigation of transformer fault identification. Various open issues and research challenges in transformer fault identification using classical methods have been highlighted. The contributions of this research work are indicated as follows.

- A BCSVM is proposed for identifying a set of five transformer faults using 70% of the oil samples for training and 30% for training the proposed model, with 30-fold cross-validation applied in the training dataset. The proposed approach yields higher accuracy than ANN and the IEC method.
- The classification accuracy of the considered DGA samples is investigated by considering various machine learning algorithms, i.e., linear SVM, quadratic SVM, cubic SVM, fine Gaussian SVM, medium Gaussian SVM, and coarse Gaussian SVM, and comparing them in terms of accuracy (in percentage), prediction speed (objects/second) and training time (in seconds).
- A case study based on 14 transformer samples is presented using practical DGA data sets supplied by a South African manufacturer to comprehend fault classification proficiency in terms of accuracy of ANN and IEC methods and practical data versus the accuracy obtained in recent works.

The novelty of the current research—The fundamental purpose of this study is to ascertain a reliable transformer fault diagnosis approach using DGA and the application of the artificial intelligence technique for fault diagnosis in power transformers. Notwithstanding that numerous research workers have worked on the application of AI to diagnose transformer faults, as shown in Table 1, very seldom has research been published on fault diagnoses using BCSVM, particularly in transformer condition assessment. A reliable fault classification and prediction algorithm are crucial criteria for developing an efficient fault identification system. Three methods are considered a benchmark of the proposed method, and it was found that the IECGR method yields a "not detectable" response to some of the set of studied case studies and ANN yields a higher fault class than the actual fault in some oil samples due to model overfitting. However after the application of the proposed BCSVM algorithm, the "not detectable" samples were effectively identified.

Many works compared the classical DGA methods in their investigations. Though there is similar work, in the current investigation, various machine learning (ML) algorithms, i.e., linear SVM, quadratic SVM, cubic SVM, fine Gaussian SVM, medium Gaussian SVM, and coarse Gaussian SVM, are compared in terms of accuracy (%), prediction Speed (objects/sec) and training time (sec). In the current study, the effect of the dissolved gas concentration levels is studied exclusively. This parameter is adopted for examining the performance of ML algorithms and two different diagnostic methods. From this investigation, it may be concluded that the proposed BCSVM algorithm is reliable in accurately diagnosing transformer faults using dissolved gas concentration levels in parts per million (ppm).

Table 1. Summary of applicable works.

Ref. No	Year	Method	Summary
[11]	2020	SVM, ANN	The transformer condition is monitored daily and yields a precision of about 81.4% and 76%. The precision can be further increase.
[12]	2021	SVM, Bat algorithm	The study considers 160 data samples. which yield an accuracy of about 93.75%
[13]	2022	SVM, seagull optimization algorithm	180 field transformers are considered, and the proposed method yields an accuracy of 91.67%
[14]	2019	Least-square SVM, grey wolf optimization	MATLAB is utilized to classify transformer faults and yields an accuracy of 97.45%. The quantity of the data has not been highlighted.
[15,16]	2018	SVM, particle swarm optimization	Transformer faults are diagnosed using 118 databases adopted from the IEC TC 10 database and yield the highest accuracy of 85.71%

The manuscript organization—This research has been structured as follows. Section 2 presents the fundamental principle of the SVM algorithm and proposed BCSVM. Section 3 presents the results and discussion of the proposed algorithm for fault diagnosis of transformers. A corroboration of the proposed algorithm with the real sample datasets from local transformer companies in South Africa is presented. Lastly, Section 4 provides a conclusion of the manuscript.

2. Materials and Methods

2.1. Transformer Fault Classification Procedure

Transformer faults can be recognized in conformity with the dissolved gases proliferated attributable to the heating of the dielectric oil and the gases that are prevailing at diverse temperatures, i.e., hydrogen (H_2), carbon monoxide (CO), methane (CH_4), ethane (C_2H_6), ethylene (C_2H_4), acetylene (C_2H_2), and carbon dioxide (CO_2). Nevertheless, five gases, i.e., H_2, CH_4, C_2H_4, C_2H_6, and C_2H_2, are considered in this research in classifying transformer faults. A set of five transformer conditions, including no fault (NF), partial discharge (PD), and thermal fault conditions, are discerned.

2.2. The Fundamental Principle of the SVM Algorithm

SVM is a vigorous supervised learning technique for developing a dataset classifier. SVM purports to establish a decision boundary among two classes of datasets that facilitate the forecasting of data labels from one or several feature vectors [11–17]. A decision boundary can be described as the area of a problem space whereupon the output dataset label of a classifier is equivocal. This decision boundary is referred to as the hyperplane, and it is positioned such that it is furthest from the nearest datasets from other classes. These nearest data points are so-called support vectors. The learning procedure of the SVM is illustrated in Figure 1. The SVM learns based on the training dataset entered. Then, the validation dataset is applied to determine the learning performance of the trained SVM algorithm. The trained SVM algorithm model is then applied to classifying samples of unknown test datasets.

Considering a tagged SVM training dataset, the latter can be expressed as follows in Equation (1).

$$(x_1,\ y_1),\ldots(x_n,\ y_n),\ x_i \in R^d \text{ and } y \in (-1,+1) \tag{1}$$

Here,

i–Training compound

x_i–Feature vector (predictors)

y_i–Class label

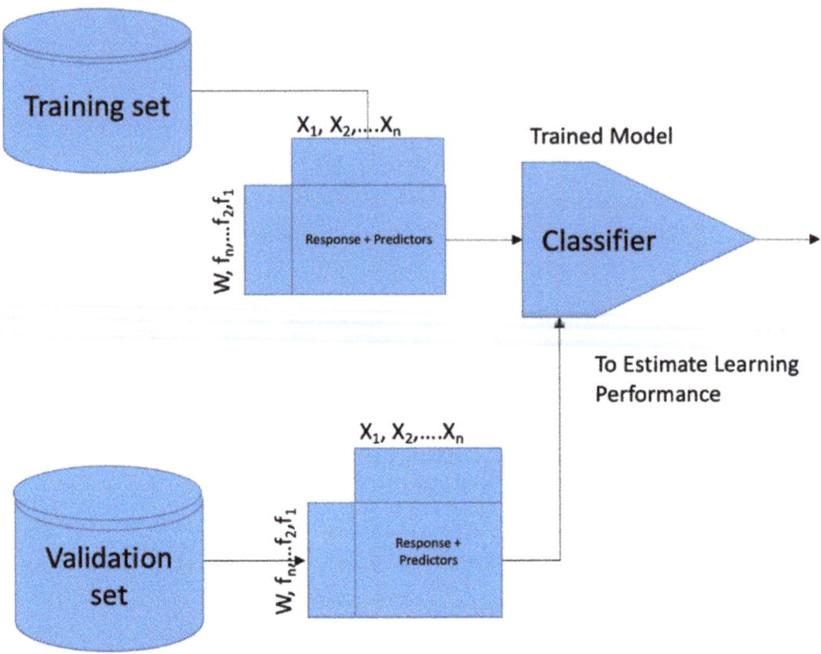

Figure 1. The learning procedure of an SVM classifier.

An optimal hyperplane can therefore be expressed as follows in Equation (2).

$$wx^d + b = 0 \tag{2}$$

Here,
w–Weight vector
x–Input feature vector,
b–is the bias.
For all elements of the training dataset, the values of w and b must fulfill the conditions of the inequalities expressed in Equations (3) and (4).

$$wx_i^T + b \geq +1 \text{ if } y_i = 1 \tag{3}$$

$$wx_i^T + b \leq -1 \text{ if } y_i = -1 \tag{4}$$

The purpose of training an SVM algorithm is to ascertain the w and b such that the hyperplane isolates the data points and makes the best use of the margin $\frac{1}{||w||^2}$. In Figure 2, the vectors x_i with the property that $|y_i| \left(wx_i^T + b\right) = 1$ will be appellate as support vectors [12].

An alternative to a linear SVM classifier for the nonlinear application of the SVM is the kernel technique, which allows the modelling of higher dimensional and nonlinear models [11–16]. For a nonlinear problem, a kernel function can be employed to append supplemental dimensions to the coarse data and therefore create a linear problem in the eventuating higher dimensional space. In a nutshell, a kernel function, which is expressed as shown in Equation (5), can facilitate carrying computations rapidly, which would in other respects necessitate high dimensional space computations.

$$K(x,y) = \; < f(x), f(y) > \tag{5}$$

Here,

K–Kernel function

x, y–n dimensional inputs

f–Map function of the input from n dimensional to m dimensional space

$< f(x), f(y) >$—Indicate the dot product

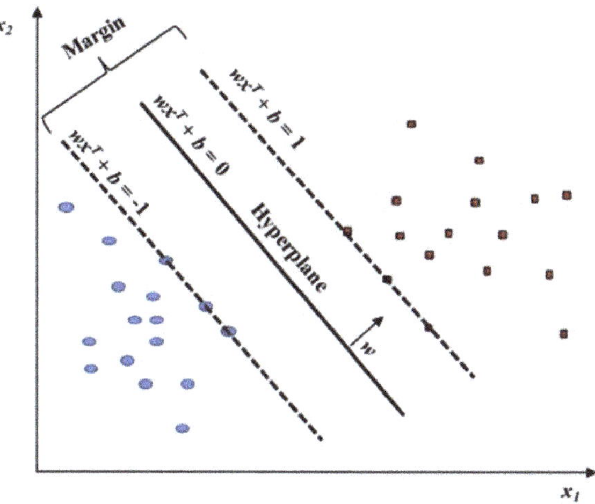

Figure 2. Linear SVM model classifying red vs. blue data points.

Using kernel functions, the computation of the scalar product of data points in a higher dimensional space except specifically evaluating the mapping from the input space to the higher dimensional space can be conducted. In many instances, calculating the kernel is straightforward whereas in the high dimensional space, calculating the inner product of feature vectors is complex. The feature vector for even straightforward kernels can inflate in dimensions, and kernels such as the radial basis function (RBF) kernel can be expressed as follows in Equation (6).

$$K_{RBF}(x, y) = e^{\left(-\gamma ||x-y||^2\right)} \tag{6}$$

The proportionate feature vector is incalculable dimensionally. Even so, calculating the kernel is nearly insignificant. Contingent on the essence of the problem, it is likely that one kernel can be a whole lot better than other kernels. An optimum kernel function can be chosen from an established set of kernels numerically in an extremely thorough and careful manner by applying cross-validation.

2.3. Proposed BCSVM Algorithm

In the current study, DGA oil samples were obtained from mineral oil-immersed transformers in the field owned by different local independent power utilities. The comprehensive flow diagram of the proposed BCSVM is illustrated in Figure 3.

- Initially, the concentration levels of five feature gases are ingested as inputs to the first SVM (SVM 1). It will distinguish whether the oil samples represent a normal or faulty condition. If the sample is in normal condition, the algorithm ends.
- Secondly, if the SVM1 output has identified a fault condition, it is further ingested as input to the second SMV (SVM 2) which then distinguishes whether the fault is a thermal fault class or an electric discharge fault.
- Thirdly, if the fault class is pigeonholed as a PD fault, then it will be ingested as input to the third SVM (SVM 3), which will then distinguish whether the fault is PD1 or PD2.

- At the same time, if the fault is categorized as a thermal (T) fault, then it will be ingested as input to the fourth SVM (SVM 4) which will distinguish whether the fault is a T1 or T2 fault as demonstrated.

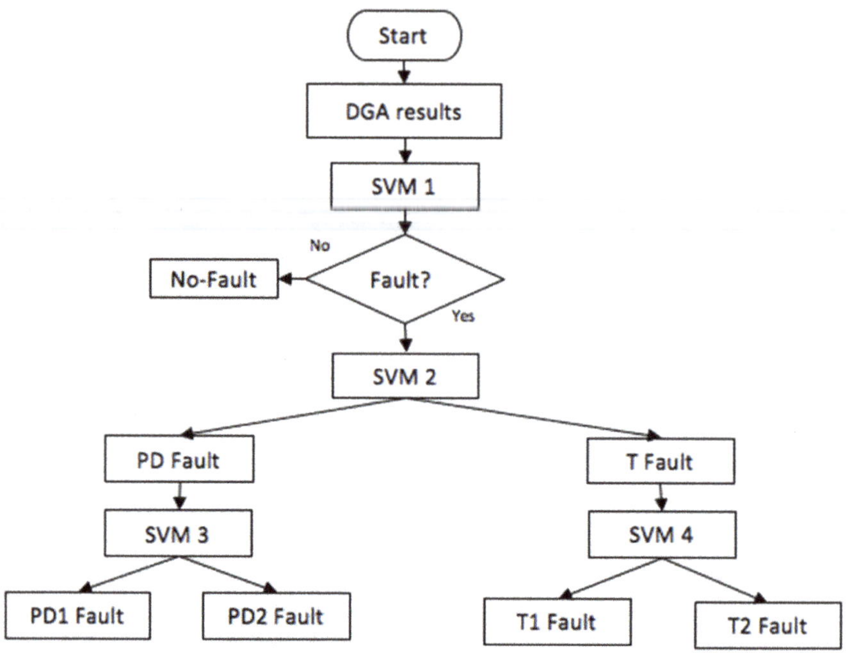

Figure 3. Proposed BCSVM classification tree.

To evaluate the statistical significance of the experimental DGA dataset, a single-factor ANOVA was employed. The results are illustrated in Table 2.

Table 2. Statistical analysis of experimental DGA dataset.

Source of Variation	SS	df	MS	F	*p*-Value	F Crit
Between Groups	4.7231×10^{11}	5	9.4463×10^{10}	5.33534763	9.1433×10^{-5}	2.23578833
Within Groups	7.3299×10^{12}	414	1.7705×10^{10}			
Total	7.8022×10^{12}	419				

It can be observed from the *p*-value of 9.1433×10^{-5} ($<\alpha = 0.05$), that the experimental DGA results are statistically significant.

The sensitivity analysis of the input gases was carried out by adopting descriptive statistics analysis (DSSA), as shown in Table 3. DSSA examines the quantitative performance of the input dataset features.

The complexity of the proposed solution design is demonstrated in Figure 4. The latter is partially based upon the related state of the problem diagnosing transformer faults and solution of DGA and artificial intelligence knowledge. Additionally, this can be illustrated in terms of a solution design complexity heatmap comprising all the main activities undertaken.

Table 3. Descriptive statistics sensitivity analysis of input gases.

Parameter	H_2	CH_4	C_2H_2	C_2H_4	C_2H_6	CO
Mean	38,978.9143	28,260.8571	10,6671.346	94,012.2143	29,719.6571	24,078.9143
Standard Error	13,315.308	10,012.6008	25,868.0788	19,828.9585	10,192.548	8590.25301
Median	3895.5	835	311.5	4908	4028.5	3269
Mode	0	744	2	11	220	300
Standard Deviation	111,403.859	83,771.4282	21,6427.875	165,900.97	85,276.9749	71,871.2131
Sample Variance	1.2411×10^{10}	7,017,652,184	4.6841×10^{10}	2.7523×10^{10}	7,272,162,445	5,165,471,270
Kurtosis	8.06695552	11.282565	1.63684628	0.37595746	9.53422061	11.585725
Skewness	3.0627171	3.35926699	1.8135374	1.47653406	3.22003396	3.52177109
Range	487,297	449,556	683,643	478,904	404,053	348,857
Minimum	0	0	0	0	0	0
Maximum	487,297	449,556	683,643	478,904	404,053	348,857
Sum	2,728,524	1,978,260	7,466,994.22	6,580,855	2,080,376	1,685,524
Count	70	70	70	70	70	70
Confidence Level (95.0%)	26,563.3126	19,974.592	51,605.4051	39,557.6899	20,333.5769	17,137.0858

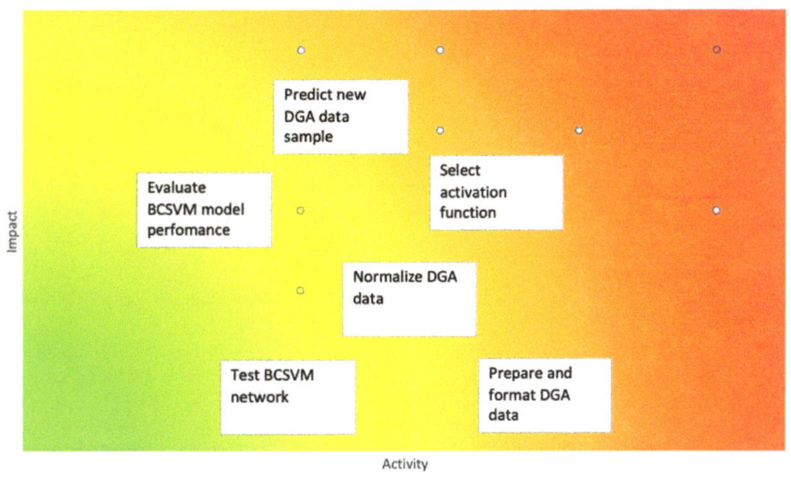

Figure 4. Complexity of the proposed BCSVM classification tree.

The red regions reveal locales of conceivably sheer complexity. The green regions denote locales of conceivably little complexity.

3. Results

In this section, the results are presented and discussed. The transformer oil testing samples based on mineral oil were abstracted from a South African transformer manufacturer after laboratory analysis. The BCSVM training phase was ingested with 70% of the oil testing samples and tested using 30% of the transformer oil samples. This investigation was conducted by employing the classification learner app in the MATLAB_R2018a software platform. Once the oil samples dataset was trained, the veracity of the distinctive SVMs was corroborated. The configuration matrix of a particular SVM that provides the highest degree of accuracy was selected and imported to predict the new testing data sample. The fault class prediction for the new data sample is carried out by utilizing Equation (7).

$$yfit = fitcecoc(Tbl, ResponseVarName) \tag{7}$$

Here,

Tbl ̄Multiclass predictor variables in table
$Response\,Var\,Name$ ̄Response variables

3.1. Training and Testing of SVM 1

The training and testing of SVM 1 are designed to categorize between the transformer's normal and faulty conditions. The training stage was carried out using 70 oil samples with fault and normal conditions. A total of 14 oil data samples were then used in testing SVM 1. The results indicated that all the new oil data samples were pigeonholed accurately. The selection response and predictors selected by SVM1 using the ingested data are demonstrated in Figure 5. A 30-fold cross-validation has been selected.

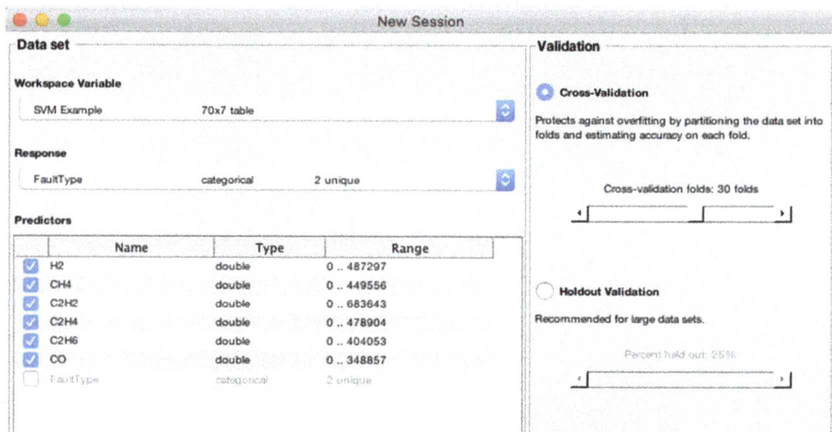

Figure 5. The dataset variables were selected by the SVM 1 classifier.

The dissolved gases were automatically placed as the predictors and the fault type as the response. In the SVM 1 algorithm H_2 and C_2H_2 was considered to acquire the desired output. The characterization of H_2 versus C_2H_2 is illustrated in Figure 6.

Further, various ML algorithms, i.e., linear SVM, quadratic SVM, cubic SVM, fine gaussian SVM, medium Gaussian SVM, and coarse Gaussian SVM, are compared in terms of accuracy (%), prediction speed (objects/sec) and training time (sec), as tabulated in Table 4. The testing and calculation of the accuracy of referenced algorithms tested on the same data were evaluated based on the percentage of accuracy to classify fault type, the prediction speed and the training time of the respective algorithm.

Table 4. SVM 1 DGA classification outcomes.

Type	Accuracy (%)	Prediction Speed	Training Time (sec)
Linear SVM	81.4	260	7.0125
Quadratic SVM	92.9	360	1.6171
Cubic SVM	58.6	370	53.793
Fine Gaussian SVM	80.0	360	1.3882
Medium Gaussian SVM	68.6	380	1.4147
Coarse Gaussian SVM	68.6	340	1.4259

In the SVM 1 algorithm, the quadratic SVM provides the highest degree of accuracy, and the corresponding configuration matrix shown in Figure 7 was imported to the workspace to predict the new testing data samples.

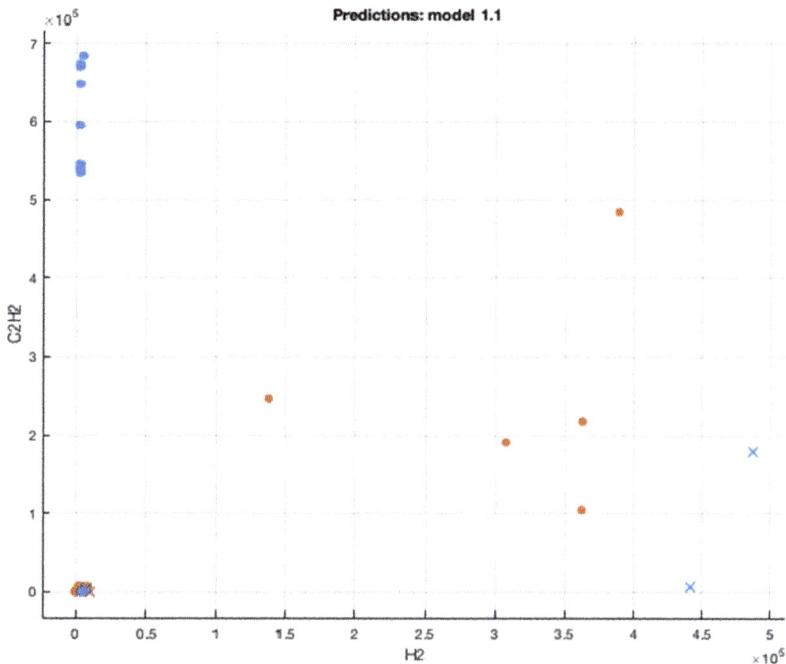

Figure 6. H_2 versus C_2H_2. Blue and orange (●)—raw data, orange and blue (×)—predicted data.

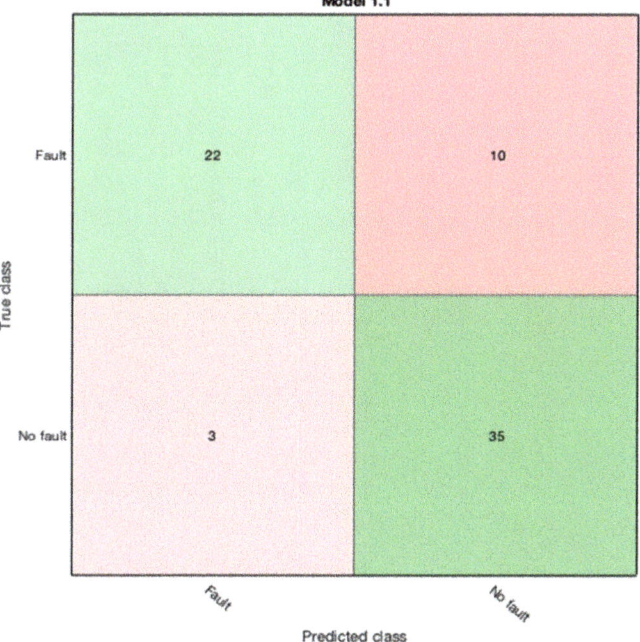

Figure 7. SVM 1 confusion matrix.

3.2. Training and Testing of SVM 2

The training and testing of SVM 2 were purposed to categorize between PD and T faults. The training stage was carried out by using a total of 15 oil data samples with T fault and with PD fault conditions respectively. Subsequently, a total of 15 oil data samples were utilized in testing SVM2. The results indicated that a total of 14 oil data samples were classified accurately. The selection of the response and predictors selected by SVM 2 using this ingested data is demonstrated in Figure 8. A 30-fold cross-validation was selected.

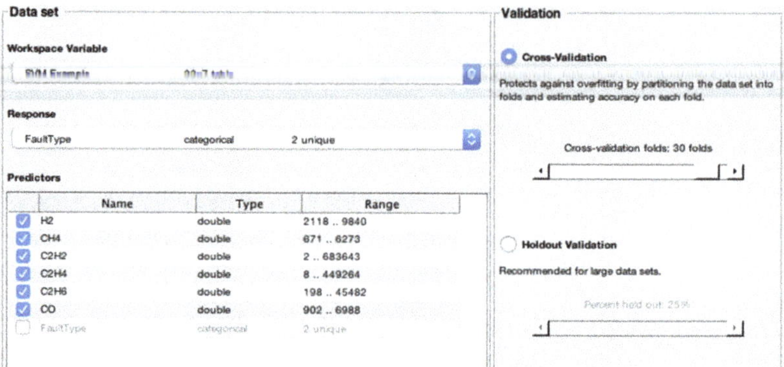

Figure 8. The dataset variables selected by the SVM 2 classifier.

The dissolved gases were automatically placed as the predictors and the fault type as the response. In the SVM 2 algorithm, CH$_4$ and C$_2$H$_2$ were considered to acquire the desired output. The characterization of CH$_4$ versus C$_2$H$_2$ is illustrated in Figure 9.

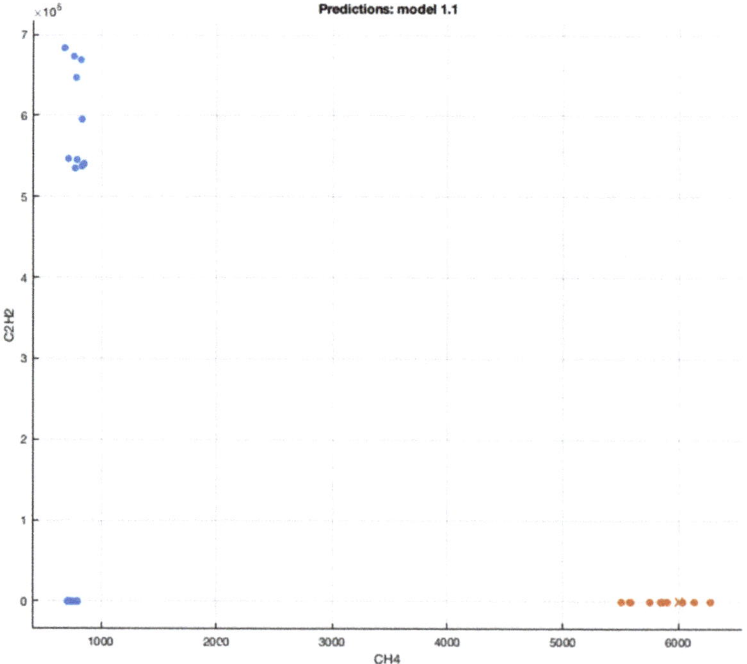

Figure 9. CH$_4$ versus C$_2$H$_2$. Blue and orange ($\bullet$)—raw data, orange and blue ($\times$)—predicted data.

Further, the various ML algorithms were compared in terms of accuracy (%), prediction speed (objects/sec) and training time (sec), as tabulated in Table 5.

Table 5. SVM 2 DGA classification outcomes.

Type	Accuracy (%)	Prediction Speed	Training Time (sec)
Linear SVM	96.9	170	2.0822
Quadratic SVM	81.2	180	1.3123
Cubic SVM	81.2	170	1.3312
Fine Gaussian SVM	96.9	170	1.3723
Medium Gaussian SVM	96.9	140	1.5766
Coarse Gaussian SVM	96.9	120	1.9193

In the SVM 2 algorithm, the linear SVM, fine Gaussian SVM, medium Gaussian SVM and coarse Gaussian SVM provided the highest degree of accuracy and the coarse corresponding configuration matrix shown in Figure 10 was imported to the workspace to predict the new testing data samples.

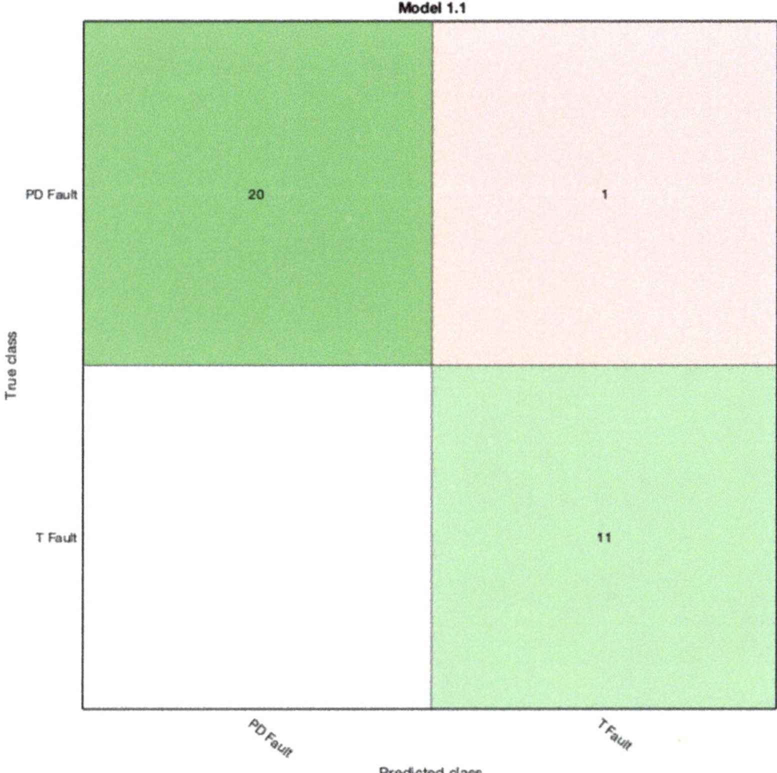

Figure 10. Confusion matrix for SVM 2 classifier.

3.3. Training and Testing of SVM 3

SVM 3 was trained and tested to classify PD1 and PD2 faults. The training stage was conducted by using a total of 8 oil data samples with a PD1 fault condition and a total of 7 data samples with a PD2 fault condition. Therefore, a total of 10 data samples were utilized in the testing of SVM. The results indicated that all the data samples were pigeonholed accurately. The selection of responses and predictors selected by SVM 3 using the ingested data is demonstrated in Figure 11.

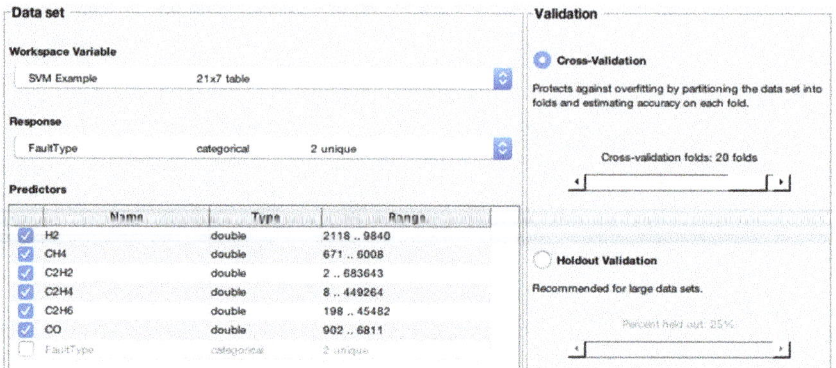

Figure 11. The dataset variables were selected by the SVM 3 classifier.

The dissolved gases were automatically placed as the predictors and the fault type as the response. In the SVM 3 algorithm, C_2H_2 and C_2H_4 were considered to acquire the desired output. The characterization of C_2H_2 versus C_2H_4 is illustrated in Figure 12.

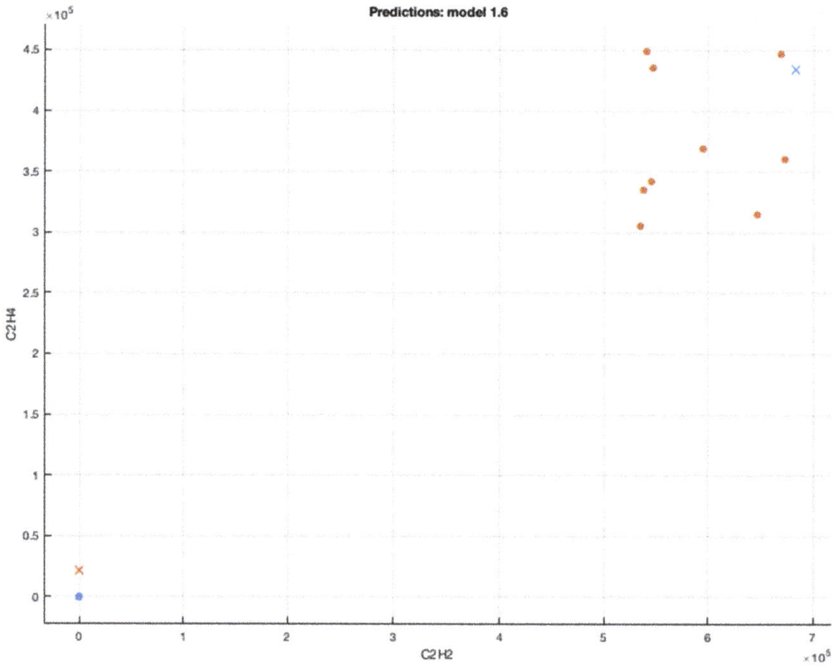

Figure 12. C_2H_2 versus C_2H_4. Blue and orange (●)—raw data, orange and blue (×)—predicted data.

Further, the ML algorithms, were compared in terms of accuracy (%), prediction speed (objects/sec) and training time (sec), as tabulated in Table 6.

Table 6. SVM 3 DGA classification outcomes.

Type	Accuracy (%)	Prediction Speed	Training Time (sec)
Linear SVM	95.2	160	1.465
Quadratic SVM	95.2	160	0.967
Cubic SVM	100	130	1.008
Fine Gaussian SVM	90.5	150	0.989
Medium Gaussian SVM	90.5	160	0.941
Coarse Gaussian SVM	90.5	140	1.055

In the SVM 3 algorithm, linear SVM, quadratic SVM, and cubic SVM provided the highest degree of accuracy, with cubic SVM yielding 100% accuracy, and the corresponding configuration matrix shown in Figure 13 was imported to the workspace to predict the new testing data samples.

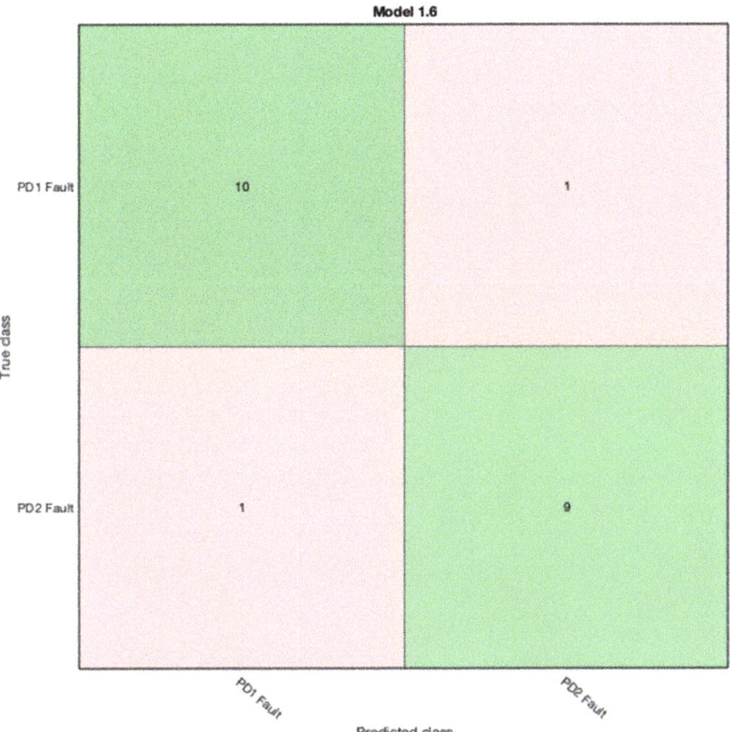

Figure 13. Confusion matrix for SVM 3 classifier.

3.4. Training and Testing of SVM 4

The training and testing of SVM 4 were purposed to categorize between T1 and T2 fault classes. SVM 4 was trained by a total of 8 data samples with a T1 fault condition and a total of 7 data samples with a T2 fault condition. Consequently, a total of 10 data samples were used in testing SVM 4 and the results showed that 9 data samples were classified fittingly. The complete BCSVM was tested using 26 DGA data samples, and 24 samples were pigeonholed rightly. The precision of the overall BCSVM was 92%. The selection of responses and predictors selected by SVM 4 using the ingested data are demonstrated in Figure 14.

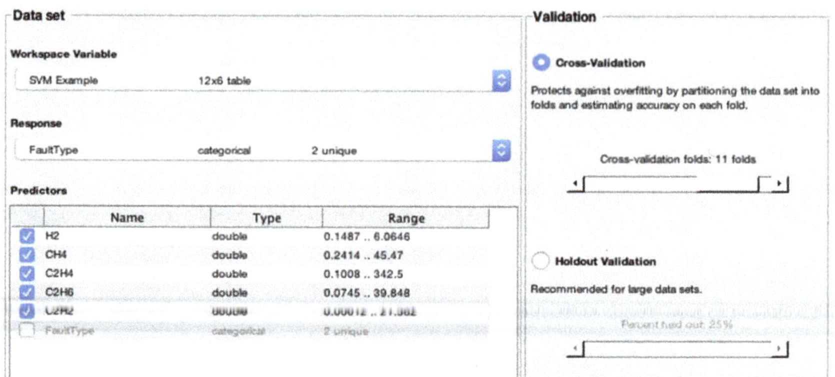

Figure 14. The dataset variables were selected by the SVM 4 classifier.

The dissolved gases were automatically placed as the predictors and the fault type as the response. In the SVM 3 algorithm, H_2 and CH_4 were considered to acquire the desired output. The characterization of H_2 versus CH_4 is illustrated in Figure 15.

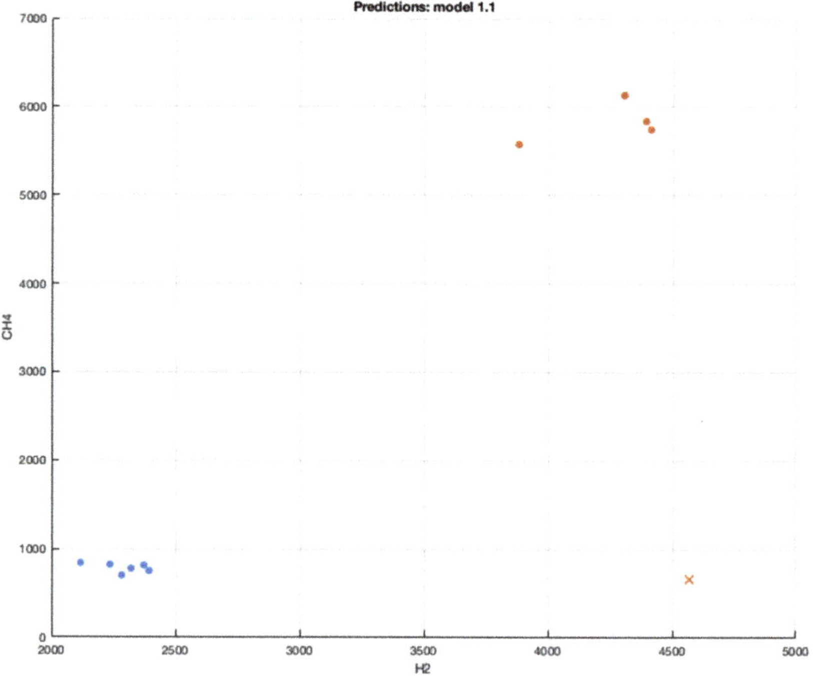

Figure 15. H_2 versus CH_4. Blue and orange (●)—raw data, orange and blue (×)—predicted data.

Further, the ML algorithms were compared in terms of accuracy (%), prediction speed (objects/sec) and training time (sec), as tabulated in Table 7.

Table 7. SVM 4 DGA classification outcomes.

Type	Accuracy (%)	Prediction Speed	Training Time (sec)
Linear SVM	90.9	470	0.629
Quadratic SVM	100	480	0.253
Cubic SVM	90.9	390	0.242
Fine Gaussian SVM	100	410	0.254
Medium Gaussian SVM	100	390	0.275
Coarse Gaussian SVM	63.6	380	0.259

In the SVM 4 algorithm, quadratic SVM, fine Gaussian SVM, and medium Gaussian SVM provided the highest degree of accuracy and the corresponding configuration matrix shown in Figure 16 was imported to the workspace to predict the new testing data samples.

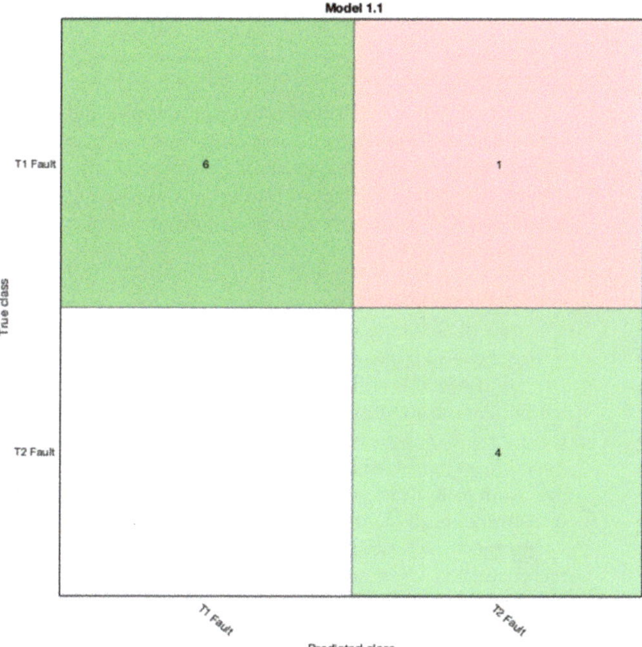

Figure 16. Confusion matrix for SVM 4.

4. Case Studies

In this section, various transformer case studies are presented to corroborate the efficacy of the proposed BCSVM algorithm in identifying faults of an unknown dataset to the proposed algorithm learning process. The training and testing of the datasets were not examined from the same utility. The training data adopted from service field oil sample data and the testing oil sample dataset were established by physical unit inspection, as shown in Table 6. The latter will be an opportunity for examination of the proposed algorithm in a more indubitable approach and realizing the capability to development of an efficacious ML algorithm for field data enactment. In the proposed approach, six ML algorithms are assessed for examining the correlation between the response and predictors.

The proposed BCSVM is evaluated in Table 8 against the actual, ANN and IECGR techniques on a set of transformers' DGA data that was not included in the training of the proposed algorithm. A major strength of BCSVM is the 30-fold cross-validation performance to circumvent overfitting problems.

Table 8. DGA fault classification case studies.

H$_2$	CH$_4$	C$_2$H$_4$	C$_2$H$_6$	C$_2$H$_2$	Actual	ANN	IECGR	BCSVM
302	490	360	182	95	T2	T2	ND *	T2
32	2	1	1	0.05	PD1	PD2	PD1	PD1
13	9	6	41	0	NF	NF	NF	NF
156	128	35	97	0	T1	T1	ND *	T1
596	81	90	10	245	PD2	PD2	PD2	PD2
23	41	7	37	2	T1	T2	T1	T1
1772	3631	8481	1071	79	T2	T2	T2	T2
87	31	36	11	30	PD2	PD2	PD2	PD2
35	40	41	10	10	PD2	PD2	PD2	PD2
143	4	9	3	2	PD2	PD2	PD2	PD2
2587.2	7.882	1.4	4.7004	0	PD1	PD1	PD1	PD1
1676	653	1006	81	418	PD2	PD2	PD2	PD2
181	262	528	210	0	T2	T2	T2	T2
181	176	51	76	5	T1	T2	T1	T1

* ND—not detectable.

From Table 6, it is observed that the fault class tag T2 has low accuracy against the actual data. Intriguingly, all other fault class tags performed well in the context of accuracy. Additionally, the IECGR was unable to conclusively diagnose some of the faults due to the limitations of the code ratios. Further, the proposed technique constitutes proof that it can be reliably applied in the prediction of unknown oil sample datasets, predicting all the case studies accurately.

5. Conclusions

The BCSVM construction for fault diagnosis of power transformers has been represented. Several faults are unpredictable by the IEC gas ratio (IECGR) method, which results in an undetectable conclusion. When using ANN, given that it is excellent at learning, the restraint of the inability of fuzzy logic to adapt the created rule base with the varying system for indivisible and irregular data is eradicated. Nonetheless, ANN has the hindrance of overfitting; therefore, it has lower generalization capability and provides circumscribed precision to fault identification. To circumvent all these challenges, in this work, power transformer fault identification was conducted by employing a binary classification support vector machine (BCSVM). The case study results demonstrate that the proposed BCSVM technique has a higher degree of diagnostic accuracy than the IECGR and ANN methods owing to its enhanced generalization capability, and it can classify indivisible DGA datasets by utilizing the kernel function.

Funding: This research received no external funding.

Institutional Review Board Statement: Not applicable.

Informed Consent Statement: Not applicable.

Data Availability Statement: Not applicable.

Conflicts of Interest: The author declares no conflict of interest.

References

1. Gouda, O.E.; El-Hoshy, S.H.; Ghoneim, S.S.M. Enhancing the Diagnostic Accuracy of DGA Techniques Based on IEC-TC10 and Related Databases. *IEEE Access* **2021**, *9*, 118031–118041. [CrossRef]
2. Draper, Z.H.; Dukarm, J.J.; Beauchemin, C. How to Improve IEEE C57.104-2019 DGA Fault Severity Interpretation. In Proceedings of the IEEE/PES Transmission and Distribution Conference and Exposition (T&D), New Orleans, LA, USA, 25–28 April 2022; pp. 1–5. [CrossRef]
3. Ward, S.A.; El-Faraskoury, A.; Badawi, M.; Ibrahim, S.A.; Mahmoud, K.; Lehtonen, M.; Darwish, M.M.F. Towards Precise Interpretation of Oil Transformers via Novel Combined Techniques Based on DGA and Partial Discharge Sensors. *Sensors* **2021**, *21*, 2223. [CrossRef] [PubMed]

4. Emara, M.M.; Peppas, G.D.; Gonos, I.F. Two Graphical Shapes Based on DGA for Power Transformer Fault Types Discrimination. *IEEE Trans. Dielectr. Electr. Insul.* **2021**, *28*, 981–987. [CrossRef]
5. Kherif, O.; Benmahamed, Y.; Teguar, M.; Boubakeur, A.; Ghoneim, S.S.M. Accuracy Improvement of Power Transformer Faults Diagnostic Using KNN Classifier with Decision Tree Principle. *IEEE Access* **2021**, *9*, 81693–81701. [CrossRef]
6. Zhang, Y.; Chen, H.C.; Du, Y.; Chen, M.; Liang, J.; Li, J.; Fan, X.; Sun, L.; Cheng, Q.S.; Yao, X. Early Warning of Incipient Faults for Power Transformer Based on DGA Using a Two-Stage Feature Extraction Technique. *IEEE Trans. Power Deliv.* **2022**, *37*, 2040–2049. [CrossRef]
7. Haque, N.; Jamshed, A.; Chatterjee, K.; Chatterjee, S. Accurate Sensing of Power Transformer Faults from Dissolved Gas Data Using Random Forest Classifier Aided by Data Clustering Method. *IEEE Sens. J.* **2022**, *22*, 5902–5910. [CrossRef]
8. Taha, I.B.M.; Ibrahim, S.; Mansour, D.E.A. Power Transformer Fault Diagnosis Based on DGA Using a Convolutional Neural Network with Noise in Measurements. *IEEE Access* **2021**, *9*, 111162–111170. [CrossRef]
9. Kaur, K.; Sharma, N.K.; Singh, J.; Bhalla, D. Performance Assessment of IEEE/IEC Method and Duval Triangle technique for Transformer Incipient Fault Diagnosis. In *2022 IOP Conference Series: Materials Science and Engineering*; IOP Publishing: Bristol, UK, 2022; Volume 1228, p. 012027. [CrossRef]
10. Ekojono, R.A.P.; Apriyani, M.E.; Rahmanto, A.N. Investigation on machine learning algorithms to support transformer dissolved gas analysis fault identification. *Electr. Eng.* **2022**, *104*, 3037–3047. [CrossRef]
11. Saravanan, D.; Hasan, A.; Singh, A.; Mansoor, H.; Shaw, R.N. Fault Prediction of Transformer Using Machine Learning and DGA. In Proceedings of the 2020 IEEE International Conference on Computing, Power and Communication Technologies (GUCON), Greater Noida, India, 2–4 October 2020; pp. 1–5. [CrossRef]
12. Benmahamed, Y.; Kherif, O.; Teguar, M.; Boubakeur, A.; Ghoneim, S.S.M. Accuracy Improvement of Transformer Faults Diagnostic Based on DGA Data Using SVM-BA Classifier. *Energies* **2021**, *14*, 2970. [CrossRef]
13. Wu, Y.; Sun, X.; Zhang, Y.; Zhong, X.; Cheng, L. A Power Transformer Fault Diagnosis Method-Based Hybrid Improved Seagull Optimization Algorithm and Support Vector Machine. *IEEE Access* **2022**, *10*, 17268–17286. [CrossRef]
14. Zeng, B.; Guo, J.; Zhu, W.; Xiao, Z.; Yuan, F.; Huang, S. A Transformer Fault Diagnosis Model Based on Hybrid Grey Wolf Optimizer and LS-SVM. *Energies* **2019**, *12*, 4170. [CrossRef]
15. Ma, H.; Zhang, W.; Wu, R.; Yang, C. A Power Transformers Fault Diagnosis Model Based on Three DGA Ratios and PSO Optimization SVM. In Proceedings of the Conference on Mechatronics and Electrical Systems (ICMES 2017), Wuhan, China, 15–17 December 2017; p. 339. [CrossRef]
16. Yu, W.; Yu, R.; Li, C. An Information Granulated Based SVM Approach for Anomaly Detection of Main Transformers in Nuclear Power Plants. *Sci. Technol. Nucl. Install.* **2022**, *2022*, 3931374. [CrossRef]
17. El-kenawy, E.S.M.; Albalawi, F.; Ward, S.A.; Ghoneim, S.S.M.; Eid, M.M.; Abdelhamid, A.A.; Bailek, N.; Ibrahim, A. Feature Selection and Classification of Transformer Faults Based on Novel Meta-Heuristic Algorithm. *Mathematics* **2022**, *10*, 3144. [CrossRef]

Article

Design, Technical and Economic Optimization of Renewable Energy-Based Electric Vehicle Charging Stations in Africa: The Case of Nigeria

Jamiu O. Oladigbolu [1,*], Asad Mujeeb [2], Amir A. Imam [1] and Ali Muhammad Rushdi [1,*]

1   Department of Electrical and Computer Engineering, Faculty of Engineering, King Abdulaziz University, Jeddah 21589, Saudi Arabia
2   Department of Electrical Engineering, Tsinghua University, Beijing 100190, China
*   Correspondence: omotayooladigbolu@gmail.com (J.O.O.); arushdi@kau.edu.sa (A.M.R.); Tel.: +234-7035634424 (J.O.O.); +20-1128247509 (A.M.R.)

Abstract: The transportation sector accounts for more than 70% of Nigeria's energy consumption. This sector has been the major consumer of fossil fuels in the past 20 years. In this study, the technical and economic feasibility of an electrical vehicle (EV) charging scheme is investigated based on the availability of renewable energy (RE) sources in six sites representing diverse geographic and climatic conditions in Nigeria. The HOMER Pro® microgrid software with the grid-search and proprietary derivative-free optimization techniques is used to assess the viability of the proposed EV charging scheme. The PV/WT/battery charging station with a quantity of two WT, 174 kW of PV panels, a quantity of 380 batteries storage, and a converter of 109 kW located in Sokoto provide the best economic metrics with the lowest NPC, electricity cost, and initial costs of USD547,717, USD0.211/kWh, and USD449,134, respectively. The optimal charging scheme is able to reliably satisfy most of the EV charging demand as it presents a small percentage of the unmet load, which is the lowest when compared with the corresponding values of the other charging stations. Moreover, the optimal charging system in all six locations is able to sufficiently meet the EV charge requirement with maximum uptime. A sensitivity analysis was conducted to check the robustness of the optimum charging scheme. This sensitivity analysis reveals that the technical and economic performance indicators of the optimum charging station are sensitive to the changes in the sensitivity variables. Furthermore, the outcomes ensure that the hybrid system of RE sources and EVs can minimize carbon and other pollutant emissions. The results and findings in this study can be implemented by all relevant parties involved to accelerate the development of EVs not only in Nigeria but also in other parts of the African continent and the rest of the world.

Keywords: transportation sector; electrical vehicle; charging scheme; renewable energy sources; sensitivity analysis

Citation: Oladigbolu, J.O.; Mujeeb, A.; Imam, A.A.; Rushdi, A.M. Design, Technical and Economic Optimization of Renewable Energy-Based Electric Vehicle Charging Stations in Africa: The Case of Nigeria. *Energies* **2023**, *16*, 397. https://doi.org/10.3390/en16010397

Academic Editors: Saad Motahhir, Najib El Ouanjli and Mustapha Errouha

Received: 25 November 2022
Revised: 11 December 2022
Accepted: 23 December 2022
Published: 29 December 2022

1. Introduction

The overall increase in population worldwide, accompanied by the contemporary technological revolution, has given rise to an excessively high rate of electricity consumption, a critical problem that necessitates sustainable energy solutions [1,2]. In the Nigerian power sector, most of the electricity supply is being secured through fossil fuels, namely natural gas, which accounts for about 80% of the total fossil sources, and oil, from which most of the remaining fossil power generation originates [3]. In fact, 80% of the country's energy mix takes the form of thermal generation sources, with the remaining 20% being derived from hydropower and other renewable energy (RE) sources. The per capita CO_2 emissions in Nigeria had dramatically risen from about 0.08 metric tons in 1960 (when Nigeria attained independence from Britain) to 0.67 metric tons in 2018 [4]. In fact, the contribution of the country to the global pollutant emissions level has increased considerably over the past few decades. Nigeria's total energy consumption in 2020 is about 164,013 kilotonne of oil equivalent (ktoe), with the transportation sector alone contributing about 16,554 ktoe [5]. The transportation sector accounted for over 70% of Nigeria's energy use [6]. This sector has been the major consumer of fossil fuels in the past 20 years. Many countries, including Nigeria, are looking toward mitigating the usage of conventional energy sources where greenhouse gas (GHG) emission poses serious climate and environmental challenges. The Nigerian government strives to promote the use of electric vehicles (EVs) in the country to mitigate noise and air pollution, and in its effort toward that end, it is incorporating Evs in its development plan for the National Automotive Industry [7]. Environmental and noise pollution concerns, unstable cost and supply of petroleum products as well as efficiency are the driving factors for the development and implementation of alternative means of running vehicles in Nigeria.

Furthermore, the EV named Kona, which was the first one assembled in Nigeria, was unveiled in 2020 [8], and it was stated [9] that it can be charged anywhere within the five-year life of its storage device. The Nigerian JET Motor firm (JMF) launched its electric van JET EV [10,11] with a moving range of more than 250 km [12], and meanwhile, the Nigerian Siltech company introduced its electric bikes with a battery swapping technology. Recently, the National Automotive Design and Development Council (NADDC) of Nigeria, under the EV pilot program, has launched two solar energy-based EV charging stations (EVCSs) in the western Nigerian cities of Lagos and Sokoto with a capacity of 86.4 kilowatts per hour each [13]. The EVCS in the southeastern Nigerian city of Enugu is expected to be unveiled soon to boost the Nigerian government's commitment to efficient, cost-effective, and eco-friendly vehicles [14,15]. The Lagos state government has given assurance towards providing EVCSs to support the full utilization of EVs in the state [16]. The partnership between JMC and GIG logistics (the country's leading logistics company) to introduce the first electric vans (Jet Mover) has resulted in the installation of EV charging station infrastructures on the premises of this latter company. A charging station is capable of charging two EVs at the same time to full capacity with two hours of charging. Each EV van has a 107.8 kWh lithium-ion battery and can go a distance of about 300 km [17]. This shows that the government and the private sectors are highly committed to the establishment of EV pilot projects that support carbon neutrality. However, insufficiency in electrical power supply, which is essential for the utilization of EVs, and the shortage in EV charging station (CS) infrastructures are obstacles to a bright future for EVs in Nigeria.

In addition, wind and solar power have become essential and common technologies for electricity production due to the recent technological advancements in power electronic storage systems as well as the price reduction of the required devices [18]. According to Aliyu et al. [19], with an average of 6 h of daily sunshine being captured over 1% of land size for the solar photovoltaic (PV) system, Nigeria has the potential to generate annual electrical energy of about 1,850,000 GWh. The coastal and offshore regions, hilly areas of the north, middle belt mountainous terrains, as well as northern fringes of the country are windy with the potential for harvesting great wind power all over the year [20]. Moreover, it was indicated by Nehrir et al. [21] that renewable energy sources (RESs) such as solar energy (SE) and wind energy (WE) have more benefits in terms of electricity generation compared to other sources in addition to being freely available in nature. The integration of solar and wind systems with storage devices, either as standalone microgrids or with a utility grid network, has proven to save a substantial amount of carbon dioxide (CO_2) and other pollutant emissions, thereby making such microgrids environment-friendly [22]. The microgrid could serve as a green solution for the EVCS energy demand. The utilization of solar and wind energies incorporated with a better end-use efficiency will almost definitely be needed to cope with the increasing energy charging demand from EVs for sustainable, eco-friendly, and affordable power supply as the use of fossil fuels is gradually destroying the atmosphere and creating adverse and alarming climatic conditions.

Additionally, to solve the problem of everyday power outages, which could affect the operational time and efficiency of EVs in the country due to insufficient generation, the

Nigerian government has unveiled plans to increase the power generation from fossil fuels to 18,200 MW [23] and targeted 30% of the gross electricity generation from renewable energy (RE) resources by the year 2030 [24]. This shows that the future energy plan still targets enormous electricity generation from resources that pollute the environment and put the ecosystem in danger [25]. The sustainable development goal of the country can be accomplished via the adequate provision of reliable and affordable electricity supply from RE sources for various applications, including EVCS systems, in addition to preserving environmental integrity [26].

2. Literature Review

Various studies have been carried out to optimally design and techno-enviro-economically investigate the performance of EVCS systems and the Vehicle-to-Grid (V2G) technology strategy using different techniques and tools. Among the works assessed in the literature, few have simultaneously considered locations with diverse geographical characteristics and climatic conditions for the design and analysis of EVCS systems. The system topology, operating mode, locations, sensitivity, and evaluation variables using different optimization strategies and methodologies for electric vehicle charging station (EVCS) design in many other parts of the world are given in Table 1. Both grid-tied, hybrid models and standalone systems for meeting EV charge demand have been effectively studied for EV charging applications, but at a smaller number in a few locations around the world. Some studies have shown the technical and economic advantages of using an off-grid system based on RESs for EV charging to mitigate the burden and protect the utility grid network [14], while others have indicated the environmental benefits of a standalone EVCS [26]. Observation of the various system designs in Table 1 for the power supply of an EVCS conforms to our belief that no study has been conducted so far in the whole of Africa, particularly in the Nigerian context, to design and conduct a detailed performance assessment of a standalone system based on available RESs for EV charging. Furthermore, a few studies (Table 1) have performed a sensitivity analysis to check the resilience of the optimized system against uncertain variables. However, none of the few studies that actually performed some sort of sensitivity evaluation has performed so as an in-depth sensitivity analysis considering variations in diverse, uncertain, and critical parameters against the technical and economic performance of the selected EV charging system.

By utilizing bio-gas sources, Karmaker et al. [27] designed a 20 kW EVCS with a detailed feasibility assessment. The technical, environmental, and financial perspectives relating to their proposed EV charging points were studied with the aid of the HOMER optimization tool. Their outcome reveals the economic and environmental benefits of the proposed EVCS system as compared to the utility grid-based charging points in Bangladesh. The design and viability study of a specialized EV charging station model was investigated using the grid analysis software tool known as HOMER (short for "Hybrid Optimization of Multiple Energy Resources"). Based on three various cases of EV charging points analyzed, a yearly profit of USD63,680 was provided in addition to recovering the charging station's installation costs in less than three years. The level-2 EVSE is the most effective, reducing greenhouse gas emissions by 104 t [28]. The investigation of the environmental effect of the carbon dioxide (CO_2) emissions point of view embedded in green off-grid power schemes and the assessment of the environmental impact of their execution in the electricity supply of EVCS was conducted by Filote and Felseghi [29]. Their outcomes show that the clean power schemes represent viable solutions for the autonomous electricity support of EVCSs, being capable of supplying electric power based on 100% availability of on-site alternative energy resources. Moreover, the mean cost of 1 kWh of energy produced by the evaluated configurations is 4.3 times the mean unit cost of the EU electric distribution network electricity. Machado et al. [30] investigated the techno-economic viability evaluation on EV and RE integration as a case study, where three possible scenarios varying between a grid-alone operation, the integration of RESs for providing electric power to the EVs and a utility grid, together with the vehicle-to-grid (V2G) strategy were presented. The V2G scheme and the RE integration outcome indicated cost benefits along with minimal power intake during the peak loading time of the design. The HOMER software was used by Boddapati and Venkatesh [31] for the design of an EVCS by utilizing hybrid power sources such as solar photovoltaic, wind, and diesel plants (DG). They indicated that the utility power network-tied EVCS is more cost-effective than the standalone charging scheme. Moreover, they also conducted a sensitivity evaluation to check the impact of some variable parameters on the EVCS.

Furthermore, another study by Ye et al. [32] assessed the feasibility of a solar-based EVCS model for application in Shenzhen City in China. The results of this study reveal prospects of the proposed model in terms of emissions minimization and the satisfaction of the huge demand needed for electric vehicles. The study also recommends that carbon pricing promotes RE only when the cost of carbon is more than USD20/t. Moreover, a technical and economic investigation of a standalone renewables-based EVCS to find the optimum system to produce the needed charging demand per day was performed [33]. The outcomes of this investigation revealed that the optimized solution for the chosen locations comprises a 250-kilowatt wind system with a 60 m hub height. Furthermore, the optimum scenarios' gross net present cost (NPC) varies between USD2.53M to USD2.92M, while the cost of energy (COE) varies between USD0.285 and USD0.329 per kWh. Based on the change in the feed-in tariff technique, a techno-economic performance investigation of the V2G system model in Indonesia's largest electricity grid network was performed [1]. The investigation indicated that the utilization of EVs can potentially reduce the peak hour supply by about 2.8% and 8.8%, respectively, for coal and gas. From the electricity company's point of view and because of fuel replacement, the annual revenue can be improved by around 3.65% with the vehicle-to-grid approach. The design and viability evaluation of a RE-based hybrid EVCS system was carried out to lessen the stress on the electric utility network system owing to the fast rise in electrical vehicles in Bangladesh [34]. The EVCS system in [32] is designed using solar PV and a biogas system. The system configuration estimates an energy cost of USD0.1302/kWh and a gross net present cost of USD56,202 at a running cost of USD2540. Moreover, the system model minimizes carbon dioxide emission by 34.68% in comparison to a traditional electric utility network-based CS.

Efficient, reliable, and cost-effective EV charging infrastructure is one of the main factors that can facilitate the utilization and adoption of electric vehicles in Africa, particularly in Nigeria, to support the decarbonization of the economy and environment. It is difficult to achieve an efficient and reliable EVCS in Nigeria with the current utility grid network because of the so-far erratic and unreliable nature of that network. Therefore, an autonomous RE-based system could provide clean, reliable, and cost-effective electricity for EV charging. Moreover, since the adequate and effective use of RE sources for power generation depends on the climatic conditions of a place and the massive and successful roll-out of EVs is a function of the availability of reliable and cost-effective EV charging infrastructures, it is important to come up with an alternative approach to designing an efficient EVCS that can generate reliable, clean and cost-effective electricity. Currently, there is no grid-independent hybrid EV charging station scheme under Nigerian conditions that are readily available in the open literature. Therefore, this study strives to carry out the design and performance assessment of a hybrid RE-based system to provide reliable, eco-friendly, cost-effective electricity with enhanced efficiency for EV charging stations. The proposed robust methodology intends to accurately investigate the establishment of EV charging stations in Nigeria. This methodology is examined by considering different optimal system configurations across the six geo-political zones with diverse geographical characteristics and climatic conditions.

The operational behavior of the charging system was tested via sensitivity analysis to check the robustness of the optimal charging scheme by varying some critical sensitivity parameters. The HOMER Pro® software has been employed for the sizing and performance assessment of the proposed charging station scheme owing to its precision, performance, and efficiency in evaluating the optimum hybrid energy system configuration [35]. The grid-search and proprietary derivative-free optimization (GPDO) techniques were used in the microgrid analysis tool to obtain the cost-effective, viable charging system.

Table 1. The EVCS system designs and sensitivity analyses conducted in previous works in the open literature.

System Configuration	Operating Mode	Country	Sensitivity Variables	Year	Evaluation Parameters	Refs
PV/Wind/Fuel cell/battery	Off grid	India	Nil	2022	NPC/OPEX/COE	[36]
PV/Grid/battery	On/Off-grid	Vietnam	Solar EVCS	2021	NPC/COE/RF	[37]
CPV/WT/Bio-Gen/FC/Battery	Stand-alone	Qatar	WT hub heights.	2021	NPC/COE/Unmet load	[33]
PV/Grid/Battery	Grid-based	India	Nil	2021	RF/COE/Prod./GHG	[28]
Wind/PV/battery	Stand-alone	Turkey	Nil	2020	NPC/Prod./COE	[38]
PV/Wind/Fuel cell/battery	Off grid	Romania	Nil	2020	COE/NPC/GHG	[29]
PV/WT/Grid/V2G	Grid-tied/V2G	Brazil	Nil	2020	LCOE/Prod./NPV	[30]
V2G technology	Grid-based	Indonesia	Nil	2020	GHG/Energy-supply/cost	[1]
PV/Biogas Gen/Grid/Battery	On/Off-grid	Bangladesh	Nil	2018	NPC/COE/GHG	[34]
DG/PV/Grid/Battery	On/Off-grid	Canada	Nil	2017	NPC/COE/GHG	[39]
PV/Grid/Battery	Grid-tied	Bulgaria	Nil	2016	COE/NPC/GHG	[40]
PV-Grid based	Grid-tied	China	Economic variables	2015	COE/GHG/NPC	[32]

3. Methodology

3.1. Microgrids Design and Optimization Tool (HOMER)

The hybrid optimization model for electric renewable (HOMER) simulation tool is an important and widely used simulation software tool. It was developed in 1993 by the National Renewable Energy Laboratory (NREL) in the USA [41]. The software is utilized for analyzing various system design alternatives for both standalone and grid-tied designs in a simplified way for different applications. The optimal evaluation simulation procedure of the charging station models in the HOMER Pro® software is illustrated in Figure 1. Three major tasks are accomplished here: simulation, optimization, and sensitivity evaluation. For the simulation, HOMER models hourly the performance of each of the system subunits to ensure the optimal possible matching between the energy demand and supply. It models various system designs in the optimization section to find the systems that meet the technical constraints as well as fulfill the charge demand at a low life-cycle cost.

Lastly, HOMER performs numerous optimization operations with various ranges of input variables to check the effects of changes in input parameters on the selected system in the sensitivity analysis section [42]. The input data required by HOMER are load profile, meteorological resources, economic constraints, and system component specifications (prices and sizes), which are used to provide a list of ranked feasible systems according to the least total NPC and energy cost [43]. The modeling and optimization of different hybrid energy systems were previously studied in the literature in terms of diverse techno-economic and environmental parameters. Such a study was accomplished with the aid of the HOMER analysis tools for both grid-connected and off-grid applications in different parts of the world are presented in Table 2. This shows that HOMER software is an important and widely used tool for the design and assessment of hybrid energy systems worldwide. Therefore, the HOMER Pro® simulation tool is employed in the present analysis for the optimal design and techno-economic viability analysis of the proposed renewable energy-based EV charging station. In order to obtain a cost-effective feasible system, the simulation tool utilizes the GPDO techniques [44].

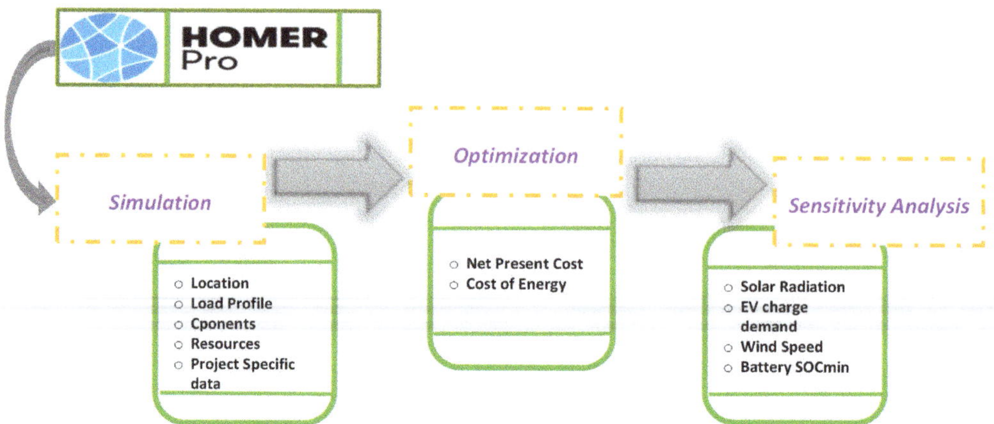

Figure 1. The evaluation process of the hybrid RE-based charging stations in HOMER Pro®.

Table 2. Hybrid renewable energy-based system models that were studied in different parts of the world with the aid of the HOMER simulation tool.

System Configuration	Application	Country	Simulation Tool/	Year	Parameters	Refs.
PV/WT/DG/BES	Rural load	Nigeria	HOMER	2021	NPC/COE/GHG	[45]
WT/DG/FC/BES	Standalone	Saudi Arabia	HOMER	2021	NPC/COE	[46]
WT/BES/PV/DG	Off-grid	Malawi	HOMER	2021	COE/RF/NPV	[47]
PV/BES/GRID	On/Off Grid	Iraq	HOMER	2020	NPC/GHG/RF	[48]
PV/BES	Off-grid	Morocco	PVsyst/HOMER	2019	LCC/RF/COE	[49]
WT/DG/BES	Off-grid	Pakistan	HOMER/MATLAB	2019	THD/GHG/COE	[50]
WT/PV/BES/DG	Off-grid	Bangladesh	HOMER	2018	NPC/COE/GHG/	[51]
GRID/PV	On-grid	Saudi Arabia	HOMER	2018	RF/NPC/COE	[52]
PV/WT/BES/FC	Off-grid	UAE	HOMER	2017	NPC/COE/GHG	[53]
WT/PV/DG/BES	Off-grid	Canada	HOMER	2016	GHGepc/COE	[54]
WT/PV/BES/DG	Off/On-grid	Sri Lanka	HOMER	2015	TNPC/LCOE	[55]
PV/DG/BES	Off-grid	India	HOMER	2014	NPC/RF	[56]
PV/DG/BES	Off-grid	Saudi Arabia	HOMER	2010	COE/RF	[57]

3.2. Sites Details and Operating Strategy

The six locations selected in this study for the set-up of the proposed EV charging station are shown on the map of Nigeria displayed in Figure 2. The considered case study sites were selected from the North-Central (Minna), North-East (Maiduguri), North-West (Sokoto), South-West (Ikeja), South-East (Enugu), and South-South (Port-Harcourt) geopolitical zones. The geographical coordinates of the case study sites are given in Table 3. The selected and investigated locations are among the sites that the National Automotive Design and Development Council (NADDC) of Nigeria is considering for the EV pilot program initiatives.

Figure 2. The map of Nigeria showing the case study sites. The star in the middle denote Abuja, the Capital of Nigeria.

Table 3. Geographical coordinates of the case study sites.

State	City	Geo-Political Zone	Latitude (°N)	Longitude (°E)	Altitude (m)
Sokoto	Sokoto	North-West	13.0059	5.2476	293.00
Lagos	Ikeja	South-West	6.6018	3.3515	46.00
Enugu	Enugu	South-East	6.4483	7.5139	200.00
Rivers	Port-Harcourt	South-South	4.8472	6.9746	13.00
Borno	Maiduguri	North-East	11.8311	13.1510	325.00
Niger	Minna	North-Central	9.5836	6.5463	446.00

The power flow diagram of the proposed hybrid RE-EV charging station is depicted in Figure 3. The wind turbines (WTs) are integrated into the hybrid charging station system via the alternating current (AC) link, while the PV panels and the batteries storage are connected through the direct current (DC) bus. The bi-directional converter keeps the flow of electric power between the AC bus and DC link devices. It converts the electrical power from DC to AC. In the inverter mode, it changes the electricity from DC to AC and from AC to DC while operating as a rectifier.

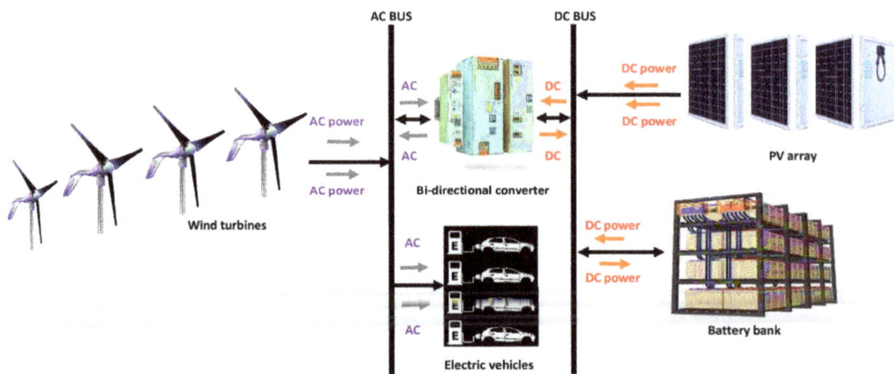

Figure 3. The energy flow schematic diagram of the suggested hybrid RE-based charging stations.

The flowchart for the assessment and the optimization evaluation utilized in the HOMER Pro® microgrid software is illustrated in Figure 4. The objective function is to find the sizes and working scenarios with the least lifespan and set up running costs among the viable systems that satisfy the charging point demand. The minimization of the objective function is based on certain constraint conditions. In this study, the following constraints are considered: the maximum yearly capacity shortage, the minimum renewable fraction, the renewable energy power output, and other standard technical constraints. The decision parameters explored in this analysis are the sizes of the integrated components, including the RE components, the battery storage device, and the converter. The values and specifications of the system control variables, economics, and constraints utilized in the analysis tool are given in Table 4. Because of the large financial cost produced by the battery device replacement, a charging control technique and control approach is needed to protect the storage device from over-charging and over-discharging [58]. The optimization evaluation and the outcomes are based on the constraints, objective function, and decision parameters. After calculating and modeling the load profile, inputting the resource data based on the geographical location of the case study sites, and identifying and specifying the predefined constraints and system components' technical and cost details, the analysis tool begins the simulation procedure. The simulation process ensured (via the optimization section) that the charging station models that satisfy the technological constraints and meet the EV charge demand at a low total net present costs are realized.

Table 4. The control variables, economics, and constraints utilized in the optimization tool [41,59,60].

Variables	Unit	Value
Initial state of charge ($SOC_{initial}$)	%	100
Minimum SOC ($SOC_{minimum}$)	%	20
Consider ambient temperature impact	Yes	Yes
Tracking system	No	No
Considered PV panel slope	Yes	Yes
Allowing system with multiple sources	Yes	Yes
System design precision	-	0.010
NPC precision	-	0.010
Economics		
Real discount rate	%	14
Project lifetime	years	25
Constraints		
Minimum renewable fraction ($RF_{minimum}$)	%	0
Load in current time step	%	10
Solar power output	%	25
Wind power output	%	50

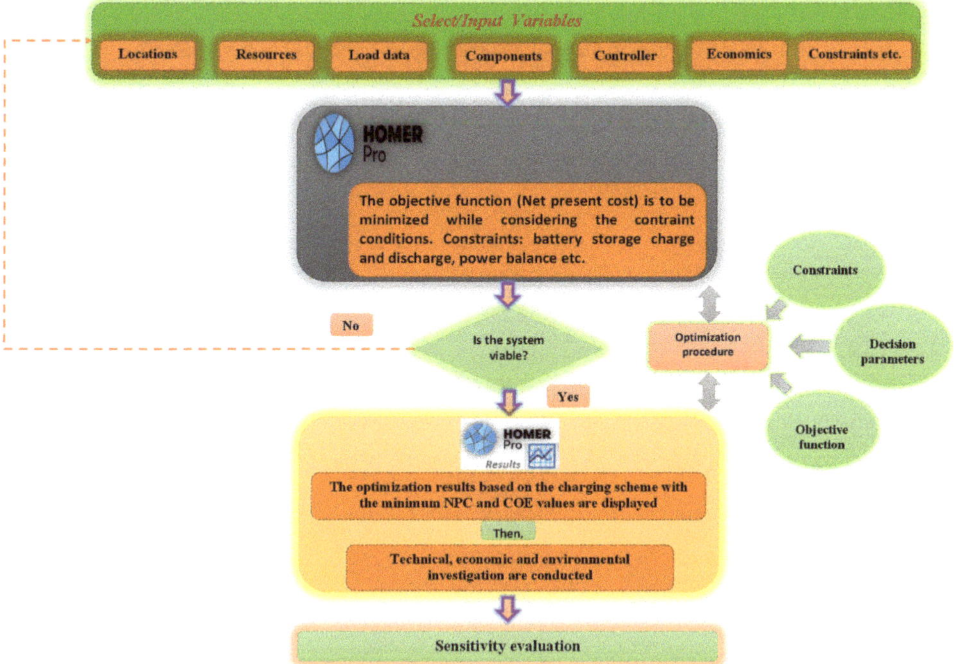

Figure 4. Assessment flowchart and optimization process utilized in the HOMER Pro® microgrid tool.

The other part of this study assesses and discusses sensitivity evaluation through the variation of some system variables. Sensitivity analysis is capable of identifying the most important variables of an investment due to the possibility of knowing in advance the effect of input parameters with uncertainty on the system cost variables and can be utilized in different contexts as well as in the assessment of investment projects [61]. The intermittent nature of RESs [62] and the change in load demand are the two major parameters that influence the reliability and consistency in the power generation of any renewable-based energy system. These variables can also impact the present and future financial aspects of the system. Hence, sensitivity analysis is a more realistic and easier technique for determining whether a specific investment is feasible or not. In the present study, the wind speed, solar radiation, battery energy storage minimum state of charge (BES SOC$_{min}$), maximum yearly capacity shortage, and EV charge demand are used as sensitivity parameters.

3.3. Renewable Sources at the EVCS Case Study Locations

3.3.1. Wind Power Sources

In Nigeria, coastal, mountainous, and offshore sites are endowed with huge wind power resources. The wind speed over the complex landscapes and the plain surface ranges between 3.60 and 5.40 m per second in the north, whereas the south is characterized by small wind resources with speeds ranging between 1.4 to 3.0 m per second. The wind speed (WS) information obtained from the NASA Prediction of Worldwide Energy Resources (POWER) database [63] over thirty years is given in Table 5. The wind data show that there is a great change in the wind speed reported from November to March of each year in the different locations. Observation of the wind speed information reveals that the minimum and maximum wind speeds are reported in different months due to distinctions in the geographical characteristics and climatic conditions of each site. The lowest wind speed recorded in Sokoto is 3.78 m/s (September), while those of Ikeja, Enugu, Port-Harcourt, Maiduguri, and Minna were obtained at 2.65 m/s (December), 2.83 m/s (November), 2.43 m/s (December), 3.96 m/s (September), and 2.75 m/s (October), respectively. Furthermore, the highest average wind speed values of 7.25 m/s (January), 4.9 m/s (August), 5.03 m/s (August), 4.0 m/s (August), 7.12 m/s (February), and 5.69 m/s (January) were reported in Sokoto, Ikeja, Enugu, Port-Harcourt, Maiduguri and Minna, respectively with a yearly average of 5.44 m/s, 3.81 m/s, 4.09 m/s, 3.15 m/s, 5.50 m/s, and 3.97 m/s, respectively.

Table 5. Wind speed data of the case study sites for EVCS (m/s) [40].

Month	Locations					
	Sokoto	Ikeja	Enugu	Port-Harcourt	Maiduguri	Minna
January	7.25	3.07	3.83	2.81	6.78	5.69
February	7.07	3.61	3.76	3.01	7.12	5.01
March	6.25	4.04	4.14	3.12	6.77	4.00
April	5.23	4.10	4.40	3.05	5.49	3.75
May	5.10	3.78	4.18	2.95	4.96	3.45
June	5.19	4.16	4.55	3.36	5.20	3.37
July	4.79	4.84	5.00	3.83	4.97	3.54
August	3.97	4.90	5.03	4.00	4.21	3.51
September	3.78	4.28	4.41	3.60	3.96	2.86
October	4.09	3.50	3.65	3.05	4.11	2.75
November	5.68	2.74	2.83	2.56	5.86	4.23
December	6.90	2.65	3.33	2.43	6.56	5.48
Average	5.44	3.81	4.09	3.15	5.50	3.97

3.3.2. Solar Power Resources

Nigeria lies between latitudes 4° and 14° N (slightly north of the equator), and longitudes 2° and 15° E (slightly east of the prime meridian). The whole country falls within an area where sunshine is plentiful. Due to its location, the country's solar radiation is relatively well distributed, and the yearly daily mean varies from about 3.5 kWh/m^2 in the coastal part to 7.0 kWh/m^2 in the far northern part [64]. The solar irradiation details of the selected sites are again retrieved from the NASA POWER database (for a period of twenty-two years) [63] by specifying the coordinates of the selected zones for the planned installation of RE-based EV charging schemes. The changes in the monthly mean solar irradiation of the different locations are presented in Table 6. The annual average values of 6.24, 4.74, 4.93, 4.13, 5.90, and 5.49 kWh/m^2/day were obtained for Sokoto, Ikeja, Enugu, Port-Harcourt, Maiduguri, and Minna, respectively. The corresponding minimum and maximum radiations of 5.25, 3.95, 3.91, 3.11, 5.14, and 4.36 kWh/m^2/day at a clearness index of 0.64, 0.394, 0.381, 0.315, 0.491, and 0.419 and 7.15, 5.49, 5.74, 5.24, 6.7, and 6.26 kWh/m^2/day at a clearness index of 0.678, 0.556, 0.58, 0.550, 0.661, and 0.611 were reported in various months. Moreover, the yearly mean air temperatures of 27.92, 25.92, 25.21, 25.59, 28.00, and 25.01 °C were reported for Sokoto, Ikeja, Enugu, Port-Harcourt, Maiduguri, and Minna for about 30 years [63] as depicted in Table 7.

Table 6. Solar radiation data of the case study sites for EVCS (kWh/m^2/day) [40].

Month	Locations					
	Sokoto	Ikeja	Enugu	Port-Harcourt	Maiduguri	Minna
January	5.47	5.28	5.68	5.24	5.61	5.72
February	6.41	5.49	5.74	5.13	6.30	6.01
March	6.87	5.46	5.57	4.73	6.70	6.26
April	7.15	5.21	5.25	4.50	6.62	6.12
May	7.03	4.76	4.94	4.09	6.36	5.73

Table 6. *Cont.*

Month	Locations					
	Sokoto	Ikeja	Enugu	Port-Harcourt	Maiduguri	Minna
June	6.91	4.04	4.54	3.45	5.97	5.17
July	6.26	3.95	4.14	3.11	5.43	4.64
August	5.73	3.98	3.91	3.42	5.14	4.36
September	6.01	4.09	4.19	3.22	5.57	4.82
October	6.03	4.55	4.57	3.60	5.89	5.42
November	5.79	4.95	5.11	4.18	5.84	5.85
December	5.25	5.17	5.46	4.88	5.35	5.73
Average	6.24	4.74	4.93	4.13	5.90	5.49

Table 7. Air temperature data of the case study sites for EVCS (degree celsius) [40].

Month	Locations					
	Sokoto	Ikeja	Enugu	Port-Harcourt	Maiduguri	Minna
January	23.03	25.90	24.02	25.40	23.35	22.90
February	25.91	26.94	25.68	26.43	26.16	25.19
March	29.71	27.22	26.60	26.73	29.98	27.20
April	32.84	26.99	26.62	26.53	32.77	27.49
May	33.00	26.53	26.16	26.19	32.77	26.52
June	31.06	25.59	25.29	25.31	30.94	25.41
July	28.48	24.65	24.60	24.61	28.17	24.48
August	26.80	24.41	24.58	24.52	26.61	24.23
September	27.09	24.92	24.78	24.83	27.23	24.82
October	27.86	25.55	25.05	25.26	28.22	25.19
November	25.91	26.27	25.27	25.77	26.16	24.04
December	23.37	26.01	23.92	25.53	23.64	22.65
Average	27.92	25.92	25.21	25.59	28.00	25.01

3.4. Mathematical Representation and Specifications of the Hybrid System Components

3.4.1. Wind Turbine System

The detailed technical information of the considered wind turbine (WT) is given in Table 8, while its cost data are presented in Table 9 [44]. The Weibull k parameter is a measure of the long-period distribution of wind speed (WS) for a year, taken herein as 2. The diurnal pattern strength, specified as 0.25, is a measure of how strong WS depends on the daytime, while the 1 h autocorrelation factor in the HOMER Pro® software is a measure of the hour–hour randomness of WS, considered as 0.85. The hour of peak WS is taken as 15. The quantities of WTs needed to reliably satisfy the EV charging requirement at a low cost are optimized. The relationship between the output power and the WS is illustrated via the WT power curve in Figure 5. The mechanical power P_m of the WT with regard to the air density ρ (1.22 kg/m^3), surface area A swept by the rotor (m^2), and velocity V are evaluated as:

$$P_m = \frac{1}{2} \times \rho \times A \times V^3 \tag{1}$$

Table 8. Technical specification of the WT.

Wind Turbine	Values
Name/Model	XANT M-21
Rated capacity	100 kW
Rotor diameter	21 m
Hub height	31.8 m
Cut-in WS	3 m/s
Rated WS	11 m/s
Lifetime	20 years

Table 9. Economic data of the hybrid energy system components.

Components	Capital Cost	Cost of Replacement	Maintenance Cost	Reference
PV panels	USD1500/kW	USD1000/kW	USD10/kW/year	[65]
Wind turbine	USD50,000/unit	USD50,000/unit	USD2500/year/unit	[44]
Converter	USD200/kW	USD200/kW	-	[50]
Batteries	USD176/unit	USD176/unit	USD8/unit/year	[50]

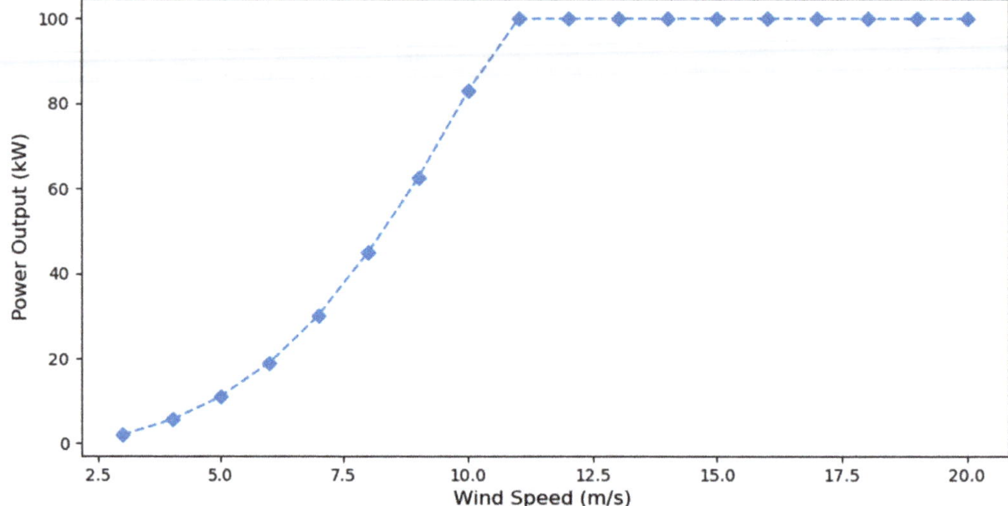

Figure 5. The WT power curve.

The electrical power P_e in terms of the power coefficient C_p is given as:

$$P_e = \frac{1}{2} \times \rho \times C_p \times A \times V^3 \times 10^{-3},\tag{2}$$

3.4.2. Solar Photovoltaic System

In this study, the SunPower X21-335-BLK PV panel was selected due to its high efficiency. The details of the cost variables associated with the PV panel are given in Table 9. The mean efficiency of the solar panel is 21%. The technical specification of the solar photovoltaic is presented in Table 10 [41]. The panel has 96 monocrystalline cells at nominal power and operating cell temperature of 0.335 kW and 43 °C, respectively. The sizes of PV panels needed to efficiently meet the EV charge demand are optimized. The output power P_{PV} of the PV module is analyzed in terms of the solar irradiation, de-rating factor, and temperature influence as follows [65]:

$$P_{PV} = Y_{PV}PV(\frac{G_T}{G_{T,STC}})[1 + \alpha_P(T_C - T_{C,STC})]\tag{3}$$

where Y_{PV} refers to the PV power output under standard test conditions (STC) in kW, PV represents the PV de-rating factor (%), G_T is the solar radiation incident on the PV panel in the current time step (kW/m^2), $G_{T,STC}$ refers to the incident radiation under standard test conditions (1 kW/m^2), α_P is the temperature coefficient of power (%/degree Celsius), T_C is the temperature of the PV cell (_C), and $T_{C,STC}$ is the PV cell temperature at STC (25 degree Celsius) [42].

Table 10. Technical data of the PV panel.

PV Panel	Values
Name/model	SunPower SPR X21
Panel type	Flat plate
Rated capacity	335 W
Temperature coefficient	−0.3
Operating temperature	43 °C
Efficiency	21%
De-rating factor	88%
Ground reflectance	20%
Tracking system	-
Lifetime	25 years

3.4.3. Battery System

The economic data and technical specifications, including the storage properties of the selected battery bank, are presented in Tables 9 and 11 respectively. The string of the battery storage consists of 20 batteries per string. The maximum capacity of the battery is 408 Ah at a capacity ratio of 0.0699. The peak charge and discharge current are 74 A and 300 A at a maximum charge rate of 1 A/Ah. The $SOC_{minimum}$ value of 20% was considered in the study. The battery capacity C_{Bat} is calculated by utilizing the daily load energy E_L and autonomy days (AD) as stated in Equation (4) below [49,66]. The battery state of charge (SOC) $SOC_B(\%)$ is determined as a percentage of the ratio of its charge q_b to its maximum charge q_{bm} using Equation (5) [67].

$$C_{Bat} = \frac{E_L\,AD}{\eta_{inv}\,DOD\,\eta_{bat}} \tag{4}$$

where η_{inv} denotes inverter efficiency, DOD is the battery's depth of discharge and η_{bat} refer to the battery efficiency.

$$SOC_B(\%) = \frac{q_b}{q_{bm}} \times 100 \tag{5}$$

Table 11. Technical details of the selected battery.

Battery Storage	Values
Nominal voltage	6 V
Nominal capacity	2.45 kWh
Roundtrip efficiency	80%
Maximum capacity	408 Ah
Lifetime (Throughput)	1958 kWh
Capacity ratio	0.0699
Rate constant (1/h)	6.01
Minimum state of charge	20%

3.4.4. Converter

The converter keeps the flow of electric power between the alternating current (AC) link and direct current (DC) link devices. It converts the electrical power from DC to AC. In the inverter mode, it changes the electricity from DC to AC and from AC to DC while operating as a rectifier. The costs of the economic variables of the power converter are given in Table 9. The technical data of the selected converter are given in Table 12 [50]. The converter capacity level is obtained using:

$$C = (3 \times L_i) + L_r \tag{6}$$

Here, L_i and L_r refer to the inductive and resistive loads.

Table 12. The technical data of the selected converter.

System Converter	Values
Inverter input	
Efficiency	95%
Lifespan	15 years
Rectifier input	
Relative capacity	100%
Efficiency	85%
Rated capacity	1 kW

3.5. Evaluation Criteria

3.5.1. The Net Present Cost (*NPC*)

The *NPC* comprises the initial cost, cost of replacement of individual devices, operation cost, maintenance cost, etc. It is an economic variable used to assess the optimum system of different combinations of system configurations. The following equation is used to analyze the *NPC* (for convenience, denoted as C_{NPC}) [68]:

$$C_{NPC} = \frac{TAC}{CRF(i, N)} \tag{7}$$

Here, the total annualized cost (USD/year) is denoted by *TAC*, *N* represents the number of years, and *i* refers to the yearly real discount rate in percent. The capital recovery factor (*CRF*) is calculated using Equation (8) below:

$$CRF(i, N) = \frac{i(1 + i)^N}{(1 + i)^N - 1} \tag{8}$$

The yearly real discount rate with regard to the anticipated inflation rate (*f*) and nominal discount rate (i') is obtained from Equation (9) below:

$$i = \frac{i' - f}{1 + f} \tag{9}$$

3.5.2. The Cost of Energy (*COE*)

The *COE* is defined as the mean cost per unit of effective electricity generated by the system configuration [69] over the system's whole lifespan. The *COE* is computed with regard to the total annualized cost (USD/year) (*TAC*) and total annual load (kWh) served by the system ($E_{anloadserved}$) as:

$$COE = \frac{TAC}{E_{anloadserved}} \tag{10}$$

3.5.3. The Renewable Fraction (*RF*)

The *RF* is the gross amount of electricity produced by sustainable resources compared to the gross energy generated from the whole hybrid EV charging scheme [70]. The *RF* is computed with regard to the output power of the sustainable resources as illustrated below:

$$RF(\%) = \left(1 - \frac{\sum P_{diesel}}{\sum Pren}\right) \times 100 \tag{11}$$

3.5.4. The Unsatisfied Load

The unsatisfied load is the electric charging load that the hybrid EV charging station system model cannot fulfill. This occurs when the load requirement is greater than the electric supply. The unfulfilled load is computed as a ratio of the annual non-served load to the gross annual demand as given below [67]:

$$\text{Unfulfilled load} = \frac{\text{Annual Non} - \text{served Load}}{\text{Annual Entire Load}} \tag{12}$$

4. Results Analysis and Discussion

In this study, an EV charging scheme based on renewable energy resources and storage devices is designed and analyzed using the *HOMER Pro®* software. The simulation software utilizes the GPDO techniques to secure the most economically feasible system configurations that can sufficiently supply the EV charging demand. Nigeria is divided into six geo-political zones with different meteorological features, and the renewable energy resources are weather dependent. Therefore, to conduct a more comprehensive investigation and to have a detailed overview of the operational performance of the proposed system, six different locations, each from one of the six geo-political regions, are considered. Furthermore, three different combinations of energy sources with storage systems were investigated for the hybrid EV charging station system in the six sites.

4.1. Load Data Estimation

In this study, the electric load analyzed is implemented under hypothetical states. The load profile of small-scale charging stations for the six locations is illustrated in Figure 6. The demand profile of the EV charging schemes is forecasted due to the small number (below 10 charging points) of EVCSs presently installed in Nigeria at the moment. According to the hypothetical daily, seasonal and annual load description of the selected six locations, about 20–30 EVs can be charged in a station. In the morning till the afternoon, between 06:00 and 16:00, up to 20 EVs can be recharged at an average load of 80 kW, whereas the hybrid charging scheme can provide energy to charge about 10 EVs averaged at 40 kW in the latter hours of the day from 16:00 until 22:00. The daily capacity of each EV was assumed to be 35 kWh of battery energy; therefore, the total average and peak load demand of 30 EVs is 1050 kWh/day and 104.99 kW at a load factor of 0.42. To establish an accurate estimation of the highest demand and depict a realistic load requirement of the proposed charging system, a time-step and day-to-day random variability of 10% and 5% were used in the EV load data analysis.

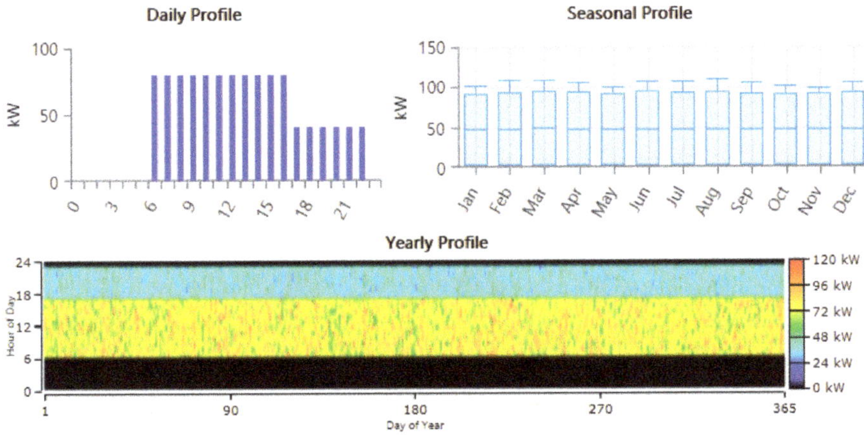

Figure 6. EV charging station system load profile.

4.2. Performance Assessment of the Proposed Charging Station Schemes

The economic and technical outcomes, including the optimum component sizes of different feasible charging station models in the six considered sites, are illustrated in Tables 13 and 14. The combination of PV and WT with battery storage is economically the best system architecture for the charging station in all six sites. It is clear from Table 13 that the PV/WT/battery charging station had the least energy cost in all the simulated sites. The COE and NPC are also very competitive, even if it is difficult to install WT in the considered locations, as seen with the PV/battery charging station scenario. However, the unavailability of a PV system in the PV/WT/battery design architecture is not economically viable, as indicated in the case of the WT/battery changing station, which has the maximum NPC and COE values of USD3,318,763 and USD1.28/kWh in the Port-Harcourt site. In general, the PV/WT/battery charging station (2 qty. of WT, 174 kW of PV panels, 380 qty. of batteries storage, and a converter of 109 kW) in Sokoto provide the best economic metrics with the lowest NPC, electricity cost, and initial costs of USD547,717, USD0.211/kWh, and USD449,134, respectively. Moreover, the charging station presented competitive annual operating and maintenance costs of USD14,344 and USD67,195.

The wind energy-based charging stations in all the case study sites had the highest operating and maintenance costs because of the large number of WTs needed to fulfill the EV charging requirement. Most of the WTs need maintenance once every two years at the minimum. Therefore, the maintenance of the key parts of the WTs by carrying out tasks such as turbine inspecting, lubricating, repairing, and cleaning also contributed to high maintenance costs. The optimal charging station (PV/WT/battery in Sokoto) model had the lowest PV Levelized cost (USD0.118/kWh) with a competitive WT Levelized cost (USD0.0594/kWh). The battery wear cost is constant at USD0.1/kWh throughout the simulated year in all the sites considered.

Furthermore, according to Table 14, the highest and lowest values of the maximum penetration of renewables were reported in Minna and Sokoto. Therefore, the maximum total annual electricity is produced at 2,204,533 kWh by the WT/battery-based charging station in Minna, whereas the minimum is generated at 495,306 kWh in Sokoto by the PV/battery charging station at a capacity shortage of only about 2%. The PV/WT/battery CS at the Sokoto site was able to reliably satisfy most of the EV charge demand as it presented a small percentage of the unmet load of 1.38%, which is the lowest when compared with the corresponding values of the other charging stations. Moreover, the optimal charging station schemes in all six locations were able to sufficiently meet the EV demand with a maximum uptime as the percentages of the unfulfilled electric load were below 2% with a capacity shortage of only approximately 2%.

Table 13. Summarized cost details of the optimal systems for the proposed charging stations.

Locations	System Design	NPC (USD)	COE (USD/kWh)	Initial Capital (USD)	Operating Cost (USD/year)	Replacement Cost (USD)	O&M Cost (USD)	Salvage (USD)	PV Levelized Cost (USD/kWh)	Battery Wear Cost (USD/kWh)	WT Levelized Cost (USD/kWh)
Sokoto	WT-Battery	950,164	0.366	689,561	37,917	56,285	222,408	(18,090)	0.000	0.100	0.0594
	PV-WT-Battery	547,717	0.211	449,134	14,344	35,383	67,195	(3995)	0.118	0.100	0.0594
	PV-Battery	601,381	0.232	504,056	14,161	52,589	48,435	(3699)	0.118	0.100	0.0000
Ikeja	WT-Battery	2,527,137	0.976	1,853,466	98,018	108,684	607,567	(42,580)	0.000	0.100	0.1570
	PV-WT-Battery	769,360	0.296	638,560	19,031	43,457	95,645	(8302)	0.155	0.100	0.1570
	PV-Battery	916,480	0.354	800,973	16,806	34,571	82,231	(1295)	0.155	0.100	0.0000
Enugu	WT-Battery	1,723,874	0.666	1,258,560	67,703	89,363	412,238	(36,287)	0.000	0.100	0.1260
	PV-WT-Battery	745,574	0.287	609,430	19,809	43,972	99,752	(7580)	0.150	0.100	0.1260
	PV-Battery	892,200	0.344	775,546	16,973	26,925	94,599	(4871)	0.150	0.100	0.0000
Port-Harcourt	WT-Battery	3,318,763	1.280	2,448,673	126,597	135,871	790,937	(56,717)	0.000	0.100	0.2920
	PV-WT-Battery	1,039,660	0.400	879,552	23,296	40,380	125,193	(5464)	0.179	0.100	0.2920
Maiduguri	PV-Battery	1,119,727	0.432	995,943	18,010	31,956	94,090	(2263)	0.179	0.100	0.0000
	WT-Battery	871,596	0.336	630,228	35,119	53,334	204,126	(16,092)	0.000	0.100	0.0579
	PV-WT-Battery	563,527	0.217	466,222	14,156	30,697	72,731	(6122)	0.125	0.100	0.0580
Minna	PV-Battery	683,409	0.264	587,219	13,995	45,806	50,910	(526)	0.125	0.100	0.0000
	WT-Battery	2,405,558	0.930	1,734,752	97,601	138,853	578,151	(46,198)	0.000	0.100	0.1370
	PV-WT-Battery	758,248	0.292	648,588	15,955	32,180	79,281	(1801)	0.133	0.100	0.1370
	PV-Battery	774,679	0.299	669,719	15,272	33,213	72,593	(845)	0.133	0.100	0.0000

Table 14. Optimal sizes and technical performance results of the optimized systems for the EV charging station.

Locations	System Design	WT (Qty.)	PV (kW)	Converter (kW)	Batteries (Qty.)	Total Renewable Production (kWh/Year)	Total EV Consumption (kWh/Year)	Capacity Shortage (%)	Unmet Electric Load (%)	Maximum Renewable Penetration (%)
Sokoto	WT-Battery	10	-	138	920	1,699,130	377,369	2.10	1.53	3310
	PV-WT-Battery	2	174	109	380	674,904	377,945	2.07	1.38	763
	PV-Battery	-	257	102	560	495,306	377,662	2.06	1.46	441
Ikeja	WT-Battery	28	-	243	2300	1,795,360	376,858	2.06	1.67	7409
	PV-WT-Battery	2	252	102	800	497,701	377,901	2.06	1.40	675
	PV-Battery	-	396	152	1000	582,115	377,110	2.09	1.60	652
Enugu	WT-Battery	19	-	170	1560	1,523,727	376,545	2.09	1.75	5574
	PV-WT-Battery	3	221	109	600	578,502	377,842	2.09	1.41	943
	PV-Battery	-	352	108	1280	537,915	377,139	2.08	1.59	597
Port-Harcourt	WT-Battery	34	-	435	3760	1,176,967	376,152	2.10	1.85	6419
	PV-WT-Battery	3	368	117	880	573,566	377,881	2.10	1.40	667
Maiduguri	PV-Battery	-	521	139	1060	665,846	377,110	2.08	1.60	849
	WT-Battery	9	-	109	900	1,569,870	377,514	2.10	1.50	3020
	PV-WT-Battery	2	174	102	480	667,997	377,894	2.10	1.40	707
Minna	PV-Battery	-	309	145	540	565,530	377,290	2.09	1.56	526
	WT-Battery	30	-	171	1140	2,204,533	376,563	2.08	1.74	8872
	PV-WT-Battery	1	296	108	760	579,666	377,570	2.10	1.48	618
	PV-Battery	-	320	137	920	548,455	377,107	2.08	1.60	550

The monthly electric generation by the PV/WT/battery-based charging station in the six case study sites is illustrated in Figure 7. Generally, the solar PV panel generated most of the electricity needed to meet the EV charging requirement in Ikeja, Port-Harcourt, and Minna as compared to the WT production. However, in Ikeja and Port-Harcourt, the WT electric production was only competitive between June and September. The overall energy production (from both PV panel and WT) in Sokoto and Maiduguri were low from April until October. The total electric production started to increase in November and maintained a continuous maximum value until March. The gross monthly electricity generated in Enugu maintained a constant value for the whole of the simulated year, with the highest production reported in July and August. Furthermore, the annual electricity production of the optimal charging station schemes (PV/WT/battery) in the case study sites is illustrated in Figure 8, where the highest excess electricity and gross electric energy is produced in Sokoto due to the enormous presence of RE resources. The surplus electricity can be sold directly to the utility grid via a CS-to-grid connection. Moreover, since the proposed charging stations are located in cities/urban areas, this will facilitate any future connection of the charging stations to the grid network to enable the buy/sell electricity approach.

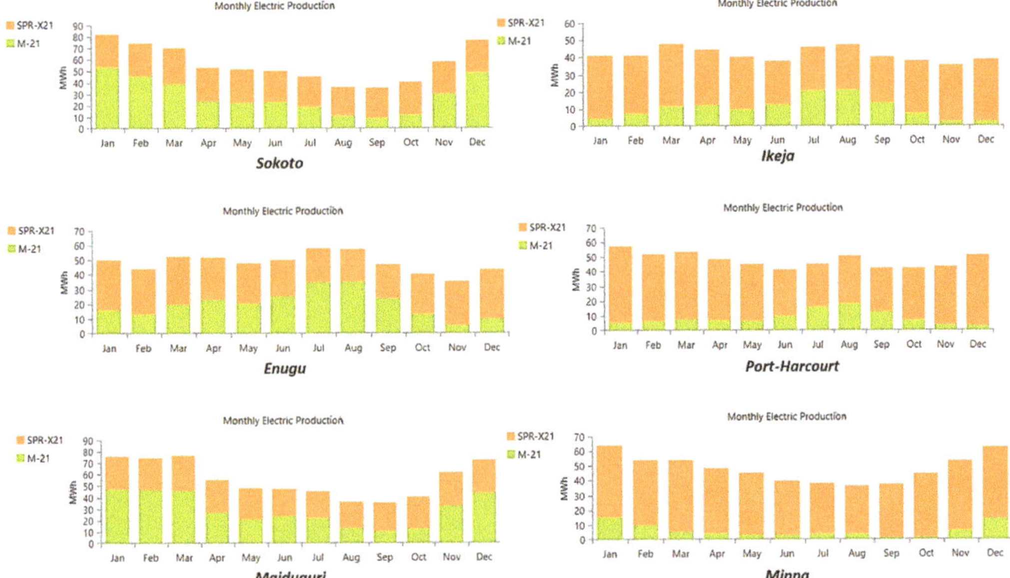

Figure 7. Monthly electricity generated by the PV/WT/battery-based charging station in the six selected sites.

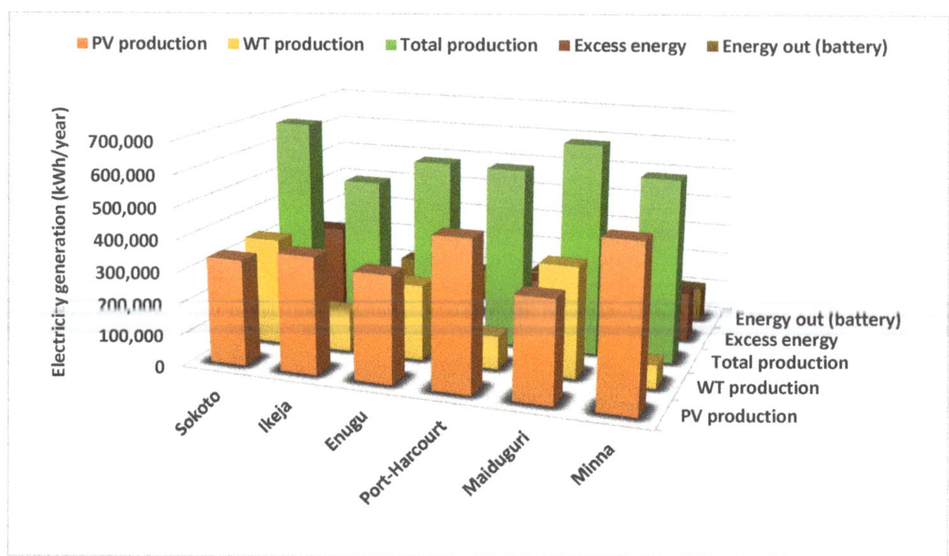

Figure 8. Yearly electric generation of the optimal charging station schemes (PV/WT/battery) in the considered locations.

In addition, Figure 8 reveals that the maximum PV production (506,181 kWh/year) is encountered in Minna, while the Maiduguri site reported the minimum PV generation at 319,137 kWh/year. The Maiduguri site recorded the highest annual electricity production from WT at 348,860 kWh, whereas a small yearly minimum value of about 73,484 kWh from the WT was reported in the Minna site. The environmental benefit of the proposed EV charging stations is that there is no carbon footprint and there is zero greenhouse gas while other harmful emissions are mitigated. This kind of system can be used to facilitate the global adoption of electric vehicles, which are often used to support economic decarbonization. Moreover, the provision of EV charge demand via the utilization of freely and readily available renewable energy resources will help to effectively promote the use of EVs as a mechanism to bring about a green solution in the transportation sector.

The optimal EVCS system designed in this study is further compared with the results of the existing EVCS designs for diverse application places using various simulation tools and approaches in the literature. The different combinations of energy resources and storage equipment and the economic and environmental results are presented in Table 15. Observation of the outcomes of the various studies revealed that the net present cost ranges approximately between USD21,000 and USD3,580,000, while the energy cost varies between about USD0.06/kWh and USD0.90/kWh. For comparison, the NPC and COE of the optimal EVCS system obtained in this study are USD547,717 and USD0.211/kWh, respectively. This provides evidence that the proposed standalone EVCS system presented herein is acceptable and possesses competitive economic metrics when compared with the previously published EVCS systems shown in Table 15. As regards the environmental benefits of the proposed standalone EV charging station, the majority of the existing works presented in Table 15 reported some non-negligible figures for greenhouse gas emissions. This highlights the drawbacks of some of the previously designed EVCS systems in terms of environmental preservation from carbon emissions. This study is claimed to be environmentally friendly indeed, as it presents no greenhouse gas emissions (i.e., no carbon footprint) whatsoever. This could facilitate the decarburization of the economy via the adoption of electric vehicles by providing fully renewable energy charging points for EVs in different parts of the country.

Table 15. Comparison of the proposed EVCS system with other existing EVCS.

Optimal System	Country	Methodology	Emissions	NPC	COE	References
Wind/CPV/FC/Bio-Gen/Battery	Qatar	HOMER Pro	xxxx	USD2.53M–USD2.92 M	USD0.285–USD0.329/kWh	[33]
PV/Battery system	Romania	iHOGA	CO_2 (430 kg/year)	USD135,524	USD 0.9//kWh	[29]
PV/Wind/Battery	China	HOMER Pro	xxxx	USD831,540	USD0.294/kWh	[44]
PV/Wind/Fuel cell/Battery	India	HOMER	Hydrogen (0.198 kg/h)	USD1,519,040	USD0.264/kWh	[36]
PV/Biogas-Gen/Battery	Bangladesh	HOMER Pro	CO_2 (222 G/kWh)	USD56,202	USD0.1302/kWh	[34]
Diesel/PV/Battery system	Canada	HOMER	Total (73,450 kg/year)	USD0.835/0.945 M	USD0.551/0.625/kWh	[39]
PV-based system	Bulgaria	Mathematical Approach	xxxx	USD21,034	0.111/kWh	[40]
PV/Grid/Battery	Vietnam	HOMER GRID	CO_2 (28,456–42,021 kg/year)	USD97,227–113,785	USD0.08–0.102/kWh	[37]
PV-based	China	HOMER	Total (463,091 kg/year)	USD3,579,236	USD0.098/kWh	[32]
Wind/PV/Battery	Turkey	HOMER Pro	xxxx	USD697,704	USD0.064/kWh	[38]
PV/WT/Battery	Nigeria	HOMER Pro	xxxx	USD547,717	USD0.211/kWh	This study

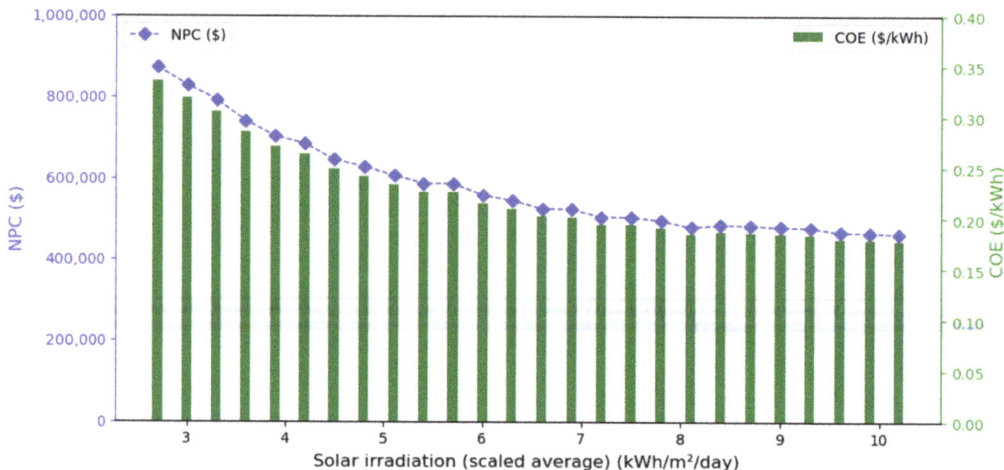

Figure 11. Impact of change in solar radiation value on the PV/WT/battery charging station costs in Sokoto.

Furthermore, the effect of varying the battery $SOC_{minimum}$ on the NPC and the electricity cost of the charging station in Sokoto has illustrated in Figure 12. The increase in the value of this sensitivity variable resulted in a rise in the values of the NPC and the COE. The NPC increases from USD541,550 to USD612,854, and the COE rises from USD0.208/kWh to USD0.236/kWh as the battery minimum state of charge rises from 5% to 60%. Increasing the battery's minimum state of charge, therefore, makes the charging station more expensive, which could create difficulties during the development and installation phase as the initial capital cost and the NPC increase due to this impact.

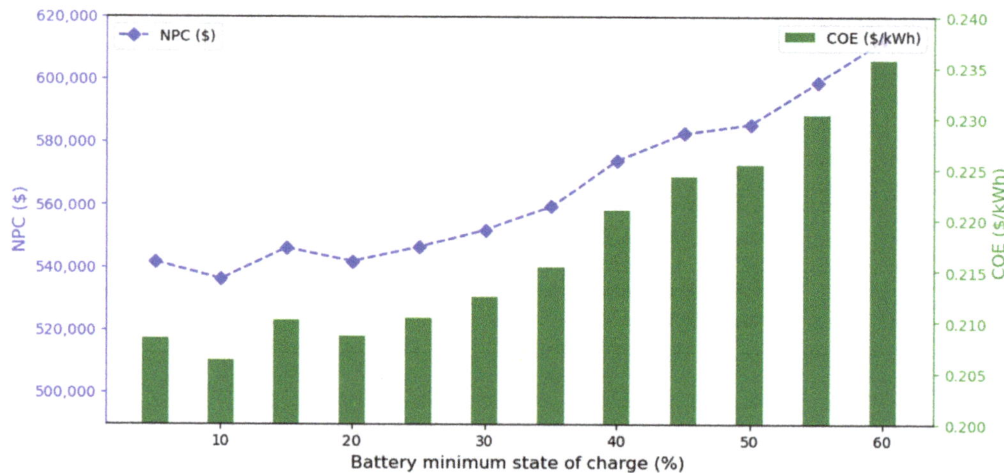

Figure 12. Effect of varying the battery $SOC_{minimum}$ on the NPC and the COE of the optimal charging system.

Finally, the change in the maximum annual capacity shortage and EV charge demand on the total NPC and the cost of electricity of the optimum EV charging scheme is depicted in Figure 13. At a particular value of the load demand, the rise in the percent value of the capacity shortage causes a reduction in the NPC and the COE values of the charging system. It is clear from the chart interface that the NPC reduces from USD730,640 to USD442,148, while the electricity cost, on the other hand, reduces from USD0.292/kWh to USD0.185/kWh as the capacity shortage increases from 0% to 8%. This has indicated that the increase in the capacity shortage can enhance the economic feasibility of the EV charging station. However, this can also create some reliability issues for the system as the charging system might not be able to adequately meet some charge demand of some EVs.

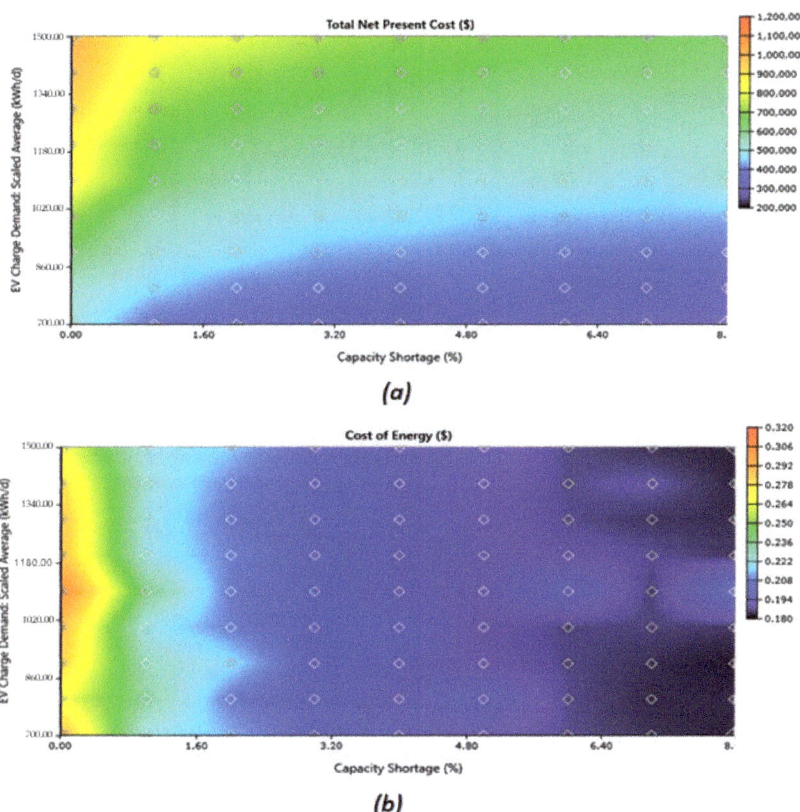

Figure 13. Effect of the sensitivity parameters on the (**a**) total NPC, and (**b**) cost of electricity of the optimal charging station.

4.3.2. Technical Impact of Sensitivity Variables

The influence of sensitivity variables on the technical performance behavior of the optimum EV charging station is investigated via the excess electricity and unmet electric load. The overall results show that the excess energy and the unmet load are sensitive to the change in wind speed, solar radiation, battery minimum state of charge, capacity shortage, and EV charge demand. The effect of variation in the EV charge demand on the excess energy and unmet load of the optimal charging station in Sokoto (Figure 14) reveals that the annual unmet load rises from 2823 kWh to 7623 kWh when the daily charging station load increases from 550 kWh to 1600 kWh. The excess electricity, on the other hand, remains constant at certain numbers of EVs before varying around 250,000 kWh/year. Its minimum value (125,123 kWh/year) was obtained at 900 kWh daily load, while its maximum value of 480,473 kWh/year was achieved when the daily EV load peaked at 1600 kWh. We can deduct from this that the greater the number of electric vehicles, the less reliable the charging system becomes.

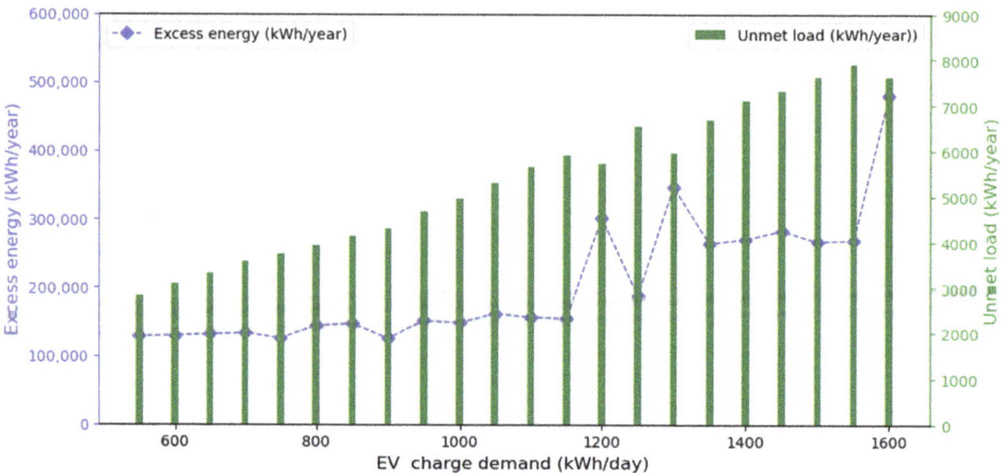

Figure 14. The effect of charge demand or EV number variation on the technical performance of the optimal charging system.

Figure 15 illustrates the effect of wind speed change on the technological performance of the optimal charging scheme. The unmet load decreases from 5588 kWh/year to 4612 kWh/year when the average wind speed rises from 2 m/s to 8.8 m/s. In the beginning, the excess energy maintains a constant minimum value before a slight fluctuation occurs around 200,000 kWh/year. The wind speed range of 2–4 m/s gives the lowest annual excess energy of 68,133 kWh, while the highest annual excess electricity of 483,326 kWh is obtained at 8.8 m/s. The excess energy and the unmet load experience fluctuation due to the impact of the solar radiation variation, as shown in Figure 16. The unmet load fluctuates around 5200 kWh/year, while the excess energy varies around 260,000 kWh/year. The annual value of the excess energy and unmet load decreases from 371,453 kWh to 132,401 kWh and from 5191 kWh to 4848 kWh when the solar irradiation rises from 2.7 to 10.2 kWh/m^2/day. This indicates an improvement in the EV charging station utility as the system becomes able to meet more EV demand.

Figure 17 illustrates the impact of changing the battery SOC$_{min}$ value on the technological performance of the optimum charging system. The figure shows that the yearly excess energy increases from 147,256 to 317,796 kWh, whereas the annual unmet load reduces from 5140 kWh to 4884 kWh when the battery SOC$_{min}$ increases from 5% to 60%. Finally, the chart interface showing the variation in the EV load and the capacity shortage (Figure 18) reveals that at a certain EV load and when the capacity shortage rises from 0 to 8%, the annual unmet load rises from 286 kWh to 18,151 kWh, while the excess electricity in this condition increases from 59,135 kWh/year to 210,334 kWh/year. It can be deduced from the outcomes that increasing the capacity shortage would lower the utility of the optimal charging scheme.

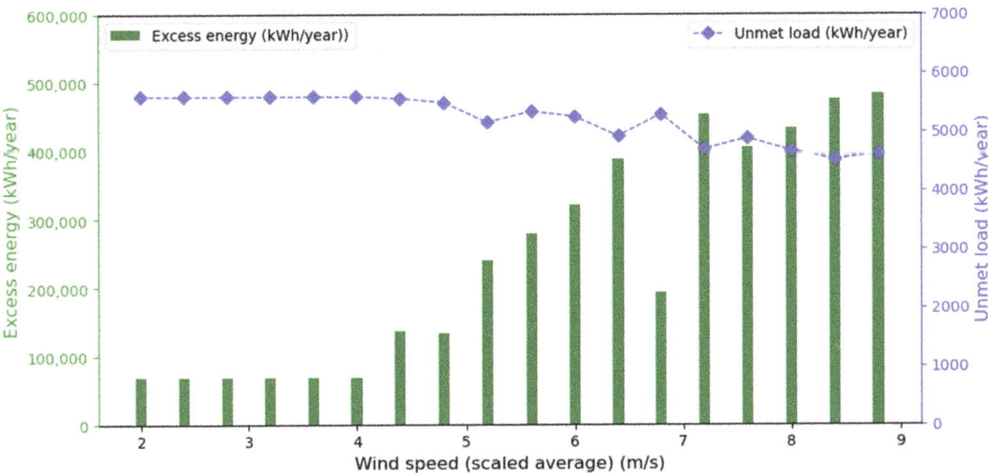

Figure 15. The effect of wind speed variation on the technical performance of the Sokoto PV/WT/battery EV charging model.

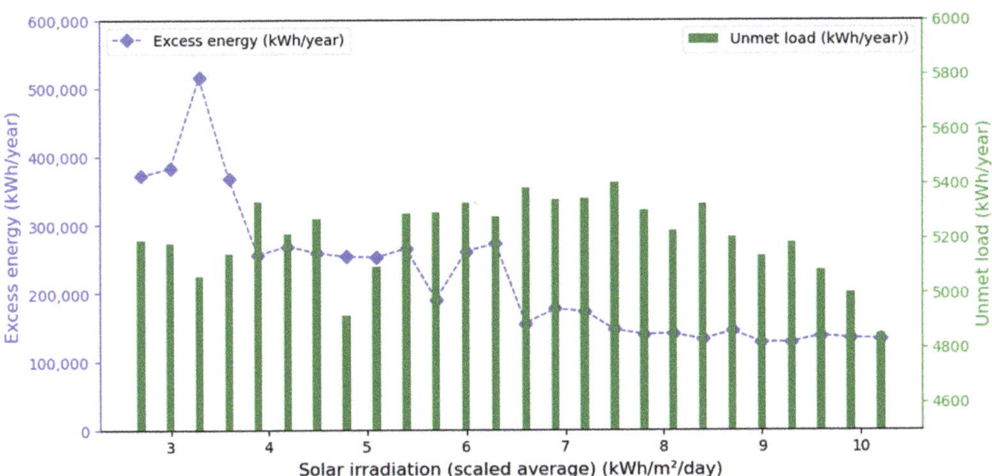

Figure 16. The impact of changing the solar radiation value on the excess energy and unfulfilled electric load of the optimal EV charging scheme.

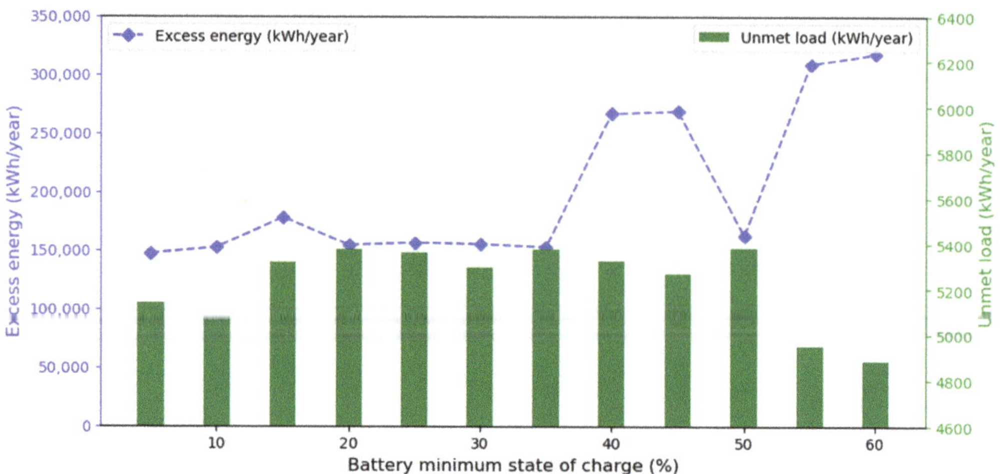

Figure 17. The effect of varying the battery $SOC_{minimum}$ percent value on the technical performance of the optimal charging system.

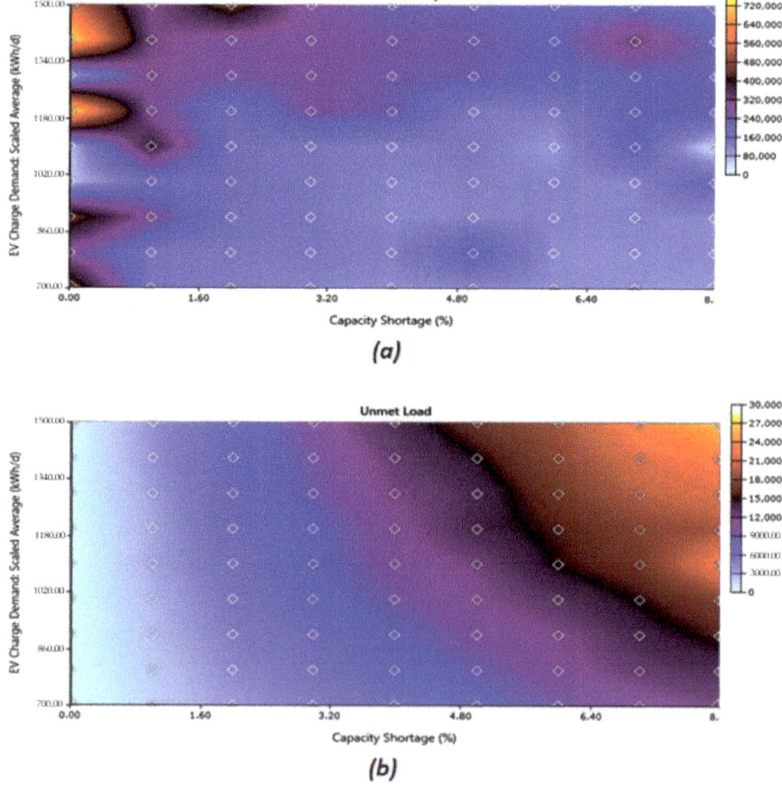

Figure 18. Impact of the sensitivity parameters on (**a**) the excess energy, and (**b**) the unmet electric load of the optimal charging station.

5. Conclusions

This paper has investigated the feasibility of EV charging stations based on RE sources in Nigeria using the HOMER optimization software by considering six different locations with diverse geographical characteristics and climatic conditions. The hybrid charging station system is configured by solar and wind resources with storage devices to charge about 20–30 EVs with a daily capacity of 35 kWh each and applied in different locations in Nigeria, namely, Sokoto, Minna, Port-Harcourt, Enugu, Maiduguri, and Ikeja. The annual average solar radiations and wind speeds used to investigate the optimum hybrid system are 6.24, 4.74, 4.93, 4.13, 5.90, and 5.49 kWh/m^2/day and 5.44, 3.81, 4.09, 3.15, 5.50 and 3.97 m/s for Sokoto, Ikeja, Enugu, Port-Harcourt, Maiduguri and Minna, respectively. The feasibility of the hybrid charging station system is assessed by using appropriate technical performance indicators, namely, unmet electric load, capacity shortage, excess electricity, monthly electric generation, individual system components electric production, battery energy out, and maximum renewable penetration, as well as pertinent economic performance indicators, namely, NPC, COE, operating cost, initial capital cost, the battery wear cost and Levelized cost of system components.

The optimization results showed that the combination of PV and WT with battery storage is economically the best system architecture for a charging station in all six sites. The PV/WT charging scheme integrated with battery storage had the least energy cost of all the simulated sites. The COE and the NPC are also very competitive even when it is difficult to install WTs in the considered locations, as seen with the PV/battery charging station scenario. However, the unavailability of a PV system in the PV/WT/battery system architecture is not economically feasible, as indicated in the case of the WT/battery charging station, which has the maximum NPC and COE values of USD3,318,763 and 1.28 USD/kWh in Port-Harcourt site. In general, the PV/WT/battery charging station (2 qty. of WT, 174 kW of PV panels, 380 qty. of batteries storage, and a converter of 109 kW) in Sokoto provides the best economic metrics with the lowest NPC, energy cost, and initial capital costs of USD547,717, USD0.211/kWh, and USD449,134, respectively. Moreover, the charging station presented competitive annual operating and maintenance costs of USD14,344 and USD67,195. The PV/WT/battery CS at the Sokoto site was able to reliably satisfy most of the EV charge demand as it presented a small percentage of the unmet load of 1.38% (In fact, the lowest when compared with corresponding values for the other charging stations). Moreover, the optimal charging station schemes in all six locations were able to sufficiently meet the EV demand with maximum uptime as the percentages of the unfulfilled electric load were below 2% with a capacity shortage of only approximately 2%. The surplus energy produced can be sold directly to the utility grid via a CS-to-grid connection. Moreover, since the proposed charging stations are located in cities/urban areas, this will facilitate any future connection of the charging stations to the grid network to enable the buying/selling electricity approach. The sensitivity analysis conducted to check the robustness of the optimal charging scheme reveals that the technical and economic performance indicators of the optimum charging station are sensitive to the changes in the sensitivity variables.

Furthermore, the outcomes ensure that the hybrid system of RE sources and EVs can minimize carbon and other pollutant emissions. As for further research, the feasibility of the hybrid charging station system can be investigated by considering distributed generation and load uncertainties. The major limitation of this study is the high initial investment cost needed to install the proposed charging system in the suggested locations. This is often the major obstacle that hinders the widespread use of a standalone renewable energy-based system in most parts of the world, particularly those parts with limited finances, such as most countries in Africa. However, with the recent technological breakthrough in renewable energy technologies as well as the numerous initiated governmental economic programs, this obstacle could be surmounted in the near future.

Author Contributions: J.O.O.: Conceptualization, Methodology, Validation, Formal analysis, Software, Investigation, Resources, Data curation, Writing—original draft preparation and editing. A.M.: Investigation, Writing—review, and editing, Validation, Resources. A.A.I.: Investigation, Writing—review, and editing, Validation, Resources. A.M.R.: Writing—review and editing, Investigation, Validation, Data curation, Visualization. All authors have read and agreed to the published version of the manuscript.

Funding: This research received no external funding.

Data Availability Statement: Not applicable.

Conflicts of Interest: The authors declare no conflict of interest.

References

1. Huda, M.; Koji, T.; Aziz, M. Techno Economic Analysis of Vehicle to Grid (V2G) Integration as Distributed Energy Resources in Indonesia Power System. *Energies* **2020**, *5*, 1162. [CrossRef]
2. Ganapaneni, S.; Pinni, S.V.; Reddy, C.R.; Aymen, F.; Alqarni, M.; Alamri, B.; Kraiem, H. Distribution System Service Restoration Using Electric Vehicles. *Energies* **2022**, *15*, 3264. [CrossRef]
3. International Energy Agency. Nigeria Energy Outlook. Available online: https://www.iea.org/articles/nigeria-energy-outlook (accessed on 28 October 2021).
4. Bank, T.W. The World Bank in Nigeria. Available online: https://data.worldbank.org/indicator/EG.ELC.ACCS.ZS?end=2019&locations=NG&start=1990&view=chart (accessed on 31 October 2021).
5. EnerData. Nigeria Energy Information. Available online: https://www.enerdata.net/estore/energy-market/nigeria/ (accessed on 17 July 2022).
6. Green Environment& Energy Conservation Initiative. Energy Consumption in Transport Sector in Nigeria. Available online: https://geeci.org/html/news/energy-transport-nigeria.html (accessed on 19 September 2022).
7. Alade, B. FG to Incorporate Electric Vehicles in Revised NAIDP. Available online: https://guardian.ng/features/fg-to-incorporate-electric-vehicles-in-revised-naidp/ (accessed on 2 November 2021).
8. Daka, T. Osinbajo Unveils First Made-in-Nigeria Electric Car. Available online: https://guardian.ng/news/osinbajo-unveils-first-made-in-nigeria-electric-car/ (accessed on 1 November 2021).
9. Ajimotokan, O. Stallion Motor Unveils Nigeria's First Electric Car in Abuja. Available online: https://www.thisdaylive.com/index.php/2021/02/06/stallion-motor-unveils-nigerias-first-electric-car-in-abuja/ (accessed on 1 November 2021).
10. JET Motor Company. Heralding the African Electric Future. Available online: https://www.jetmotorcompany.com/jet-mover-ev/ (accessed on 3 November 2021).
11. Onu, E. Chinese Parts Help JET Motor Assemble First EV Vans in Nigeria. Available online: https://www.bloomberg.com/news/articles/2021-07-29/chinese-parts-help-jet-motors-assemble-first-ev-vans-in-nigeria (accessed on 3 November 2021).
12. Idajili, A. Meet the JET EV, Nigeria's First Electric-Powered van by Jet Motor Company. 2021. Available online: https://www.techcityng.com/meet-jet-ev-nigerias-first-electric-powered-van-by-jet-motor-company/ (accessed on 3 November 2021).
13. Echewofun, S.; Aliyu, A.; Ugwuezuoha, Y. FG Commissions Solar-Powered Charging Station for Electric Vehicles. Available online: https://dailytrust.com/fg-commissions-solar-powered-charging-station-for-electric-vehicles (accessed on 3 November 2021).
14. Vanguard. NADDC Inaugurates 1st Nigeria Electric Vehicle Charging Station in Sokoto. Available online: https://www.vanguardngr.com/2021/04/naddc-inaugurates-1st-nigeria-electric-vehicle-charging-station-in-sokoto/ (accessed on 1 November 2021).
15. Odunsi, W. Nigerian Govt Commissions Electric Vehicle Charging Station in Lagos. Available online: https://dailypost.ng/2021/07/13/nigerian-govt-commissions-electric-vehicle-charging-station-in-lagos/ (accessed on 1 November 2021).
16. Olisah, C. Sanwo-Olu Launches Nigeria's First Electric Car, to Complete Lagos-Badagry Expressway. Available online: https://nairametrics.com/2020/11/13/sanwo-olu-launches-nigerias-first-electric-car-to-complete-lagos-badagry-expressway/ (accessed on 1 November 2021).
17. Alade, B. GIG Logistics, Jet Motor Partner on Electric Vehicles. Available online: https://guardian.ng/features/gig-logistics-jet-motor-partner-on-electric-vehicles/ (accessed on 3 November 2021).
18. Singh, S.; Chauhan, P.; Singh, N.J. Feasibility of Grid-Connected Solar-Wind Hybrid System with Electric Vehicle. *J. Mod. Power Syst. Clean Energy* **2021**, *9*, 295–306. [CrossRef]
19. Aliyu, A.K.; Modu, B.; Tan, C.W. A Review of Renewable Energy Development in Africa: A Focus in South Africa, Egypt and Nigeria. *Renew. Sustain. Energy Rev.* **2018**, *81*, 2502–2518. [CrossRef]
20. Ajayi, O.O. Assessment of Utilization of Wind Energy Resources in Nigeria. *Energy Policy* **2009**, *37*, 750–753. [CrossRef]
21. Nehrir, M.H.; Wang, C.; Strunz, K.; Aki, H.; Ramakumar, R.; Bing, J.; Miao, Z.; Salameh, Z. A Review of Hybrid Renewable/Alternative Energy Systems for Electric Power Generation: Configurations, Control, and Applications. *IEEE Trans. Sustain. Energy* **2011**, *2*, 392–403. [CrossRef]
22. Ma, T.; Yang, H.; Lu, L. A Feasibility Study of a Stand-Alone Hybrid Solar–Wind–Battery System for a Remote Island. *Appl. Energy* **2014**, *121*, 149–158. [CrossRef]

23. Okere, R. Government Plans 45,101 MW Electricity Generation by 2030. Available online: https://guardian.ng/business-services/government-plans-45101mw-electricity-generation-by-2030/ (accessed on 30 October 2021).

24. Nigeria-Punch Newspaper. Renewable Energy Mix. Available online: https://punchng.com/fg-targets-30-per-cent-in-renewable-energy-mix-by-2030-fashola/ (accessed on 30 October 2019).

25. Akikur, R.K.; Saidur, R.; Ping, H.W.; Ullah, K.R. Comparative Study of Stand-Alone and Hybrid Solar Energy Systems Suitable for off-Grid Rural Electrification: A Review. *Renew. Sustain. Energy Rev.* **2013**, *27*, 738–752. [CrossRef]

26. Oyedepo, S.O. Energy and Sustainable Development in Nigeria: The Way Forward. *Energy Sustain. Soc.* **2012**, *2*, 15. [CrossRef]

27. Karmaker, A.K.; Hossain, A.; Kumar, N.M. Analysis of Using Biogas Resources for Electric Vehicle Charging in Bangladesh: A Techno-Economic-Environmental Perspective. *Sustainability* **2020**, *12*, 2579. [CrossRef]

28. Nigam, K. Feasibility Analysis of a Solar-Assisted Electric Vehicle Charging Station Model Considering Differential Pricing. *Energy Storage* **2021**, *3*, e237. [CrossRef]

29. Filote, C.; Felseghi, R. Environmental Impact Assessment of Green Energy Systems for Power Supply of Electric Vehicle Charging Station. *Int. J. Energy Res.* **2020**, *44*, 10471–10494. [CrossRef]

30. Machado, A.; Daniel, S.; Nogueira, H.; Soares, B.; César, M.; Diogo, G. Techno-Economic Feasibility Study on Electric Vehicle and Renewable Energy Integration: A Case Study. *Energy Storage* **2020**, *2*, e197. [CrossRef]

31. Venkatesh, B.; Daniel, S.A. Design and Feasibility Analysis of Hybrid Energy-Based Electric Vehicle Charging Station. *Distrib. Gener. Altern. Energy J.* **2022**, *37*, 41–72. [CrossRef]

32. Ye, B.; Jiang, J.; Miao, L.; Yang, P.; Li, J.; Shen, B. Feasibility Study of a Solar-Powered Electric Vehicle Charging Station Model. *Energies* **2015**, *8*, 13265–13283. [CrossRef]

33. Al Wahedi, A.; Bicer, Y. Techno-Economic Optimization of Novel Stand-Alone Renewables-Based Electric Vehicle Charging Stations in Qatar. *Energy* **2022**, *243*, 123008. [CrossRef]

34. Kumar, A.; Ahmed, R.; Hossain, A.; Sikder, M. Feasibility Assessment & Design of Hybrid Renewable Energy Based Electric Vehicle Charging Station in Bangladesh. *Sustain. Cities Soc.* **2018**, *39*, 189–202. [CrossRef]

35. Manwell, J.M.; McGowan, J.G. Lead Acid Battery Storage Model for Hybrid Energy Systems. *Sol. Energy* **2003**, *50*, 399–405. [CrossRef]

36. Syed Mohammed, A.; Anuj; Lodhi, A.S.; Murtaza, Q. Techno-Economic Feasibility of Hydrogen Based Electric Vehicle Charging Station: A Case Study. *Int. J. Energy Res.* **2022**, *46*, 14145–14160. [CrossRef]

37. Minh, P.V.; Le Quang, S.; Pham, M. Technical Economic Analysis of Photovoltaic-Powered Electric Vehicle Charging Stations under Different Solar Irradiation Conditions in Vietnam. *Sustainability* **2021**, *13*, 3528. [CrossRef]

38. Ekren, O.; Canbaz, C.H.; Güvel, Ç.B. Sizing of a Solar-Wind Hybrid Electric Vehicle Charging Station by Using HOMER Software. *J. Clean. Prod.* **2021**, *279*, 123615. [CrossRef]

39. Hafez, O.; Bhattacharya, K. Optimal Design of Electric Vehicle Charging Stations Considering Various Energy Resources. *Renew. Energy* **2017**, *107*, 576–589. [CrossRef]

40. Ilieva, L.M.; Iliev, S.P. Feasibility Assessment of a Solar-Powered Charging Station for Electric Vehicles in the North Central Region of Bulgaria. *Renew. Energy Environ. Sustain.* **2016**, *1*, 12. [CrossRef]

41. HOMER Pro Software. Available online: www.homerenergy.com (accessed on 3 September 2022).

42. Oladigbolu, J.O. Economic Evaluation and Determination of Optimal Hybrid Energy Supply Systems for Residential and Healthcare Facilities in Rural and Urban Areas. Ph.D. Thesis, King Abdulaziz University, Jeddah, Saudi Arabia, 2020.

43. Sinha, S.; Chandel, S.S. Review of Software Tools for Hybrid Renewable Energy Systems. *Renew. Sustain. Energy Rev.* **2014**, *32*, 192–205. [CrossRef]

44. Li, C.; Shan, Y.; Zhang, L.; Zhang, L.; Fu, R. Techno-Economic Evaluation of Electric Vehicle Charging Stations Based on Hybrid Renewable Energy in China. *Energy Strateg. Rev.* **2022**, *41*, 100850. [CrossRef]

45. Oladigbolu, J.O.; Al-turki, Y.A.; Olatomiwa, L. Comparative Study and Sensitivity Analysis of a Standalone Hybrid Energy System for Electrification of Rural Healthcare Facility in Nigeria. *Alex. Eng. J.* **2021**, *60*, 5547–5565. [CrossRef]

46. Seedahmed, M.M.A.; Ramli, M.A.M.; Bouchekara, H.R.E.H.; Shahriar, M.S.; Milyani, A.H.; Rawa, M. A Techno-Economic Analysis of a Hybrid Energy System for the Electrification of a Remote Cluster in Western Saudi Arabia. *Alex. Eng. J.* **2021**, *61*, 5183–5202. [CrossRef]

47. Malanda, C.; Makokha, A.B.; Nzila, C.; Zalengera, C. Techno-Economic Optimization of Hybrid Renewable Electrification Systems for Malawi's Rural Villages. *Cogent Eng.* **2021**, *8*, 1910112. [CrossRef]

48. Aziz, A.S.; Tajuddin, M.F.N.; Adzman, M.R.; Mohammed, M.F.; Ramli, M.A.M. Feasibility Analysis of Grid-Connected and Islanded Operation of a Solar PV Microgrid System: A Case Study of Iraq. *Energy* **2020**, *191*, 116591. [CrossRef]

49. El-houari, H.; Allouhi, A.; Rehman, S.; Buker, M.S. Design, Simulation, and Economic Optimization of an Off-Grid Photovoltaic System for Rural Electrification. *Energies* **2019**, *12*, 4735. [CrossRef]

50. Ur, H.; Habib, R.; Wang, S.; Elkadeem, M.R.; Elmorshedy, M.F. Design Optimization and Model Predictive Control of a Standalone Hybrid Renewable Energy System: A Case Study on a Small Residential Load in Pakistan. *IEEE Access* **2019**, *7*, 117369–117390.

51. Mandal, S.; Das, B.K.; Hoque, N. Optimum Sizing of a Stand-Alone Hybrid Energy System for Rural Electrification in Bangladesh. *J. Clean. Prod.* **2018**, *200*, 12–27. [CrossRef]

52. Al Garni, H.Z.; Awasthi, A.; Ramli, M.A.M. Optimal Design and Analysis of Grid-Connected Photovoltaic under Different Tracking Systems Using HOMER. *Energy Convers. Manag.* **2018**, *155*, 42–57. [CrossRef]

53. Aziz, A.S. Techno-Economic Analysis Using Different Types of Hybrid Energy Generation for Desert Safari Camps in UAE. *Turk. J. Electr. Eng. Comput. Sci.* **2017**, *25*, 2122–2135. [CrossRef]
54. Rahman, M.M.; Khan, M.M.U.H.; Ullah, M.A.; Zhang, X.; Kumar, A. A Hybrid Renewable Energy System for a North American Off-Grid Community. *Energy* **2016**, *97*, 151–160. [CrossRef]
55. Kolhe, M.L.; Ranaweera, K.M.I.U.; Gunawardana, A.G.B.S. Techno-Economic Sizing of off-Grid Hybrid Renewable Energy System for Rural Electrification in Sri Lanka. *Sustain. Energy Technol. Assess.* **2015**, *11*, 53–64. [CrossRef]
56. Suresh Kumar, U.; Manoharan, P.S. Economic Analysis of Hybrid Power Systems (PV/Diesel) in Different Climatic Zones of Tamil Nadu. *Energy Convers. Manag.* **2014**, *80*, 469–476. [CrossRef]
57. Rehman, S.; Al-Hadhrami, L.M. Study of a Solar PV-Diesel-Battery Hybrid Power System for a Remotely Located Population near Rafha, Saudi Arabia. *Energy* **2010**, *35*, 4986–4995. [CrossRef]
58. Banguero, E.; Correcher, A.; Pérez-Navarro, Á.; Morant, F.; Aristizabal, A. A Review on Battery Charging and Discharging Control Strategies: Application to Renewable Energy Systems. *Energies* **2018**, *11*, 1021. [CrossRef]
59. Trading-Economics. Economic Rates. Available online: https://tradingeconomics.com (accessed on 27 September 2022).
60. Dalton, G.J.; Lockington, D.A.; Baldock, T.E. Feasibility Analysis of Stand-Alone Renewable Energy Supply Options for a Large Hotel. *Renew. Energy* **2008**, *33*, 1475–1490. [CrossRef]
61. Orrico, A.C.A.; Schwingel, A.W.; Gimenes, R.M.T.; Souza, S.V.; Orrico, M.A.P.; Maciel, T.T.; Borquis, R.R.A.; de Vargas, F.M. Can Adding Liquid Hatchery Waste to Sheep Manure Potentialize Methane Production and Add Value to Sheep Farming? *Environ. Technol. Innov.* **2021**, *24*, 101866. [CrossRef]
62. Adaramola, M.S.; Paul, S.S.; Oyewola, O.M. Assessment of Decentralized Hybrid PV Solar-Diesel Power System for Applications in Northern Part of Nigeria. *Energy Sustain. Dev.* **2014**, *19*, 72–82. [CrossRef]
63. NASA/SSE. Surface Meteorology and Solar Energy. Available online: https://power.larc.nasa.gov/ (accessed on 15 December 2021).
64. Ohunakin, O.S.; Adaramola, M.S.; Oyewola, O.M.; Fagbenle, R.O. Solar Energy Applications and Development in Nigeria: Drivers and Barriers. *Renew. Sustain. Energy Rev.* **2014**, *32*, 294–301. [CrossRef]
65. Oladigbolu, J.O.; Ramli, M.A.M.; Al-turki, Y.A. Techno-Economic and Sensitivity Analyses for an Optimal Hybrid Power System Which Is Adaptable and Effective for Rural Electrification: A Case Study of Nigeria. *Sustainability* **2019**, *11*, 4959. [CrossRef]
66. Agarwal, N.; Kumar, A.; Varun. Optimization of Grid Independent Hybrid PV-Diesel-Battery System for Power Generation in Remote Villages of Uttar Pradesh, India. *Energy Sustain. Dev.* **2013**, *17*, 210–219. [CrossRef]
67. Khan, F.A.; Pal, N.; Saeed, S.H. Optimization and Sizing of SPV/Wind Hybrid Renewable Energy System: A Techno-Economic and Social Perspective Hybrid Optimization of Multiple Energy Resources. *Energy* **2021**, *233*, 121114. [CrossRef]
68. Oladigbolu, J.O.; Ramli, M.A.M.; Al-Turki, Y.A. Optimal Design of a Hybrid PV Solar/Micro-Hydro/Diesel/Battery Energy System for a Remote Rural Village under Tropical Climate Conditions. *Electronics* **2020**, *9*, 1491. [CrossRef]
69. Imam, A.A.; Oladigbolu, J.O.; Mustafa, M.A.S.; Imam, T.A. The Effect of Solar Photovoltaic Technologies Cost Drop on the Economic Viability of Residential Grid-Tied Solar PV Systems in Saudi Arabia. In Proceedings of the 2022 International Conference on Electrical, Computer and Energy Technologies (ICECET), Prague, Czech Republic, 20–22 July 2022; pp. 1–5. [CrossRef]
70. Lambert, T.; Gilman, P.; Lilienthal, P. Micropower System Modeling. *Integr. Altern. Sources Energy* **2006**, *1*, 379–418.

Article

Modeling and Simulation of Modified MPPT Techniques under Varying Operating Climatic Conditions

Doaa Khodair [1], Saad Motahhir [2], Hazem H. Mostafa [3], Ahmed Shaker [4], Hossam Abd El Munim [5], Mohamed Abouelatta [6] and Ahmed Saeed [7,*]

1   Faculty of Engineering and Technology, Future University in Egypt, Cairo 11835, Egypt
2   Engineering, Systems and Applications Laboratory, ENSA, SMBA University, Fez 30000, Morocco
3   Energy and Renewable Energy Department, Faculty of Engineering, Egyptian Chinese University, Cairo 11724, Egypt
4   Engineering Physics and Mathematics Department, Faculty of Engineering, Ain Shams University, Cairo 11517, Egypt
5   Computer and Systems Engineering Department, Faculty of Engineering, Ain Shams University, Cairo 11517, Egypt
6   Electronics and Electrical Communications Department, Faculty of Engineering, Ain Shams University, Cairo 11517, Egypt
7   Electrical Engineering Department, Future University in Egypt, Cairo 11835, Egypt
*   Correspondence: asaeed@fue.edu.eg

Abstract: Enhancing the performance of photovoltaic (PV) systems has recently become a key concern because of the market demand for green energy. To obtain the most possible power from the solar module, it is imperative to allow the PV system to operate at its maximum power point (MPP) regardless of the climatic conditions. In this study, a comparison of distinctive Maximum Power-Point Tracking (MPPT) techniques is provided, which are Perturb and Observe (P&O) and Modified Variable Step-Size P&O, as well as Incremental Conductance (INC) and Modified Variable Step-Size INC, using a boost converter for two types of solar panels. Using MATLAB software, simulations have been performed to assess the efficiency of the solar module under several environmental conditions, standard test conditions (STCs), and sudden and ramp variations in both solar irradiance and temperature. The output power efficiency, time response, and steady-state power oscillations have all been taken into account in this study. The simulation results of the improved algorithms demonstrate an enhancement in the PV module performance over conventional algorithms in many factors including steady-state conditions, tracking time, and converter efficiency. Furthermore, a boost in the dynamic response in monitoring the MPP is observed in a variety of climatical circumstances. Moreover, the proposed P&O MPPT algorithm is implemented in a hardware system and the experimental results verified the effectiveness, regarding both fast-tracking speed and lower oscillations, of the proposed Variable Step-Size P&O algorithm and its superiority over the conventional P&O technique.

Keywords: MPPT; perturbation and observation; incremental conductance; variable step size; MATLAB-Simulink

Citation: Khodair, D.; Motahhir, S.; Mostafa, H.H.; Shaker, A.; Munim, H.A.E.; Abouelatta, M.; Saeed, A. Modeling and Simulation of Modified MPPT Techniques under Varying Operating Climatic Conditions. *Energies* **2023**, *16*, 549. https://doi.org/10.3390/en16010549

Academic Editor: James M. Gardner

Received: 17 October 2022
Revised: 12 December 2022
Accepted: 29 December 2022
Published: 3 January 2023

1. Introduction

Undoubtedly, finding sustainable and renewable energy sources alternative to fossil fuel resources is a necessity. Amongst the various resources, solar energy is considered the most demanding thanks to its reliability, abundance, and effectiveness in alleviating global warming problems [1,2]. One significant advantage of using photovoltaic (PV) systems is that they can produce clean energy and are free of pollution [3]. Furthermore, PV panels are low-cost and require minimal maintenance [4]. Recently, photovoltaic panels are used in numerous applications, including water-pumping systems, battery chargers, and aeronautical applications [5]. Notably, the output energy of the panel is influenced by both ambient temperature and solar radiation, which consequently results in the nonlinear behavior of the PV panel [1,6]. In addition, the output energy is affected by the internal parameters of the panel, which, in turn, causes the load to impose its characteristics on the output power [7,8]. Hence, acquiring the maximum accessible power from a PV panel, particularly under fluctuating weather conditions, while maintaining high reliability and lower cost is a considerable concern in research. Therefore, much effort has been devoted to obtaining solutions [9–13]. The primary proposed key solution to conquer this issue is to add an MPPT controller to the photovoltaic system to boost the obtained power in any conditions [14,15].

Various techniques have been proposed for MPPT [13,16], including Perturbation and Observation (P&O) [13], Incremental Conductance (INC) [17], Fractional Open Circuit Voltage (FOCV) [18], Fractional Short Circuit Current (FSCC) [19], Fuzzy Logic [20], and Neural Networks [21]. These MPPT algorithms vary in many aspects, including the steady-state-oscillations, the response time of tracking the MPP, cost, effectiveness, implementation complexity, and the accurate tracking of the peak point regardless of unpredictable abrupt changes of solar radiation or temperature under the conditions of partial shading [15,16]. Of the MPPT algorithms, P&O and INC techniques are considered the most common commercialized ones that are utilized in tackling the MPP [22–25], attributed to the algorithms' medium complexity and their simple implementation due to the low hardware requirements. Despite these advantages, P&O techniques suffer shortcomings that are mainly attributed to the oscillations around the MPP because the perturbation size added to the control signal in both directions is continually changing to remain in the MPP position. Furthermore, when the level of incident irradiance changes rapidly, the direction of perturbation may be incorrectly determined by the algorithm, resulting in increased power losses [26,27].

To overcome the drawbacks and shortcomings of the traditional P&O, various developments and implementations regarding the P&O technique have been addressed in the literature, such as the implementation of P&O, by using a variable duty cycle instead of a classical constant one as in [26] and proposing an adaptive P&O control algorithm that has a rapid dynamic response as presented in [27]. Another implementation of the conventional P&O was suggested in [28,29] to enhance the effectiveness of the system by employing an embedded microcontroller-based real-time algorithm with a combination of hardware-in-the-loop (HIL) along with embedded C language. Conversely, by obtaining the slope of the power–voltage curve identified and comparing it against zero at the MPP, the INC technique that has a fixed step size was introduced to minimize the oscillations of the targeted MPP [30]. On the other hand, the INC technique turns out to be much more complex when compared to P&O since it utilizes a set of divisional computations that require a calculation procedure and a stronger microcontroller [25]. Thus, upon utilizing a fixed step size, the algorithm has a low-speed response. Furthermore, in P&O and INC techniques, the tracing of the MPP may fail when solar radiation is abruptly changed or when partial shadowing occurs [15]. Consequently, modifying the INC algorithm was necessary [5,31].

Although some proposed modifications were conducted for both conventional algorithms in the literature, most of them do not achieve a fast and precise response to the first step change in the duty cycle when subjected to abrupt environmental fluctuations [25]. Moreover, to ensure the effectiveness of any of these MPPT algorithms, testing them under various operating circumstances is required. However, it is challenging to achieve the optimal test case since it is hard to control weather variations [32]. Researchers use MATLAB-Simulink [33] and other simulation environments to implement and ensure the accuracy of various MPPT algorithms before their hardware implementation because of the problems associated with the hardware components [34,35].

In the current study, modified variable step-size-based Perturb and Observe and Incremental Conductance (abbreviated as M-VSS-P&O and M-VSS-INC) algorithms are proposed to overcome the limitations of their corresponding traditional algorithms. MATLAB-

Simulink software has been applied on two kinds of PV modules, namely, the polycrystalline MSX60 and the thin-film ST40 to investigate the effectiveness of the two modified MPPT algorithms in tracing the MPP under Standard Test Conditions (STC) and three different variation profiles of climatic variations regardless of the PV panel category.

The paper is organized as follows. After the introduction, Section 2 includes the PV system configuration while the model specifications details and the modified algorithms are presented in Sections 3 and 4, respectively. Section 5 is devoted to the results and discussion. Finally, Section 6 represents the conclusions of this simulation study.

2. System Configuration and Models

2.1. Main System Configuration

The main objective of the system configuration illustrated in Figure 1 is to control the PV panel's MPP and force the system to work at this point [16]. This point changes according to the variation of the atmosphere status, including solar radiation and temperature. The PV array's output is linked to the DC-DC converter, which plays a fundamental function in the MPPT process. The circuit component design and the parameter values are given in Table 1. The MPPT controller controls the converter's duty cycle, which in turn controls the array voltage from which the maximum power is acquired and maintained. The output of the systems is connected to a 30 Ω resistive load.

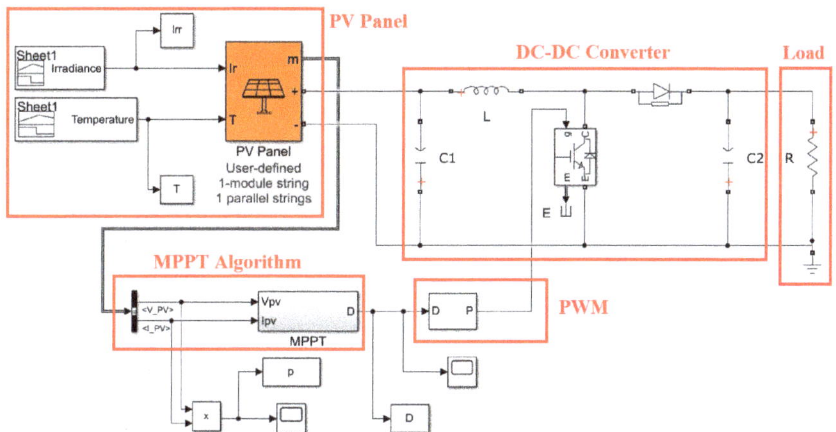

Figure 1. System General Configuration illustrated by Simulink model.

Table 1. Boost converter design parameters [36].

Parameter	Definition	Value
L	Inductor	3 mH
C_1	Input capacitor	100 μF
C_2	Output capacitor	100 μF

2.2. Models Specifications

The simulations performed in this work are linked to the manufacturer's specifications specified at STC for the two cell modules, as mentioned in Table 2. The Current-Voltage and Power-Voltage curves for both cells are presented in Figure 2. As is apparent in *P-V* characteristics, the required MPPs where the maximum power is extracted from the panel are positioned in the curve's peaks.

Table 2. PV panels' electrical specifications under standard test conditions (STC).

Parameter	Definition	Polycrystalline MSX60	Thin Film ST40
V_{oc} (V)	Open-Circuit voltage	21.1	23.3
I_{sc} (A)	Short-Circuit current	3.80	2.68
V_{mp} (V)	Voltage at MPP	17.1	16.6
I_{mp} (A)	Current at MPP	3.5	2.41
K_v (V/°C)	Temperature coefficient of the V_{oc}	-0.08	-0.1
K_i (A/°C)	Temperature coefficient of the I_{sc}	0.00300	0.00035
N_s	Number of cells per module	36	36

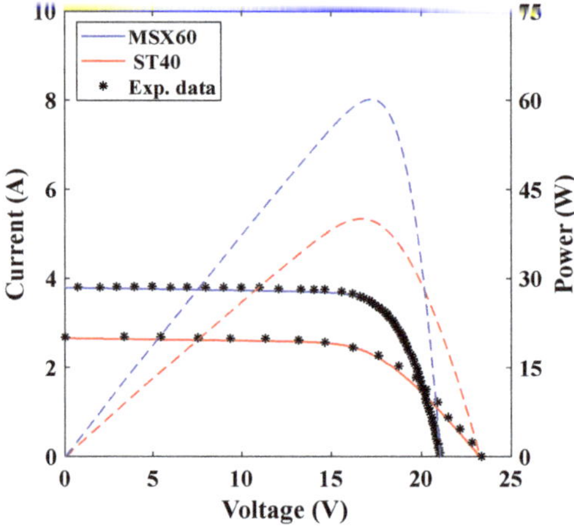

Figure 2. *I-V* (Solid lines) and *P-V* (Dashed lines) characteristics for both MSX60 and ST40 modules.

The incident solar irradiance and temperature influence these characteristic curves. Consequently, a non-linear PV characteristic is observed due to the variation in climatic conditions. This, in turn, causes a considerable change in the MPP position. Notably, the load also influences the MPP. Hence, an MPPT algorithm is highly required to be implemented in the PV system to trace the MPP under varying conditions. Further, to examine the effect of the temperature and solar irradiance fluctuations, the two PV modules are tested and simulated for distinctive temperature and irradiance values. The photocurrent (I_{ph}) and the reverse saturation current (I_s) are formulated by Equations (1) and (2):

$$I_{ph} = \frac{G}{G_{STC}}\left(I_{sc} + K_i\left(T - T_{STC}\right)\right) \tag{1}$$

$$I_s = \frac{I_{sc} + K_i\left(T - T_{STC}\right)}{\exp\left(\frac{V_{oc} + K_V\left(T - T_{STC}\right)}{A\,V_T}\right) - 1} \tag{2}$$

where A is the diode ideality factor while V_T is the thermal voltage and T is the temperature in Kelvin. The constants K_i and K_v are the temperature coefficients of the short-circuit current I_{sc} and the open circuit voltage V_{oc}, respectively. The solar radiation (W/m^2) and the Standard Test Condition (1000 W/m^2 and 25 °C) are denoted by G and STC, respectively. G_{STC} is the solar irradiance under STC (1000 W/m^2) and T_{STC} is the temperature under STC (25 °C). It can be deduced from the above equations that the photocurrent primarily depends on both the temperature and incident irradiance. On the other hand, the reverse saturation current only depends on temperature. Thus, the variation in irradiance and temperature strongly impacts the current and voltage levels. These implications are displayed in Figures S1 and S2, respectively (see Supplementary Materials).

3. MPPT Algorithms

The crucial concern for a good MPPT is to guarantee that the DC-DC converter's input and output power does not change and remains the same even if the load changes. Many MPPT strategies have been proposed in an attempt to best determine the MPP. As mentioned earlier, because of their medium complexity and simple implementation, the traditional INC and P&O techniques are the most widely utilized algorithms.

3.1. Issues Related to Conventional P&O and INC Algorithms

The P&O and INC techniques utilize the (P-V) characteristics of the PV panel in the tracking procedure, fulfilling the requirement $dP/dV = 0$. Essentially, it is difficult to determine the zero point on the P-V curve's slope as in P&O and the truncation error in digital processing as in the INC. Therefore, both approaches turned out to be entirely inaccurate in tracing the MPP owing to the resulting steady-state oscillations in the vicinity of the MPP and higher response time. The perception of the P&O algorithm is centered on modulating the duty cycle. If the peak of the P-V characteristic is detected, there will be no more additional perturbations to the duty cycle. This process, however, causes the system operation to be adjacent to the MPP, but not at the point itself. Therefore, contentious duty cycle modifications must be employed to maintain the MPP position resulting in oscillations that are proportional to the step size. The larger step size is translated to higher oscillations while slower tracking is the result of the small step size.

In contrast, the concept of the INC algorithm depends on determining the slope of the power curve by employing the incremental conductance of the PV panel. When the incremental conductance has the same value as its instantaneous one, the maximum power can be successfully traced. Unlike the P&O algorithm, the INC process requires a powerful microcontroller which increases the system cost [5]. As a result of these issues raised for both P&O and INC algorithms, they are not the best choice when sudden changes occur in solar irradiance and/or temperature.

3.2. Modified MPPT Algorithms

Based on the discussion mentioned above, in this subsection, we present modified algorithms to address the issues related to the traditional P&O and INC techniques. The modification is based on the variable-step size as an alternative to using a fixed-step size. Additionally, a comptonization between fast response and steady-state oscillations will be met. The approach of variable step size adjustment was previously outlined in [37,38]. It used a scaling factor reliant on the variation in power (ΔP) and voltage (ΔV). Nevertheless, this modification may not display good tracing capabilities in abrupt irradiance variation [5]. So, to swamp this problem, we follow the modified algorithms presented in [5] using Matlab-Simulink, depending on an enhanced variable step size that relies solely on the power change (ΔP) with a scaling factor accustomed to the settlement of the response time and decreasing the steady-state oscillations in such a way as to reduce the oscillations of the output PV power as given in Figures 3 and 4, which illustrate the flow chart of M-VSS-P&O and M-VSS-INC techniques, respectively. The power (P) is calculated from $V \times I$, while the change in power is calculated as the difference between the new and old values (i.e., $\Delta P = P_{new} - P_{old}$).

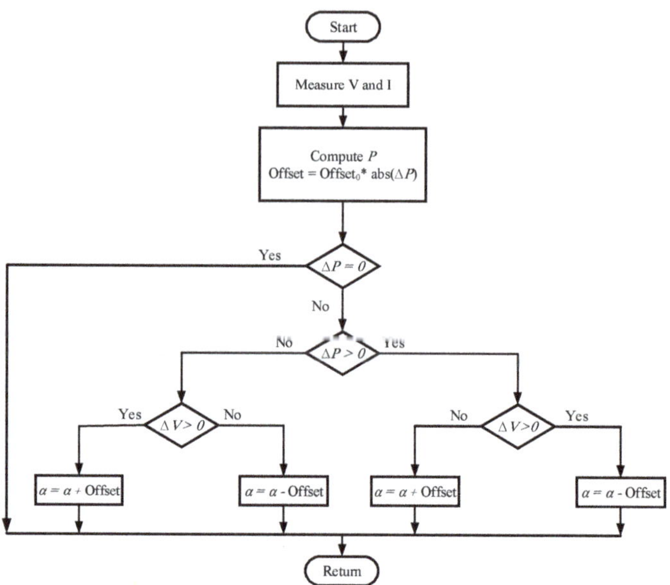

Figure 3. Flowchart of modified P&O algorithm.

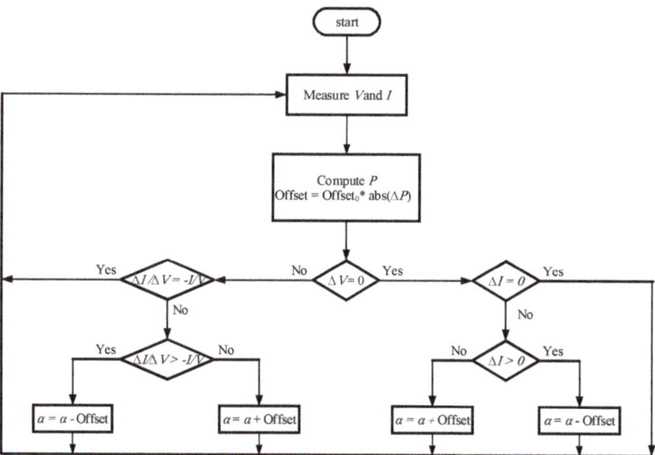

Figure 4. Flowchart of Modified INC algorithm.

It is well known, for the conventional P&O algorithm, that when irradiance increases, a drift problem occurs, which causes a delay in tracing the MPP. This is indicated in Figure 5, which displays the *P-V* curves for two cases of irradiance. If the operating point is at A, and there is an abrupt increase in insolation and the point will settle to point B. At point B, $\Delta P > 0$ and $\Delta V > 0$, so the algorithm imposes a lower duty cycle resulting in moving the operating point to C, which is separate from the MPP in the new curve. On the other hand, when using our modified P&O technique, it is noted when $\Delta P > 0$ and $\Delta V > 0$, ΔV will be negative because the offset is positive (see the flowchart in Figure 3) causing the duty cycle to decrease and, consequently, the voltage to decrease as indicated in Figure 5, which shows the new operating point at D implying a faster response toward the new MPP. The previous discussion demonstrates that our modified P&O algorithm is a drift-free technique.

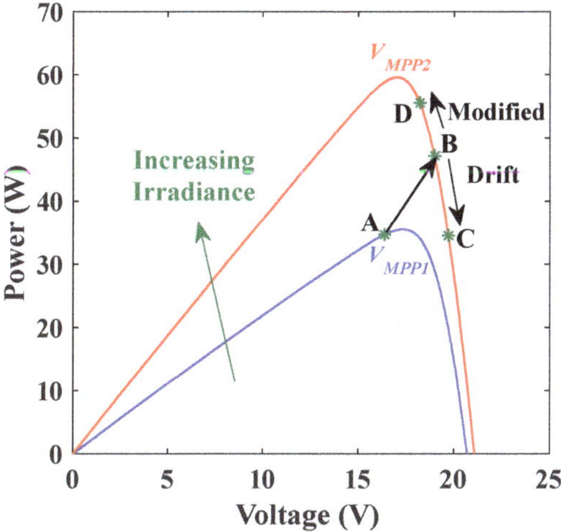

Figure 5. Drift analysis of conventional P&O versus modified P&O algorithms when considering a rapid increase in irradiance.

4. Results and Discussion

This section is concerned with presenting the tests of traditional and enhanced techniques: P&O, INC, and the modified P&O and INC, using MATLAB-Simulink to explore the steady-state and dynamic performance results by utilizing two distinct kinds of solar panels, MSX60 and ST40, whose types are polycrystalline thin films, respectively. Simulations are carried out and presented for standard test conditions and varying operating circumstances in ambient temperature and incident solar irradiation to investigate and compare the algorithm's efficiency regarding the response time desired to trace the peak power point and steady-state power oscillations and converter tracking efficiency.

4.1. Test under STC

Figure 6 shows the polycrystalline cell MSX60 performance using the four MPPT algorithms. All simulations are performed at an irradiance of 1000 W/m² and 25 °C. According to the simulations observed in the figure, the MPP was reached using both traditional P&O and INC algorithms, but the output power had a high percentage of oscillations (see Figure 6a). In contrast, the efficiency of the cell performance can be enhanced using the modified versions of the classical methods, M-VSS-P&O and M-VSS-INC. As observed in the figure, the M-VSS-P&O algorithm (58% duty cycle) was the faster algorithm among the four algorithms to obtain the MPP with the same o/p power of P&O and INC techniques (see Figure 6b). While the Modified Variable Step Size INC algorithm (59% duty cycle) reached the MPP with a quicker response time (18.22 ms) than the classical methods, it was the only algorithm among the four algorithms that achieved 100% power efficiency with negligible steady-state oscillations.

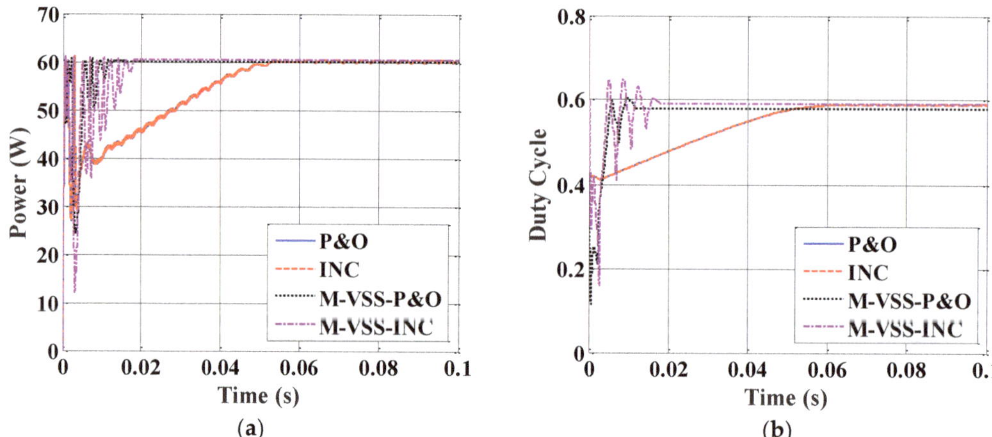

Figure 6. STC test of MSX60 solar cell performance: (**a**) Output power and (**b**) duty cycle variation.

In the same manner, and according to the results presented in Figure 7, another solar cell (thin-film ST40) was used to test the four algorithms' performance. A comparison between the performances of the four MPPT algorithms using both cells is listed in Table 3. Accordingly, the performance efficiency of the traditional P&O in both cells can be noted as the lowest efficiency compared to the other algorithms. The algorithm also presents a higher level of oscillations to arrive at the MPP. Furthermore, the INC algorithm typically operates in the same way as P&O in the MSX60 panel in tracking the MPP. Nevertheless, corresponding to the ST40 cell simulations, it is obvious that all trackers could achieve a conversion efficiency of approximately 100%; both traditional techniques require a similar duration to operate at the same point (~34.0 ms). In contrast, the updated modified algorithms in MSX60 cell simulations were able to improve the efficiency of the solar cell to achieve the panel's full available power to hit the MPP with negligible oscillations quicker than traditional high oscillation algorithms. In addition, the M-VSS-INC improved the performance of the MPPT controller to 100% in 18.22 ms.

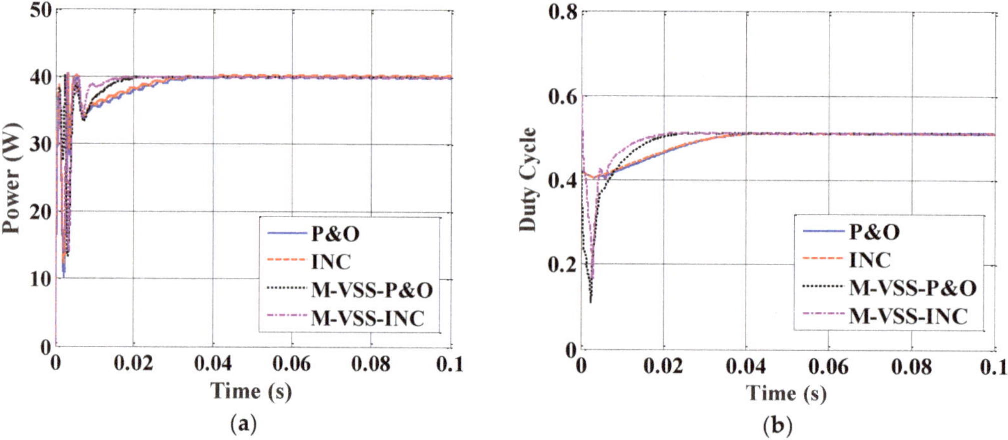

Figure 7. STC test of ST40 solar cell performance: (**a**) Output power and (**b**) duty cycle variation.

Table 3. Performance of the MPPT algorithms under STC conditions.

Module	MPPT Algorithm	MPP (W)	Power (W)	Tracking Time (ms)	Oscillations (W)	Eff. (%)
	P&O	60.53	60.15	54.76	0.52	99.37
	INC	60.53	60.15	54.76	0.52	99.37
MSX60	M-VSS-P&O	60.53	60.15	16.34	Neglected	99.37
	M-VSS-INC	60.53	60.53	18.22	Neglected	100
	P&O	40.006	34.19	34.19	0.21	99.78
	INC	40.006	33.71	33.71	0.15	100
ST40	M-VSS-P&O	40.006	26.21	26.21	0.1	100
	M-VSS-INC	40.006	23.02	23.02	0.1	100

In summary, under STCs, the adjustments performed for both traditional P&O and INC algorithms increased the steady-state efficiency of both solar cells to achieve the full usable o/p power from the panel and grasp the MPP quicker than traditional algorithms that have low oscillations for both cells. However, according to the observations of both cells, the M-VSS-INC was proven to be the most efficient algorithm among the three algorithms.

4.2. Test under an Abrupt Variation in Irradiance with Constant STC Temperature

To extend our analysis to the four mentioned MPPT techniques under a constant STC temperature, a sudden variation in irradiance was examined for both solar cells as follows (Figure 8): The irradiance was initially 1000 W/m^2 but was unexpectedly reduced to 600 W/m^2 at $t = 0.35$ s; then, upon reaching 0.65 s, an extra variation from 600 W/m^2 to 1000 W/m^2 was abruptly applied; eventually, the radiation remained steady at 1000 W/m^2 until the completion of the simulation time at $t = 1$ s.

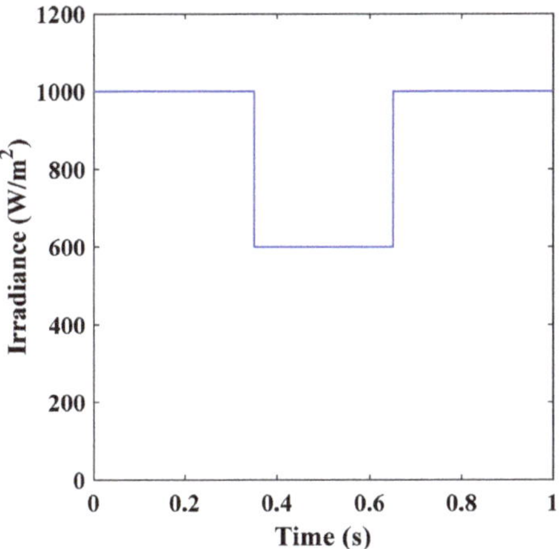

Figure 8. Sudden Irradiance change at 25 °C.

In conjunction with [39], simulations were carried out to ensure the performance and efficiency of the algorithms, in particular, the updated algorithms to meet the MPP. Checking the MSX60 cell, as seen in Figure 9a, the four MPPTs were initially performed in the STC case addressed above. When the irradiance was abruptly reduced to 600 w/m^2, both classical algorithms were able to detect the MPP but many oscillations occurred, exhibiting the same phenomena with a response time of 379.6 ms and an output power of 37.13 W, displaying an approximate error of 26.9 ms.

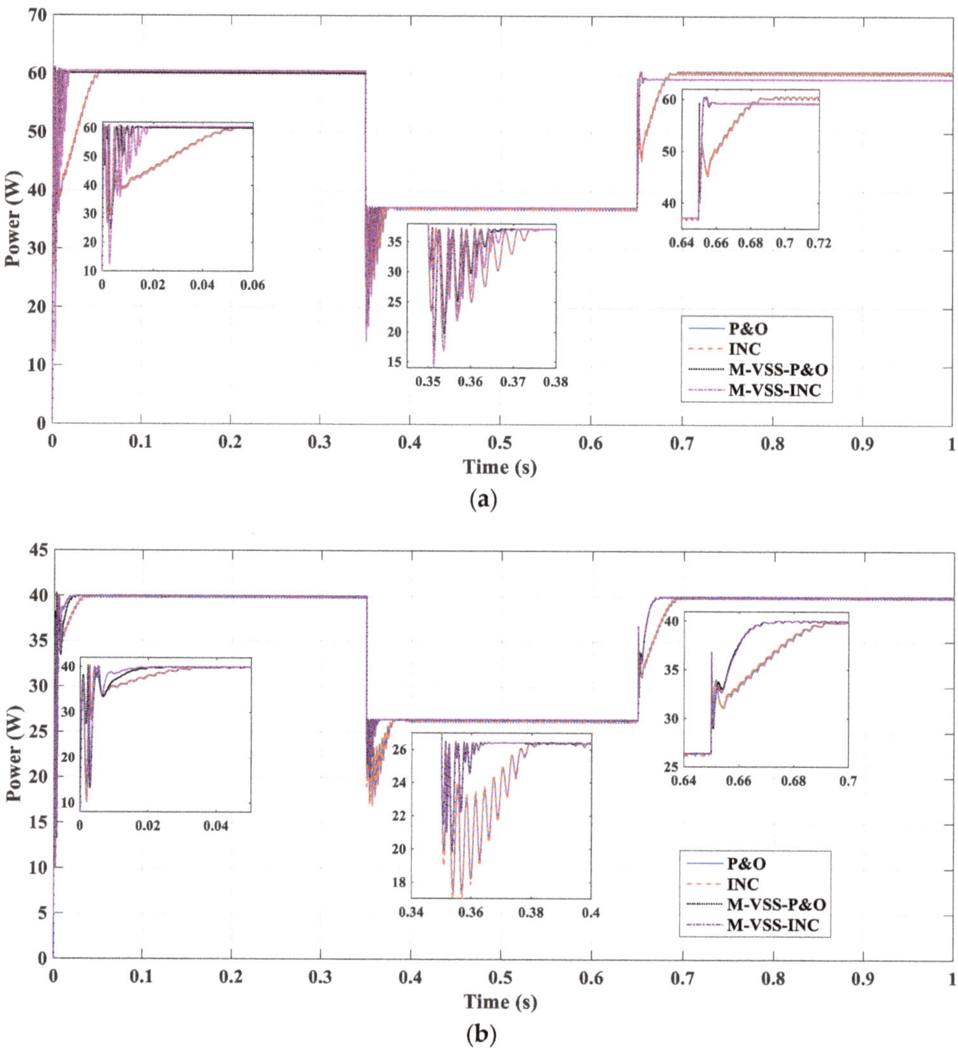

Figure 9. Cell performance under irradiance sudden variation for (**a**) MSX60 and (**b**) ST40.

In contrast, with a faster response time of 367.5 ms and a minor error of 17.5 ms, both modified strategies could congregate to 37.14 W, almost the equivalent MPP of the conventional algorithms. Moreover, the enhanced algorithms succeeded in minimizing the steady-state oscillations around the MPP. In comparison to both P&O and INC (60.14 W), both modified MPPTs can perceive the abrupt spike in solar irradiance even better and faster, with an inaccuracy of just 7.2 ms and a minor reduction in power (59.52 W) in 690.4 ms.

A similar evaluation of the four controllers on the ST40 solar cell is indicated in Figure 9b. Both P&O and INC algorithms were able to monitor and detect the MPP for solar radiation unexpectedly reduced to 600 W/m^2, but with a high percentage of oscillations, demonstrating identical behavior with a performance power of 26.4 W in a response time of 381.8 ms, with almost 34.8 ms error. In comparison, both modified algorithms can converge to an approximately identical highest power of 26.43 W but quicker than the classical algorithms with a response time of 363.4 ms and a smaller error of 13.4 ms.

Furthermore, the improved algorithms minimized steady-state oscillations all over the MPP. Furthermore, with a quicker reaction time, the enhanced algorithms could also recognize the abrupt rise in solar radiation with just a 25.2 ms error with an almost equivalent o/p power of 40.02 W related to P&O and INC with a time of 45.2 ms. Similar to the performance in the polycrystalline cell, the output power curves of the improved techniques showed better performance and fewer oscillation levels than the output power of the other conventional MPPT approaches. However, in the thin-film solar cell simulations, both modified algorithms were better in the MPP tracking process with a sudden increase and sudden reduction in irradiance. In contrast, the polycrystalline solar cell simulations showed different behavior in tracking as M-VSS-P&O and M-VSS-INC prospered in tracking the sudden increase in the irradiance, but they gave slightly less power than the conventional algorithms; however, they were very fast with almost no oscillations around the MPP.

4.3. Test under an Abrupt Variation in Temperature with Constant STC Irradiance

Extending our study, to ensure the validity of the MPPTs in reaching the MPP, especially by the modified ones, another operating condition was inspired by [40] and performed on both solar cells using the four MPP tracking algorithms. Specifically, a sudden variation in temperature with a constant STC irradiance (1000 W/m^2) was tested, as shown in Figure 10.

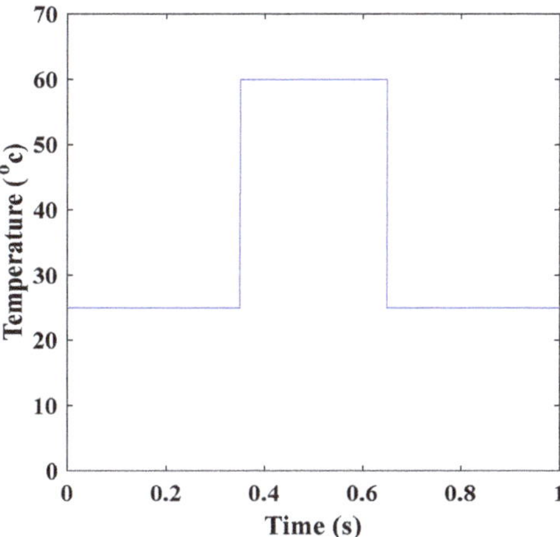

Figure 10. Sudden temperature change at 1000 W/m^2.

With the same value of irradiation (1000 W/m^2), the temperature was initially at 25 °C and then it suddenly increased to 60 °C at t = 0.35 s; then, at 0.65 s, a drop in temperature from 60 °C to 25 °C was applied abruptly; eventually, the temperature was kept steady at 25 °C until the simulation time ended at 1 s. In the MSX60 solar cell test, as seen in Figure 11a, first, before 0.35 s, the algorithms completed the STC case as analyzed herein. At 0.35 s, when temperature increased to 60 °C, P&O and INC algorithms could track the MPP, with identical reactions in 362 ms with 53.88 W output power, with an approximate error of 12 ms. Furthermore, both modified algorithms converged to a higher MPP of 54.22 W and responded faster than the classical methods, with a similar response time of 356.5 ms and a slight error of 6.5 ms. Moreover, with negligible steady-state oscillations, the updated algorithms can also detect the abrupt drop in temperature that occurred at 0.65 s with a 7.5 ms error and a minor reduction in the power (60.18 W) relative to P&O and INC (60.48 W) in 656.5 ms.

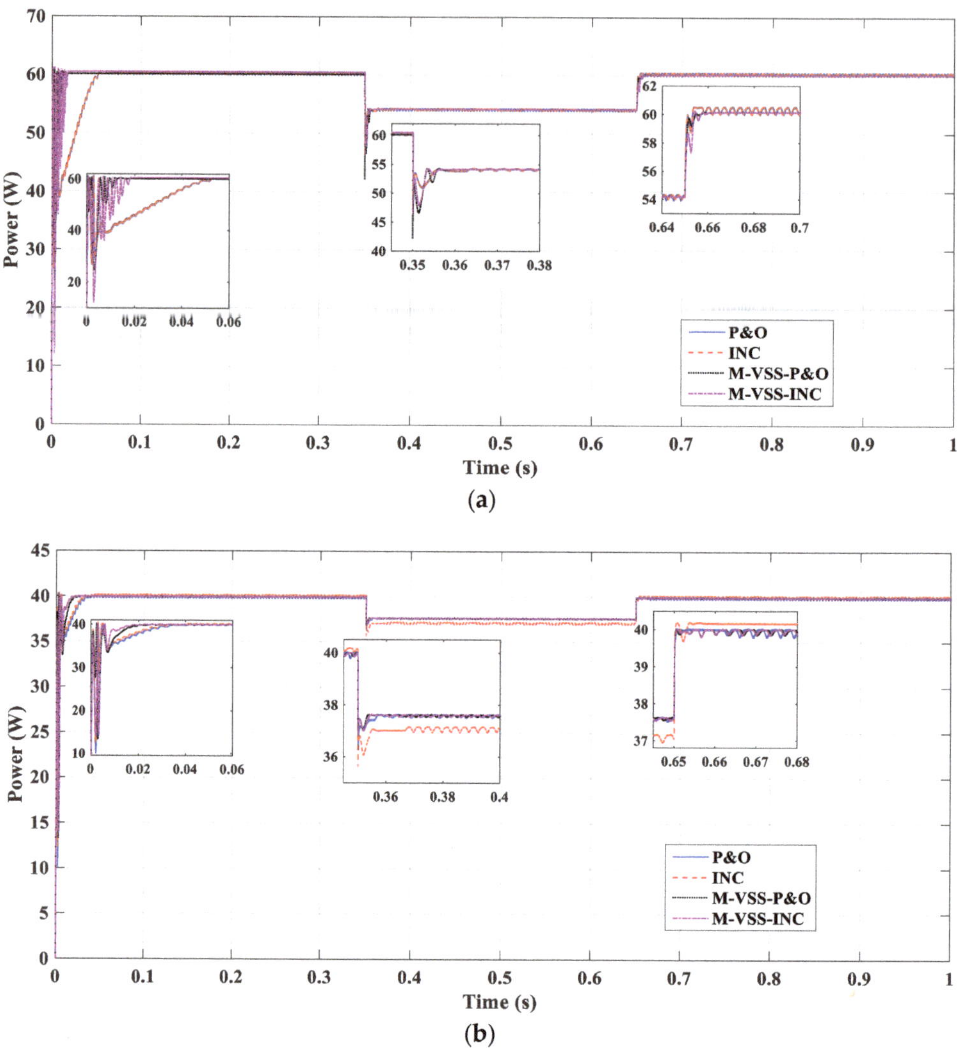

Figure 11. Cell performance under temperature sudden variation for (**a**) MSX60 and (**b**) ST40.

Moreover, as presented in the simulation results related to the ST40 panel as seen in Figure 11b, as the temperature increased abruptly to 60 °C, all algorithms succeeded in tracking the peak point. The P&O algorithm could reach the MPP of 37.62 W in 357.7 ms. However, the INC algorithm succeeded in tracking the MPP slightly quicker than the P&O algorithm with a tracking time of 357.3 ms but with lower power of 37.03 W. In contrast, both improved techniques can converge faster than the traditional algorithms to the same MPP as P&O had the same response in 354.2 ms and a small error of only 4.2 ms and could minimize the oscillations in the vicinity of the MPP.

Furthermore, the enhanced techniques could also reveal approximately the same power of 40.02 W with a very fast response and a sudden decrease in the temperature with only 0.6 ms. In addition, P&O could reach its MPP of 39.97 W with a response time of 652.5 ms, whereas INC could achieve an MPP of 40.1 W but slower with a time of 654.7 ms. P&O and INC algorithms were able to monitor the peak point, showing the same behavior in the MSX60 cell, while both M-VSS-P&O and M-VSS-INC functioned superiorly with approximately the same response. However, in the ST40 solar cell, according to the observations, both M-VSS-P&O and M-VSS-INC algorithms showed better performance than both P&O and INC in the two sudden temperature changes; but in the case of the temperature increment, the INC algorithm gave slightly less power than the other three algorithms and slightly higher power than the other algorithms when a decrement in the temperature occurred.

4.4. Test under Ramp Variation in Both Irradiance and Temperature

Solar radiation and ambient temperature changed simultaneously with random changes during the 1 s period, as presented in Figure 12, to examine the MPPTs process. Before 0.15 s, the irradiation and temperature rose almost simultaneously. After that, at 0.25 s, the irradiation value was maintained at 1000 W/m^2 and the temperature increased from 40 °C until it reached 60 °C. Reaching 0.54 s, the radiation started to decline until it reached 500 W/m^2 at 0.55 s and the system operated at that irradiance until the end of the simulation. At 0.7 s, the temperature started to change and decreased from 60 °C to reach 35 °C at 0.85 s and remained constant until the end of the simulation time.

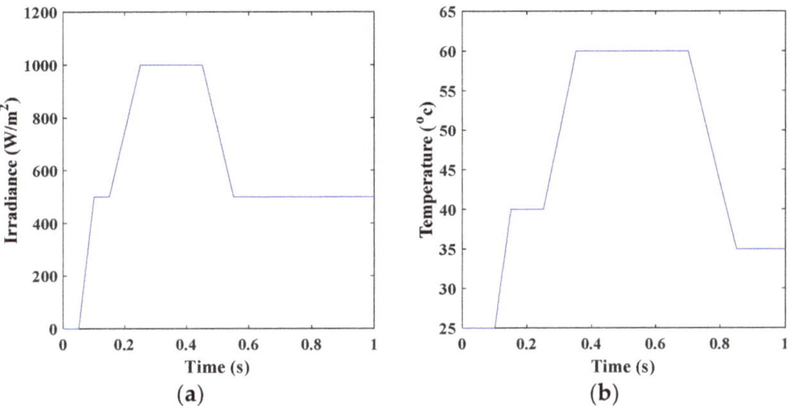

Figure 12. Simultaneous ramp changes in both (**a**) solar irradiance and (**b**) temperature.

In Figure 13, the performances associated with these changes in irradiance and temperature are presented for both cells. From the observations of the output power curve for both cells, it is observed that the P&O algorithm was less efficient in dealing with these simultaneous ramp changes among the two cells. At the same time, the modified approaches were able to track the MPP through these operating conditions, showing a better dynamic response performance than both traditional P&O and INC algorithms and consequently improving the system efficiency.

4.5. Test under Real Solar Radiation Measurements

Finally, a test using practical measurement during a relatively dusty or cloudy day was studied. The actual radiation, extracted from REF [41], is shown in Figure 14a, while a focused scaling down of this radiation is shown in Figure 14b, which is used as an input to the simulation. The scaling down is performed due to the limitation of the simulation time. Figure 15 shows the performance of the modified P&O algorithm versus the conventional P&O for the MSX60 module. Furthermore, the theoretical maximum power is shown for comparison. The figure demonstrates the effectiveness of the modified technique, especially for critical times of a sudden increase in irradiation.

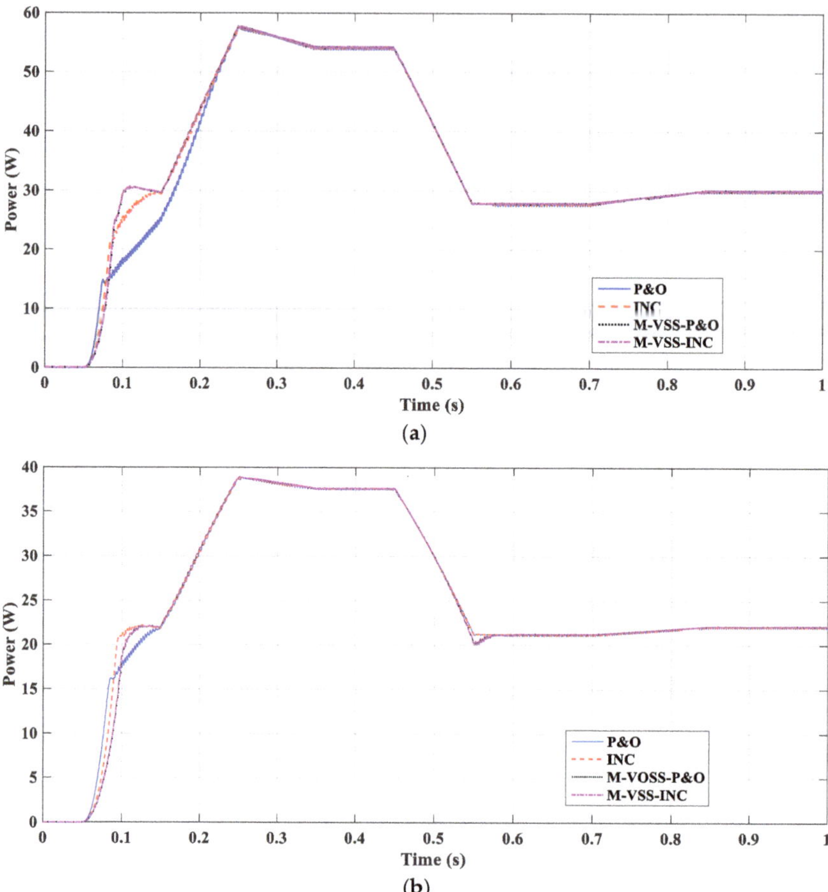

Figure 13. Performance under simultaneous ramp changes in both solar irradiance and temperature: (**a**) MSX60 (**b**) ST40.

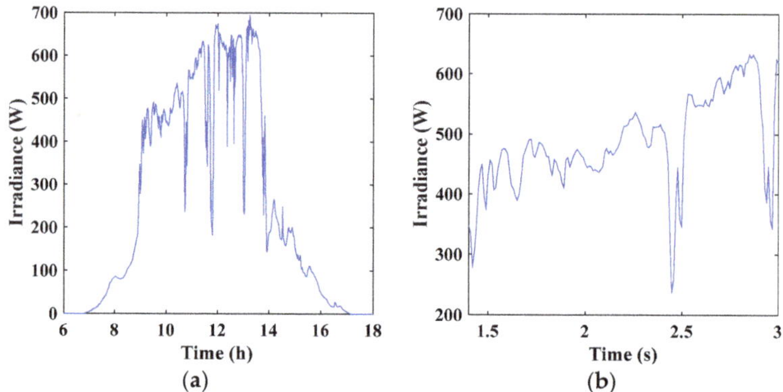

Figure 14. Measured solar irradiance during a cloudy weather day: (**a**) Whole day and (**b**) scaling down of solar irradiance to cope with simulation.

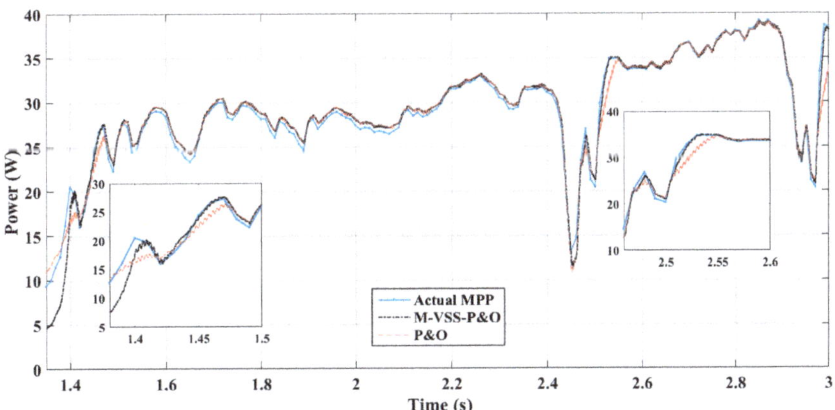

Figure 15. Performance under influence of scaled measured solar radiation for MSX60. The actual theoretical maximum power is also shown.

Moreover, we performed a comparative study of our modified algorithms and other techniques published in the literature. Table 4 summarizes the comparison in terms of the oscillation level, efficiency, response time during an abrupt increase in irradiation, complexity of implementation, and cost. As illustrated, the proposed methodologies show very fast tracking speeds, higher efficiencies, and neglected oscillations around the MPP. Other techniques may provide a faster response and possible tracking capabilities for partial shading but at the expense of complex implementation and high cost. Furthermore, our proposed techniques can be extended to include the tracking of partial shading as has been proposed in [42]. Regarding future work, adding this feature will be considered to enhance the capabilities of the presented MPPT techniques.

Table 4. Comparison of the proposed algorithms with other algorithms proposed in the literature.

MPPT Technique	PV Panel	Oscillation Level (W)	Eff. (%)	Response Time during Sudden Changes in Irradiance	Complexity	Cost	REF
M-VSS-P&O	MSX60	Neglected	99.37	Very fast	Simple	Medium	Our Work
M-VSS-INC	MSX60	Neglected	100	Very fast	Simple	Medium	Our Work
Modified INC	MSX60	1	96.50	Fast	Simple	Medium	[43]
Fractional Short-Circuit and P&O	PB-115	Neglected	97.56	Fast	Medium	Medium	[44]
Fuzzy INC	NA	1	97.50	Medium	Complex	Expensive	[45]
PSO	SM55	Neglected	99.90	Very fast	Complex	Very Expensive	[46]

5. Hardware Implementation

The complexity and cost of the proposed algorithms are promising factors that pave the way for real implementation. Therefore, as a proof-of-concept, the M-VSS-P&O algorithm is implemented on hardware to validate the usefulness of the proposed algorithm. The proposed M-VSS-P&O algorithm is implemented for a low-voltage PV module using Arduino Due, which is equipped with a Cortex-M3 CPU. To analyze and visualize the output voltage and current of the PV module, the output data are exported using GDM Digital Multi Meter (DMM) software through the serial port. DMM software receives the data from the DMM and saves it in an MS Excel sheet. The experimental setup of the MPPT technique is demonstrated in Figure 16a. The setup includes Arduino Due, a voltage and current sensor (MAX471), a boost converter, and a 30-Ω load. The PV module used in the experimental study is shown in Figure 16b while the STC outputs of the module are shown in the datasheet in Figure S3 in the Supplementary Materials.

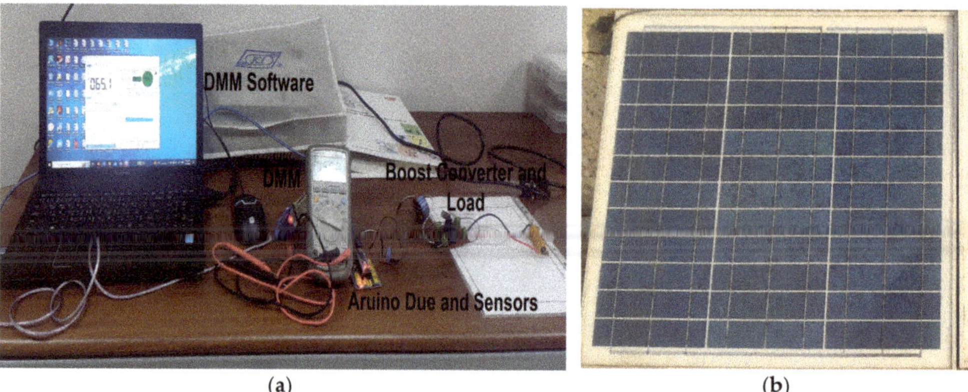

(a)   (b)

Figure 16. (**a**) Hardware Implementation and (**b**) PV module.

Figure 17 shows the characteristic curves of the PV module used, as it was tested before using the MPPT algorithm in order to acquire the *I-V* and *P-V* curves. The PV module is tested by recording the current and voltage using DMM software. The maximum power that PV modules can deliver is measured as 15.24 W at a voltage of 15.4 V and a current of 0.98 A as is depicted in Figure 17.

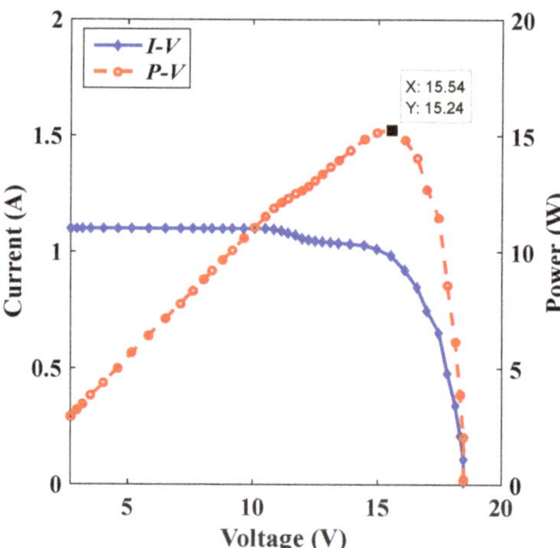

Figure 17. Practical *I-V* and *P-V* curves of the PV module.

The response of the conventional P&O is shown in Figure 18 where the PV voltage is plotted in Figure 18a while the PV power is plotted in Figure 18b. As can be inferred from the figure, there is a certain delay in tracking the maximum power in addition to the obvious oscillations. On the other hand, the modified P&O algorithm tracked the maximum power on the *P-V* curve of the solar panel and continued to operate the PV modules at that point as evident in Figure 19. Figure 19a displays the PV voltage while Figure 19b shows the PV power waveforms of the modified algorithm. The figure clearly indicates the improvement in the M-VSS-P&O algorithm as there are no oscillations around the MPP in addition to the response time being reduced. The modified technique tracks the power of the PV module to 14 W, which exhibits an efficiency of 92%. The main reason for the difference in efficiency between the simulation and practical implementation can be attributed to the resolution of sensors.

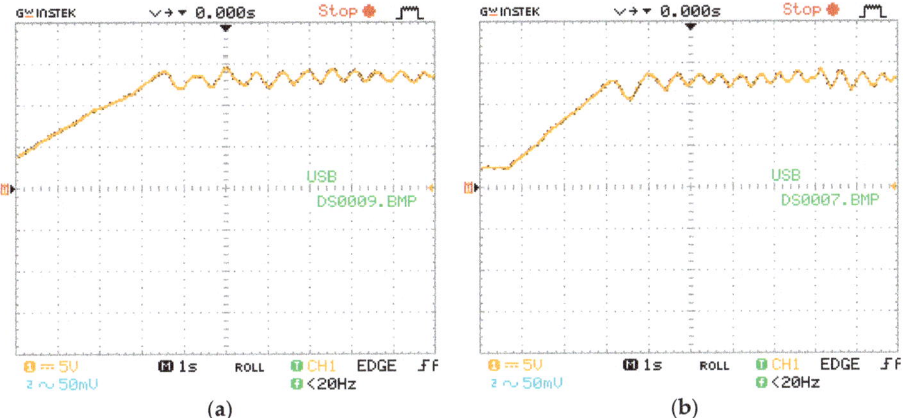

Figure 18. (**a**) PV Voltage and (**b**) PV Power of conventional P&O algorithm.

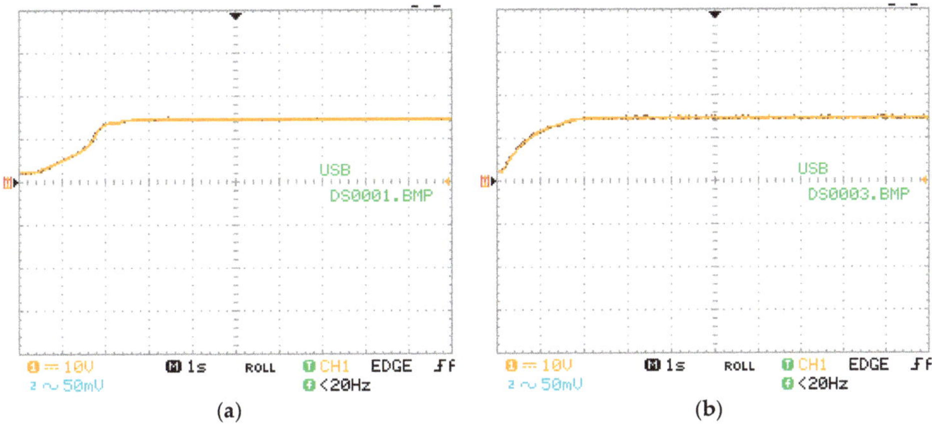

Figure 19. (**a**) PV Voltage and (**b**) PV Power of modified P&O algorithm.

6. Conclusions

The performance analysis of two MPPT Algorithms, namely the P&O and INC, against their enhanced versions, the Modified-Variable-Step Size P&O and Modified-Variable-Step Size INC (M-VSS-P&O and M-VSS-INC), is discussed within this simulation study. Within the context of this paper, the algorithms were implemented and simulated on two distinct PV modules, namely, the polycrystalline MSX60 and the ST40 thin coupled with a boost converter, to examine their performances under varying environmental statuses. All the simulations were carried out utilizing MATLAB-Simulink software. The tracker's efficiency was analyzed in light of the STCs, an abrupt change in solar irradiance with a stable temperature, an abrupt change in temperature with stable solar irradiance, under simultaneous ramp changes in both irradiance and temperature, and finally, under real solar radiation measurements. The simulation results generally prove that both modified MPPTs controllers enhance steady-state and dynamic PV system performances regarding efficiency, oscillations, and tracking speed MPP compared to the traditional MPPT techniques (P&O and INC). Generally, the same case studies presented in this work could be applied to other algorithms as a measure of their effectiveness. Moreover, the hardware implementation of conventional and modified P&O was performed. A substantial improvement of the modified P&O algorithm over the conventional algorithm was experimentally observed in terms of lower oscillations around the MPP in addition to a faster tracking response.

Supplementary Materials: The following supporting information can be downloaded at: https://www.mdpi.com/article/10.3390/en16010549/s1, Figure S1: Solar Irradiance variation (at fixed $T = 25\ °C$) impact on (a) V_{oc} and I_{sc} and (b) voltage and current at MPP. Figure S2: Temperature variation (at fixed $G = 1000\ W/m^2$) impact on (a) V_{oc} and I_{sc} and (b) voltage and current at MPP. Figure S3: Datasheet of the PV module used in the hardware implementation.

Author Contributions: Conceptualization, S.M., A.S. (Ahmed Saeed), A.S. (Ahmed Shaker) and H.A.E.M.; methodology, D.K., S.M., A.S. (Ahmed Shaker) and M.A.; software, H.H.M., D.K., S.M and A.S. (Ahmed Shaker); validation, D.K. and S.M.; formal analysis, D.K.; investigation, D.K., S.M. and H.H.M.; resources, A.S. (Ahmed Saeed), H.H.M., H.A.E.M. and M.A.; writing—original draft preparation, D.K. and A.S. (Ahmed Shaker); writing—review and editing, D.K., S.M., H.H.M., A.S. (Ahmed Saeed), A.S. (Ahmed Shaker), H.A.E.M. and M.A.; supervision, A.S. (Ahmed Saeed), A.S. (Ahmed Shaker), H.A.E.M. and M.A. All authors have read and agreed to the published version of the manuscript.

Funding: This research received no external funding.

Data Availability Statement: Not applicable.

Conflicts of Interest: The authors declare no conflict of interest.

References

1. Pervez, I.; Antoniadis, C.; Massoud, Y. Advanced Limited Search Strategy for Enhancing the Performance of MPPT Algorithms. *Energies* **2022**, *15*, 5650. [CrossRef]
2. Martinez Lopez, V.A.; Žindžiūtė, U.; Ziar, H.; Zeman, M.; Isabella, O. Study on the Effect of Irradiance Variability on the Efficiency of the Perturb-and-Observe Maximum Power Point Tracking Algorithm. *Energies* **2022**, *15*, 7562. [CrossRef]
3. Akram, N.; Khan, L.; Agha, S.; Hafeez, K. Global Maximum Power Point Tracking of Partially Shaded PV System Using Advanced Optimization Techniques. *Energies* **2022**, *15*, 4055. [CrossRef]
4. Marinic-Kragic, I.; Nižetic, S.; Grubišic-Cabo, F.; Papadopoulos, A.M. Analysis of flow separation effect in the case of the free-standing photovoltaic panel exposed to various operating conditions. *J. Clean. Prod.* **2018**, *174*, 53–64. [CrossRef]
5. Motahhir, S.; El Ghzizal, A.; Sebti, S.; Derouich, A. MIL and SIL and PIL tests for MPPT algorithm. *Cogent Eng.* **2017**, *1*, 1378475. [CrossRef]
6. El Hammoumi, A.; Motahhir, S.; Chalh, A.; El Ghzizal, A.; Derouich, A. Low-cost virtual instrumentation of PV panel characteristics using Excel and Arduino in comparison with traditional instrumentation. *Renew. Wind WaterSol.* **2018**, *5*, 3. [CrossRef]
7. Patel, H.; Agarwal, V. MATLAB-Based Modeling to Study the Effects of Partial Shading on PV Array Characteristics. *IEEE Trans. Energy Convers.* **2008**, *23*, 302–310. [CrossRef]
8. Motahhir, S.; El Ghzizal, A.; Sebti, S.; Derouich, A. Proposal and implementation of a novel perturb and observe algorithm using embedded software. In Proceedings of the 3rd International Renewable and Sustainable Energy Conference (IRSEC), Marrakech, Morocco, 10–13 December 2015; pp. 1–5.
9. Barker, L.; Neber, M.; Lee, H. Design of a low-profile two-axis solar tracker. *Sol. Energy* **2013**, *97*, 569–576. [CrossRef]
10. Rambhowan, Y.; Oree, V. Improving the Dual Axis Solar Tracking System Efficiency via Drive Power Consumption Optimization1. *Appl. Sol. Energy* **2014**, *50*, 74–80. [CrossRef]
11. Visconti, P.; Costantini, P.; Orlando, C.; Lay-Ekuakille, A.; Cavalera, G. Software solution implemented on hardware system to manage and drive multiple bi-axial solar trackers by PC in photovoltaic solar plants. *Measurement* **2015**, *76*, 80–92.
12. Kawamoto, H.; Shibata, T. Electrostatic cleaning system for removal of sand from solar panels. *J. Electrost.* **2015**, *73*, 65–70. [CrossRef]
13. Hafeez, M.A.; Naeem, A.; Akram, M.; Javed, M.Y.; Asghar, A.B.; Wang, Y. A Novel Hybrid MPPT Technique Based on Harris Hawk Optimization (HHO) and Perturb and Observer (P&O) under Partial and Complex Partial Shading Conditions. *Energies* **2022**, *15*, 5550. [CrossRef]
14. Saharia, B.J.; Manas, M.; Talukdar, B.K. Comparative evaluation of photovoltaic MPP trackers: A simulated approach. *Cogent Eng.* **2016**, *3*, 1137206. [CrossRef]
15. Verma, D.; Nema, S.; Shandilya, A.M.; Dash, S.K. Maximum power point tracking (MPPT) techniques: Recapitulation in solar photovoltaic systems. *Renew. Sustain. Energy Rev.* **2016**, *54*, 1018–1034. [CrossRef]
16. El-Khozondar, H.J.; El-Khozondar, R.J.; Matter, K.; Suntio, T. A review study of photovoltaic array maximum power tracking algorithms. *Renew. Wind. Water Sol.* **2016**, *3*, 1. [CrossRef]

17. Gupta, A.; Chauhan, Y.K.; Pachauri, R.K.; Yin, X.; Pickert, V. A comparative investigation of maximum power point tracking methods for solar PV system. *Sol. Energy* **2016**, *136*, 236–253. [CrossRef]
18. Ahmed, J. A fractional open circuit voltage based maximum power point tracker for photovoltaic arrays. In Proceedings of the 2nd International Conference on Software Technology and Engineering (ICSTE), San Juan, PR, USA, 3–5 October 2010; Volume 1, pp. V1–V247.
19. Sher, H.A.; Murtaza, A.F.; Noman, A.; Addoweesh, K.E.; Chiaberge, M. An intelligent control strategy of fractional short circuit current maximum power point tracking technique for photovoltaic applications. *J. Renew. Sustain. Energy* **2015**, *7*, 013114. [CrossRef]
20. Kottas, T.L.; Boutalis, Y.S.; Karlis, A.D. New maximum power point tracker for PV arrays using fuzzy controller in close cooperation with fuzzy cognitive networks. *IEEE Trans. Energy Convers.* **2006**, *21*, 793–803. [CrossRef]
21. Miloudi, L.; Acheli, D.; Kesraoui, M. Application of Artificial Neural Networks for Forecasting Photovoltaic System Parameters. *Appl. Sol. Energy* **2017**, *53*, 85–91. [CrossRef]
22. Sellami, A.; Kandoussi, K.; El Otmani, R.; Eljouad, M.; Mesbahi, O.; Hajjaji, A. A Novel Auto-Scaling MPPT Algorithm based on Perturb and Observe Method for Photovoltaic Modules under Partial Shading Conditions. *Appl. Sol. Energy* **2018**, *54*, 149–158. [CrossRef]
23. Elgendy, M.A.; Zahawi, B.; Atkinson, D.J. Assessment of Perturb and Observe MPPT Algorithm Implementation Techniques for PV Pumping Applications. *IEEE Trans. Sustain. Energy* **2012**, *3*, 21–33. [CrossRef]
24. Elgendy, M.A.; Zahawi, B.; Atkinson, D.J. Assessment of the Incremental Conductance Maximum Power Point Tracking Algorithm. *IEEE Trans. Sustain. Energy* **2013**, *4*, 108–117. [CrossRef]
25. Khodair, D.; Shaker, A.; El Munim, A.H.E.; Saeed, A.; Abouelatta, M. A Comparative Study Between Modified MPPT Algorithms Using Different Types of Solar Cells. In Proceedings of the 2020 2nd International Conference on Smart Power & Internet Energy Systems (SPIES), Bangkok, Thailand, 15–18 September 2020; pp. 215–218. [CrossRef]
26. Femia, N.; Granozio, D.; Petrone, G.; Spagnuolo, G.; Vitelli, M. Predictive & Adaptive MPPT Perturb and Observe Method. *IEEE Trans. Aerosp. Electron. Syst.* **2007**, *43*, 934–950. [CrossRef]
27. Piegari, L.; Rizzo, R. Adaptive perturb and observe algorithm for photovoltaic maximum power point tracking. *IET Renew. Power Gener.* **2010**, *4*, 317–328. [CrossRef]
28. Elbaset, A.A.; Ali, H.; Sattar, M.; Khaled, M. Implementation of a modified perturb and observe maximum power point tracking algorithm for photovoltaic system using an embedded microcontroller. *IET Renew. Power Gener.* **2016**, *4*, 551–560. [CrossRef]
29. Motahhir, S.; El Ghzizal, A.; Sebti, S.; Derouich, A. Shading effect to energy withdrawn from the photovoltaic panel and implementation of DMPPT using C language. *Int. Rev. Autom. Control. (IREACO)* **2016**, *2*, 88–94. [CrossRef]
30. Ishaque, K.; Salam, Z.; Lauss, G. The performance of perturb and observe and incremental conductance maximum power point tracking method under dynamic weather conditions. *Appl. Energy* **2014**, *119*, 228–236. [CrossRef]
31. Farayola, A.M.; Hasan, A.N.; Ali, A. Implementation of Modified Incremental Conductance and Fuzzy Logic MPPT Techniques Using MCUK Converter under Various Environmental Condition. *Appl. Sol. Energy* **2017**, *53*, 173–184. [CrossRef]
32. Elbreki, A.M.; Alghoul, M.A.; Al-Shamani, A.N.; Ammar, A.A.; Yegani, B.; Aboghrara, A.M.; Rusaln, M.H.; Sopian, K. The role of climatic-design-operational parameters on combined PV/T collector performance: A critical review. *Renew. Sustain. Energy Rev.* **2016**, *57*, 602–647. [CrossRef]
33. Matlab. *Version 9.11 (R2021b)*; The MathWorks Inc.: Natick, MA, USA, 2021.
34. Motahhir, S.; Chalh, A.; Ghzizal, A.; Sebti, S.; Derouich, A. Modeling of photovoltaic panel by using proteus. *J. Eng. Sci. Technol. Rev.* **2017**, *10*, 8–13. [CrossRef]
35. Mukti, R.; Islam, A. Modeling and Performance Analysis of PV Module with Maximum Power Point Tracking in Matlab/Simulink. *Appl. Sol. Energy* **2015**, *51*, 245–252. [CrossRef]
36. Ayop, R.; Tan, C.W. Design of boost converter based on maximum power point resistance for photovoltaic applications. *Sol. Energy* **2018**, *160*, 322–335. [CrossRef]
37. Al-Diab, A.; Sourkounis, S. Variable step size P&O MPPT algorithm for PV systems. In Proceedings of the International Conference on Optimization of Electrical and electronic Equipment, Brasov, Romania, 20–22 May 2010.
38. Zakzouk, N.E.; Elsaharty, M.A.; Abdelsalam, A.K.; Helal, A.A.; Williams, B.W. Improved performance low-cost incremental conductance PV MPPT technique. *IET Renew. Power Gener.* **2016**, *10*, 561–574. [CrossRef]
39. Zhang, L.; Yu, S.S.; Fernando, T.; Ho-Ching IU, H.; Wong, K.P. An online maximum power point capturing technique for high efficiency power generation of solar photovoltaic systems. *J. Mod. Power Syst. Clean Energy* **2019**, *2*, 357–368. [CrossRef]
40. Wang, Y.; Yang, Y.; Fang, G.; Zhang, B.; Wen, H.; Tang, H.; Fu, L.; Chen, X. An advanced Maximum Power Point Tracking Method for Photovoltaic Systems by Using Variable Universe Fuzzy Logic Control Considering Temperature Variability. *Electronics* **2018**, *7*, 355. [CrossRef]
41. Hassan, M.A.; Khalil, A.; Kaseb, S.; Kassem, M.A. *Machine Learning Models for Generating Synthetic Solar Radiation Data at Cairo, Egypt*; Mendeley Data: Amsterdam, The Netherlands, 2017. [CrossRef]
42. Mahmod, A.N.; Mohd, M.A.; Azis, N.; Shafie, S.; Atiqi, M.A. An enhanced adaptive perturb and observe technique for efficient maximum power point tracking under partial shading conditions. *Appl. Sci.* **2020**, *10*, 3912. [CrossRef]
43. Belkaid, A.; Colak, I.; Isik, O. Photovoltaic maximum power point tracking under fast varying of solar radiation. *Appl. Energy* **2016**, *179*, 523–530. [CrossRef]

44. Sher, H.A.; Murtaza, A.F.; Noman, A.; Addoweesh, K.E.; Al-Haddad, K.; Chiaberge, M. A new sensorless hybrid MPPT algorithm based on fractional short-circuit current measurement and P&O MPPT. *IEEE Trans. Sustain. Energy* **2015**, *6*, 1426–1434.
45. Sekhar, P.C.; Mishra, S. Takagi–Sugeno fuzzy-based incremental conductance algorithm for maximum power point tracking of a photovoltaic generating system. *IET Renew. Power Gener.* **2014**, *8*, 900–914. [CrossRef]
46. Kaced, K.; Larbes, C.; Ramzan, N.; Bounabi, M.; Dahmane, Z. Bat algorithm based maximum power point tracking for photovoltaic system under partial shading conditions. *Sol. Energy* **2017**, *158*, 490–503. [CrossRef]

*energies*

Review

Comprehensive Review of Recent Advancements in Battery Technology, Propulsion, Power Interfaces, and Vehicle Network Systems for Intelligent Autonomous and Connected Electric Vehicles

Ghulam E Mustafa Abro [1,2,*], Saiful Azrin B. M. Zulkifli [1,2], Kundan Kumar [1,2], Najib El Ouanjli [3,*], Vijanth Sagayan Asirvadam [1] and Mahmoud A. Mossa [4]

[1] Centre for Automotive Research & Electric Mobility (CAREM), Universiti Teknologi PETRONAS, Seri Iskandar 32610, Perak, Malaysia
[2] Electrical and Electronic Engineering Department, Universiti Teknologi PETRONAS, Seri Iskandar 32610, Perak, Malaysia
[3] Laboratory of Mechanical, Computer, Electronics and Telecommunications, Faculty of Sciences and Technology, Hassan First University, Settat 26000, Morocco
[4] Electrical Engineering Department, Faculty of Engineering, Minia University, Minia 61111, Egypt
* Correspondence: mustafa.abro@ieee.org (G.E.M.A.); najib.elouanjli@uhp.ac.ma (N.E.O.)

Citation: Abro, G.E.M.; Zulkifli, S.A.B.M.; Kumar, K.; El Ouanjli, N.; Asirvadam, V.S.; Mossa, M.A. Comprehensive Review of Recent Advancements in Battery Technology, Propulsion, Power Interfaces, and Vehicle Network Systems for Intelligent Autonomous and Connected Electric Vehicles. *Energies* **2023**, *16*, 2925. https://doi.org/10.3390/en16062925

Academic Editors: Ahmed Abu-Siada and Byoung Kuk Lee

Received: 26 February 2023
Revised: 13 March 2023
Accepted: 21 March 2023
Published: 22 March 2023

Abstract: Numerous recent innovations have been achieved with the goal of enhancing electric vehicles and the parts that go into them, particularly in the areas of managing energy, battery design and optimization, and autonomous driving. This promotes a more effective and sustainable eco-system and helps to build the next generation of electric car technology. This study offers insights into the most recent research and advancements in electric vehicles (EVs), as well as new, innovative, and promising technologies based on scientific data and facts associated with e-mobility from a technological standpoint, which may be achievable by 2030. Appropriate modeling and design strategies, including digital twins with connected Internet of Things (IoT), are discussed in this study. Vehicles with autonomous features have the potential to increase safety on roads, increase driving economy, and provide drivers more time to focus on other duties thanks to the Internet of Things idea. The enabling technology that entails a car moving out of a parking spot, traveling along a long highway, and then parking at the destination is also covered in this article. The development of autonomous vehicles depends on the data obtained for deployment in actual road conditions. There are also research gaps and proposals for autonomous, intelligent vehicles. One of the many social concerns that are described is the cause of an accident with an autonomous car. A smart device that can spot strange driving behavior and prevent accidents is briefly discussed. In addition, all EV-related fields are covered, including the likely technical challenges and knowledge gaps in each one, from in-depth battery material sciences through power electronics and powertrain engineering to market assessments and environmental assessments.

Keywords: electric vehicles; autonomous vehicles; digital twins; EV components; e-mobility; IoT

1. Introduction

To acquire a proper pace to meet the increasing demands of sustainable transportation one may come across many technical hurdles that currently exist in the area of electric vehicles and their electrification mobility plan [1]. As per the literature survey [2–4], it has been observed that the price of current electric vehicles, range, and the facilities to charge them are the major concerns in the present market of electric vehicles. With time, these three challenging issues have gone through tremendous evolution. For instance, the price has been decreased by almost 90 percent and it is predicted that it may reduce more in future by 2050. Its range comparatively has been increased from 80–160 km to 320+ km [5–7], whereas the charging infrastructures have been observed as still a major concern. These electric vehicles are no doubt environmentally friendly [8,9] and have the great boom of renewable sources in the electrical grid, as witnessed in [10]. This review manuscript will try to address these issues in more detail along with emphasizing the current cutting-edge advancements in this e-mobility [11–13].

Looking at 2020, one may see that the battery's specific energy has been increased from almost 110 Wh/kg to 250 Wh/kg. Hence, looking at this advancement one may predict that it may reach up to 450 Wh/kg by the year of 2030. Moreover, the energy density also increased from 300 Wh/L to 550 Wh/L from 2010 to 2020 (in merely 10 years). Thus, one may also predict for 2030 that it may increase up to 1100 Wh/L. Where the prices of batteries are concerned, there is reduction in the cost as well. The battery price that used to be EUR 1200 per kWh has been reduced to EUR 120 per kWh and will likely to go down to EUR 50 per kWh in the near future. This review paper therefore shares the current state of the battery technology as well [11–13].

Discussing further the traction inverter power density, one may see that it is also increased up to 35 kW/L and likely to be increased up to 60 kW/L in the upcoming 10 years [11]. Researchers have so far incorporated wide band gap strategies to increase the efficiency up to 98% with an increment in driving range by 8%. The method utilized to generate electricity has the biggest impact on how environmentally friendly electric vehicles are. The European Energy Mix estimates that CO_2 emissions in 2010 were at 300 CO_2 g/kWh. The deployment of sustainable energy sources and the potential retirement of nuclear power plants are expected to reduce CO_2 emissions to below 200 g/kWh by 2030, and maybe even lower. The CO_2 emissions per vehicle will drop from 66 CO_2 g/km in 2010 to under 30 CO_2 g/km in 2030 when the consumption and emissions of the electric vehicles used to generate the electricity are taken into account [11–14].

Autonomous vehicles (AVs) are anticipated to be used by 2030. They will probably be electric and shared. Some of the commercial used vehicles have been incorporated with automation level 1, as mentioned by the SAE in 2010. Now, the latest cutting-edge vehicular technologies have reached level 3 already, and some of them have level 4 automation. They have artificial intelligence features already incorporated with advanced communication systems that enable their use for brand new mobility services right now. Such development is also present in this article. In this present era, thousands of things cannot only be sensed but processed and actuated using the Internet of Things feature. This will also enable the smooth collaboration and makes sharing of data easy [14–16]. Such platforms are used in the automotive mobility, automation, as well as in smart cities. These platforms are not only used to identify threats but also used to tackle them [14–20]. While driving one may see that driver has to perform a series of actions, such as accelerating and de-accelerating the vehicle, taking care of directions, and using indicators and changing lanes accordingly [21]. An autonomous vehicle must consider its surroundings in order to operate [22]; the five core processes of perception and planning, along with localization, vehicle control system, and system administration, are needed for this.

Estimating the position of the vehicle is the responsibility of the localization module, while the perception module uses data from several sensors to build a representation of the driving environment. Thus far, the module related to planning is concerned; its main task is to make decisions for maneuvering the EV based on safer localization and mapping. This is all possible because of the perception data only. Moreover, the acceleration, steering, and braking mechanisms all are controlled by the vehicle control system [23]. Thus, taking all factors on the road into account, such as pedestrians, cyclists, other vehicles, etc., the procedure becomes a bit complex. Therefore, the communication system module plays a vital role in the autonomous electric vehicles, which allows the vehicle to take care of such factors while being driven on roads. A frequent name for this type of communication is "vehicle-to-everything" (V2X) communication, which encompasses a number of situations including "vehicle-to-vehicle", "vehicle-to-infrastructure", "vehicle-to-pedestrian", and "vehicle-to-network" (V2N) communication [24,25].

Thus far, it has been observed and studied in the literature [25,26] that two vehicles can communicate with each other, known as vehicle-to-vehicle communication. This enables fewer collisions and enables road departure with nominal speed and acceleration by letting other side vehicles know about each other [27]. Instead, V2I communication allows the car to link to the infrastructure on the side of the road to spread information widely [28]. Among the advanced services, one may find all relevant information related to safe distances from surrounding cars, speed limits, safety, roadblocks, and accidental warnings, and it also helps in assisting lane tracking as well [29]. In order to reduce accidents, the term "V2P" refers to the idea of information being exchanged between a vehicle and a pedestrian utilizing sensors and intelligent technology [30–32]. The server that provides centralized control and data on traffic, roads, and services is connected by V2N, which connects car user equipment [33]. As a result, the deployment of V2X communications in conjunction with already existing vehicle-sensing capabilities serves as the basis for complex applications intended at enhancing vehicle traffic, passenger infotainment, manufacturer services, and road safety [34,35].

If such systems are to be successful when implemented in a real-life scenario, they will ultimately depend on the data gathered from actual contact [36]. For instance, machine vision makes use of image processing to keep an eye on the back cars [37] and trajectory analysis of the vehicles in particular jurisdiction [38]. The best control parameters for maximizing fuel efficiency and saving fuel are also determined using historical data [39]. The chance of drunk or sleepy driving is decreased by using data gathered by in-car sensors to examine driver behavior even when the vehicle is not fully autonomous [40].

There is literature that examines V2X communication and focuses on the connectivity and security of networks [41–45]. There have also been reviews that concentrate on various aspects of the autonomous vehicle. The state of the art for connected automobiles was outlined by Siegel et al. [40] starting with the obstacles, applications, and requirements for vehicle data. It was endorsed in [46] that the communication between transport infrastructures with cooperative traffic management solves half of the problems. They focus on non-signalized junctions in their study while also including approaches for signalized intersections. The detailed review on autonomous overtaking was published in [47]. The authors demonstrated that the dynamics of vehicles and restrictions associated with the environment, as well as proper understanding of the environment and nearby obstructions, are the two key components of high-speed overtaking. Bresson et al. [48] conducted an assessment on localization methods for autonomous cars equipped with on-board sensor-based systems along with the combination of a communication network, either V2V or V2I, or in some of the cases both are equipped.

Development of Vehicular Networks and Cloud Options

Furthermore, one may see several research contributions revolving around cloud computing [49–58]. In these research manuscripts, authors have investigated the vehicular computing based on cloud things and associated its extension to mobile-based cloud computing and vehicular networks. In addition to this, one may find additional information related to privacy, security topics, and concern areas such as cloud applications and their formation along with the communication system design. The difficulties of vehicle cloud networks were discussed in [50,51]. Last but not least, one may see similar discussions on vehicular cloud options, including traffic models, services, and applications that can make vehicular clouds possible in a more dynamic setting [52,53]. Reviews written by various authors have concentrated on a certain topic. Although review paper [40] focused on a more general topic, network connectivity was highlighted. The application had very few details. However, the researcher in [40] only highlighted applications that may use gathered data to check drivers, hence lowering the danger of sluggish drinking, as an example of driver monitoring. As per our best efforts on studying the literature, one may be unable to find any study related to the cutting-edge trends in autonomous vehicular technology. Hence, the aim to include this in our review manuscript is to comment whether the above-stated jobs will be performed by the modern autonomous vehicular technology or not. Fuzzy, modern predictive control, and a number of other techniques are used to enhance fuel efficiency and energy consumption without sacrificing the vehicle's performance. Some of these provide information on the EMS's real-time development process and its calibration parameters, which are utilized to improve the vehicle output characteristics. These parameters include SOC, vehicle speed, power-split, etc. The primary goal of this type of research is to investigate a comprehensive strategy for creating the control architecture of an EMS using multiple control techniques. Along with a brief suggestion and debate regarding the improvement of future EMS research, constraints and difficulties relating to EMS breakthroughs are also suitably emphasized. The significance and potential effects of real-time EMSs with different control systems were finally revealed by an interpretative analysis [39]. For the transportation of the future, which is anticipated to be sustainable in terms of energy generation, consumption, and vehicle emissions, these methods are all put forth. Vehicle electrification, autonomy, and implementation all heavily rely on embedded intelligent systems. Despite the fact that electric vehicle technology is anticipated to dominate the automotive powertrain design in the next decades, a number of obstacles currently prevent their widespread adoption in the automotive industry. These obstacles can generally be divided into four categories: consumer behavior, charging infrastructure, car performance, and governmental backing. Hence, a thorough understanding of these obstacles is a matter of concern. Based on the importance of each barrier to be found and removed, this article studies them and deduces the relative order of their removal [40].

This paper starts with information about the digital-twin-based vehicle propulsion system (DTVPS) and its revolutionary benefits associated with the wide band gap (WBG)-based semiconductor trend utilized in power converters. Later, one may read about the rapid charger technology in detail with respect to vehicle-to-grid (V2G) and vehicle-to-device (V2D) communication systems [59,60]. In addition to this, one may see the overall investigation provided by these modern strategies that make modern autonomous cars more powerful. At the same time, there is a recommendation of resolving the issues as stated in this manuscript. Thus, the main objective of this research article is to propose an overall picture of this topic though comprehensive literature work which includes related areas but no discussion on algorithms at this moment.

2. Future Electric Vehicle Propulsion Systems

In terms of the concept related to electric vehicle propulsion systems, one may understand it in an easy way, as it requires a power converter, a battery, an electric motor, and, last but certainly not least, a fixed transmission. Moreover, there is no need for a gearbox, and it is also free from clutch and oil filters. This cuts down the costs as well as improves the driving comfort [61–64]. One may find the information associated with the upcoming market trends with EV propulsion in this section and will be able to draw the new directions for research as well.

2.1. Development of Digital Twins for EVs and Their Perks

For many years, automotive researchers and engineers have created analytical and simulation models of the individual parts of EVs as well as complete EVs. With time, these models have advanced and become more precise. With the advent of sensor technologies and the powerful IoT-like feature, all offline models have been turned into digital models that provide liberty of monitoring, rescheduling maintenance, predictive maintenance, and fault endurance and recognition for their lifetime. This results in reducing the costs over the intermediate steps during manufacturing, such as system design verification and validation. These are the reasons that digital twin has been initiated based on the advanced strategies such as AI, IoT, and cloud computing [65,66].

The entire idea about digital twin is illustrated in Figure 1, where one may see an electric vehicle containing all essential components, such as a power converter, battery, and then a suitable number of sensors equipped with a motor. The representational model of the simulation platform is housed in the virtual environment. Thus, using a multi-physics framework, one of the highly accurate models has been developed. The exchange of information and data links the physical and digital worlds. The vehicle designer can develop a virtual process that runs concurrently with the real one and functions as a source that helps in analyzing the model in terms of dynamic and static perspectives.

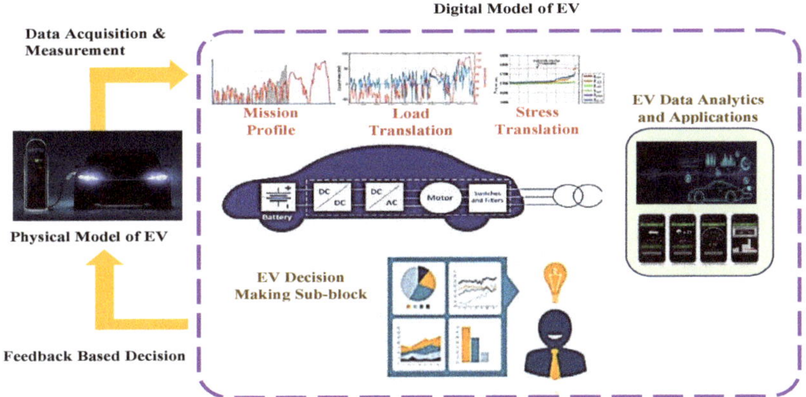

Figure 1. The creation of a digital twin implementation concept.

The EV's digital twin model and tool can provide the following advantages:

- Ensuring that future EVs' functionality, energy efficiency, and user comfort all advance significantly: These criteria can gauge the usability and attributes of the vehicle, for instance, price, driving distance, range forecast, overall journey time, appropriateness of a long-distance trip, comfort in all environmental circumstances, and comfort in traffic situations.
- Stress modeling and study using multi-physics: This analysis predicts breakdowns in advance, allowing for the prevention of failures and a reduction in downtime.
- Predictive maintenance using reliability analysis based on mission profiles: Battery-based electric vehicle drivetrain deterioration of the components crucial to system dependability can be identified via the mission-profile-oriented accelerated lifetime testing. As a result, product developers will be better equipped to innovate quickly and consistently, testing several combinations of various variations of drivetrain components and experimenting with novel ideas.

Additionally, maintenance protocols and schedules can be created utilizing the data collected from the digital twins of the cars to ensure that the parts are available before they are anticipated to fail in the EV and reduce inventory stocks. The use of the digital twin in control design, designing powertrain, and the dependability of innovative new powertrains is also one of the major themes of the future. This shares three significant domains, such as the digital-twin-based reliability, its design itself, and digital-twin-based control design. These areas are all crucial for creating future vehicle generations that are more dependable and cost effective.

2.2. Power Electronic Interfaces

Power electronic converters are no doubt important components of any electric vehicle propulsion [11]. Numerous studies have been undertaken in the area of semiconductor-based materials designed as switches for these power converters. These switches are currently proposed based on silicon (Si) materials or silicon carbide (SiC). A few of them use gallium nitride (GaN), as referred to in [67–70]. The only limitation or constraint with such switches is their switching frequency. As per the user requirement, it is seen that the silicon-based IGBT traction inverter designs restrict the switching frequency [71]. These wide band gap materials need one- or two-electron volt energy to transfer their electron to the valence band in order to execute the conduction [67–73]. Such properties are illustrated in Figure 2.

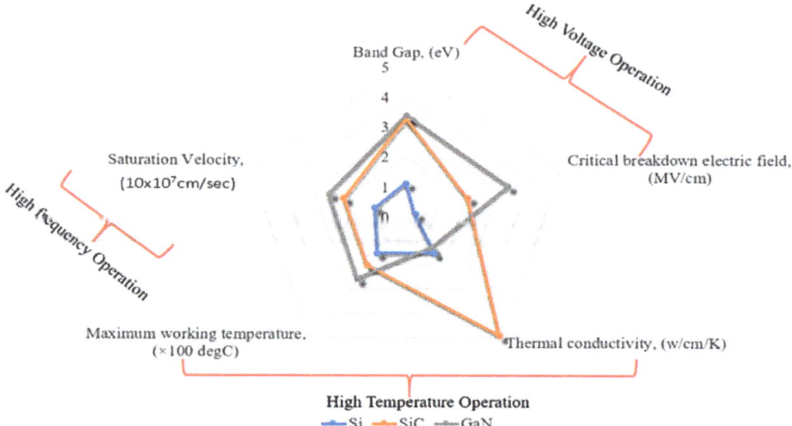

Figure 2. Comparison between silicon (Si), silicon carbide (SiC), and gallium nitride (GaN) [73].

For Si-based OBCs, the switching frequency for MOSFET-based on-board chargers (OBCs) must be less than 100 kHz [72]. The WBG semiconductors, in contrast to typical Si semiconductors, have intriguing characteristics and advanced material features, such as the capacity to function at higher voltages and with less leakage current, as well as greater switching frequencies and thermal conductivity. Thus, for low-voltage applications, the high-frequency-based WBG semiconductor provides good efficiency along with better power density. This results as a reduction in the overall weight of the converter and efficiency of the electric powertrain. Moreover, these high frequencies that are in between 40 kHz–100 kHz for active front-end inverters and 200 kHz to 500 kHz for OBC systems will allow to even operate on high temperature readings as well. It has been observed that there is much less focus dedicated to the thermal control of GaN-based semiconductor devices. Thus, precise models for GaN-based power electronic converters are required so that one may suggest the outcome based on their parametric as well as non-parametric representations.

Additionally, for power electronic converters, the most failure-prone devices are the semiconductor modules. This is because of their high thermal stress property. Many of these failures are time-dependent dielectric breakdowns [74]. WBG-based power converters are most reliable because of their higher activation and are currently preferable due to their cost-effective packaging devices. These reliability operations have so far not been discussed in previous research contributions. Although there are some research manuscripts that may share the reliability analysis of silicon- and silicon-carbide-based converters, one may find very little research on GaN-based power converters. These GaN devices in the EV power industry enable higher range efficiency, but still one of the key constraints in this area to tackle is the range of voltages. Regarding the predictive maintenance and reliability, no detailed stress study is available. Therefore, in accordance with these WBG technologies, one may see them being integrated with electric motors as well as with the battery systems.

3. The Future of Solid Batteries

Because of the dominating properties, lithium batteries are currently the most common type of batteries used in electric vehicles, according to the literature [75–78]. This battery's role in EVs is crucial, since it determines the vehicle's energy, cycle life, power performance, safety, and, most importantly, its driving range. Numerous unique scientific advancements in battery chemistry, composition, and manufacturing have improved the performance of EVs while also lowering their overall cost [79].

3.1. Previous Technology Developments Associated with Lithium Batteries

According to the cathode material utilized, Li-ion batteries are frequently categorized [80,81]. The common iron and phosphate are used to create LFP (lithium iron phosphate) batteries, on the other hand. Due to the material's hard olivine structure, these batteries have a very long lifespan and are capable of producing very high power. Unfortunately, this technology's intrinsic low potential compared to Li+ and particular capacitance make it less suitable for high-energy applications. LFP is still a good option for power applications (hybrid cars, power equipment, etc.) or situations requiring a lot of cycles, whereas both lithium nickel cobalt aluminum oxide (NCA) and lithium nickel manganese cobalt oxide (NMC) are energy-dense technologies, and electric cars frequently employ them. The amount of cobalt is being decreased in favor of the amount of nickel in both technologies, which is a definite trend. This guarantees a better energy density and lessens reliance on pricey cobalt. Studying different stoichiometric proportions enables us to see several types of NMC which are now available for commercial usage. Considering NMC111, NMC532, and NMC622, each of the three components are present in the same quantity. NMC111 is better suited for higher power applications, since it has less nickel and more manganese, but NCA, NMC-532, and NMC-622 are considered to be cutting-edge cathode materials, as seen in Figure 3.

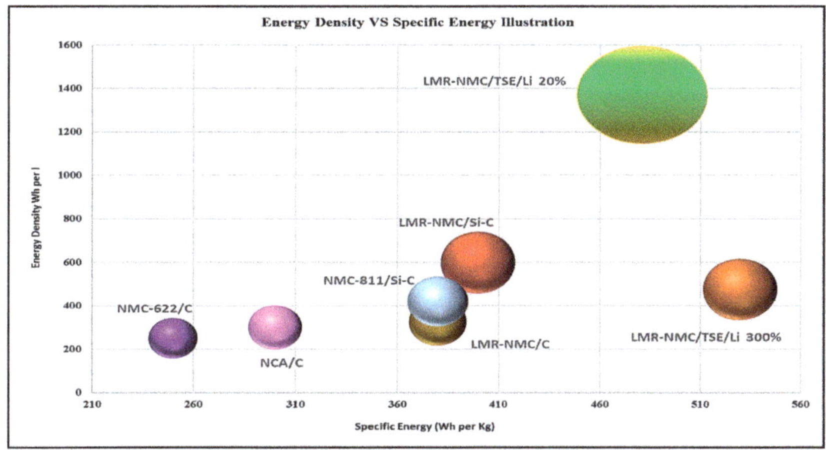

Figure 3. Illustration of Energy density vs. specific energy.

There are many constraints of negative electrodes in terms of their availability for commercial usage. Because of their low potential when compared with Li+, it is observed that high specific capacitance and carbon anodes have outclassed them since 1991. In the year 2016, almost 90% of commercial batteries were based on graphite, whereas only 7% of them had amorphous carbon and only 2% had lithium titanate oxide. These materials have the ability to charge the batteries quickly but at the same time the raw material seems quite expensive, along with the low energy density [82]. There is a plethora of perks of the electrode material available today in the market, and this is because of the recent research and developments occurring in the field of Li-ion batteries, as illustrated in Figure 3. In the near future, silicium's contribution to this will be crucial. It has been observed while going through the literature that silicon is one of the alternate solutions for the next-generation materials for anodes [83], and this leads toward the reduction in the low prices: almost eight to ten percent less than graphite. In addition to this, the life cycle of silicon-based batteries is still short, even with the amalgamation of graphite electrodes in lower amounts.

A few examples of this amalgamation have occurred already, for example, the 5 percent incorporation in Panasonic cells which were later utilized in Tesla X. We all expect that the technology that we have today will surely expand in upcoming years, and with this expansion there will surely be a boost in nickel-based cathodes that may lead to lessen the silicon's content and immediately will result in an increment in energy density. This is something that is expected until 2025, along with the anticipation of technologies based on lithium-sulfur-oxygen and solid-state batteries. In the coming eight to ten years, the current market is predicted to head toward the next generation of technology based on the lithium-ion battery. It is concluded that with an inclusion of cobalt or nickel the energy densities can be increased at the cell level as well as at the pack level, as shown in [84,85]. In order to fulfil this prediction, extensive research is required on these solid-state electrolytes. These are the thicker electrolytes with higher energy densities. They are not flammable and do not have any impact on concentration polarization voltage losses like the other liquid electrolytes. These are found to have brilliant dendritic growth resistance that allow us to use Li-based anodes easily [86].

Specifically, for a solid electrolyte to be used for the applications associated with an electric vehicle, it must have a fast charging capability. In addition to this, the maximum current density over which the battery becomes short circuited due to the Li-dendrite penetration phenomenon is one of the key things to be kept in mind. The modern parametric reading for critical current density values should be below the value of 5 mA/cm^2 [87,88], but in actuality they are even below 0.12 mA/cm^2. Furthermore, variable current densities are needed for charging and discharging. Recently, it was discovered that critical current densities are higher during charging than during discharging [86].

Investigating the interfaces such as electrode to electrolyte in solid-state batteries is one of the significant aspects in acquiring high specific energy with a longer life cycle. Moreover, one of the reasons due to which cells usually fail is the electro-chemical interfacial instability. For example, solid-state electrolytes presently have a stability window up to 6 V as compared to Li+/Li batteries. The cell impedance may eventually rise as a result of the breakdown of the solid electrolyte–solid electrode contact. The purpose of some techniques, such as liquid–solid hybrid electrolytes, is to explain the interface instabilities [87]. It is seen that the polymers and their composite electrolytes are majorly centered on solid-state batteries; this is because of their well-oriented nature towards the field of energy storage. In addition to providing improved mechanical flexibility, processability, and scaling up, they are less flammable than liquid electrolytes. Due to its broad ionic conductivity ranges, poly (ethylene oxide) (PEO) and its derivatives have fascinating solid-state battery possibilities. Compared to the conventional organic liquid electrolytes, ion conduction is still less efficient and more challenging [89–91]. Solid-state battery production comprises distinct lines for anode, cathode, and electrolyte sheets, much like standard Li-ion battery assembly. The production of battery parts and the methods by which they are assembled, however, differ. In contrast to conventional Li-ion batteries, electrodes must be added after the electrolyte has been created. Additionally, the creation of extremely poisonous H_2S (in the case of sulphides such as Li6PS5Cl) and somewhat high temperatures (over 1000 C in the case of $Li_7La_3Zr_2O_{12}$) are required for the production of solid electrolytes [92].

3.2. Problems Confronting the Solid-State Battery and Potential Solutions

It is a fact that the cost per cell or cost per pack is very high for EV applications, and the same goes for the energy per cell and pack as well. This goal is quite sincere, even though the expected level of safety is compromised. Here is a list of some of the issues that scientists are now working to resolve:

- Due to the weak wetting between lithium and the solid electrolyte, an interfacial resistance forms. Solid electrolytes, especially those made of ceramic, exhibit very high interfacial resistance due to insufficient wetting of Li. Li is therefore ineligible for use in solid-state batteries. It was discovered that solid electrolytes made of polymers exhibit improved Li wetting while having less ionic conductivity than their ceramic counterparts. In light of this, polymer/ceramic composites can be used as electrolytes to address the Li wetting problem [86].

- Li metal used in high-power applications has substantial dendrite development and spread problems. The critical current densities for solid-state batteries are quite a distance from the intended value of 5 mA/cm^2 [87,88]. Additionally, since plating (charging) and stripping (discharging) differ from one another, the critical current density must be decreased. Dendrite propagation is severely constrained in dense microstructures, hence great attention has been taken to create the electrolytes as tightly as possible, despite the fact that the exact reason and therapeutic treatments are yet unknown [86].
- Solid electrolytes with high ionic conductivity are challenging to make, store, and handle. They are expensive to use and necessitate specific methods and oxygen-free environments. In this field, there is a constant need to reduce production costs and facilitate handling of solid electrolytes.

During the process of developing a cell based on ceramic material, there are a lot of heat pressing techniques being utilized, as illustrated in Figure 4. Moreover, this step is performed to ensure the proper but smooth relation or contact between the electrode and its electrolyte. Today, the design engineering may easily perform this process, whereas it is observed that the bulk type of solid-state batteries may produce enough retention capacity [93]. On the other hand, the scalability is also one of the main constraints of bulk-type batteries. Thus, one may opt for polymer composites as a true solution during the massive production of these products. Furthermore, Li metal creeps while being operated on at high temperatures. Therefore, in modern methods it is suggested to engage Li metal in a process that actually reduces the creep tendency [94,95].

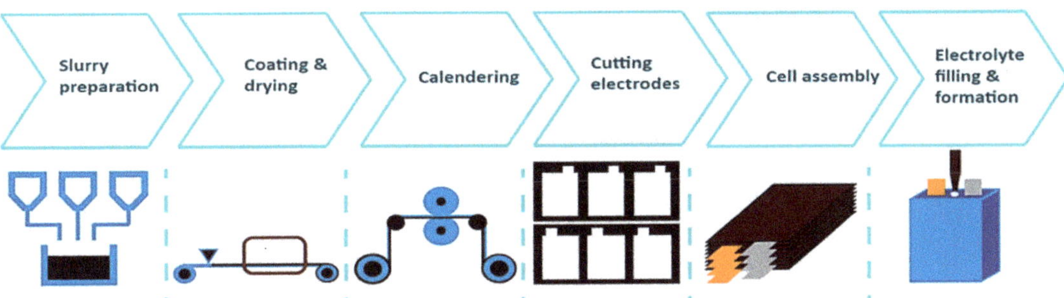

Figure 4. Battery production for process parameter optimization [94].

3.3. Batteries with Embedded Sensors

Over time, battery performance changes substantially. This may be the result of a multitude of undesirable material side reactions that eventually lead to capacity fading and impedance growth, which may raise safety concerns about dendritic short circuits. It is essential to properly manage and monitor batteries when they are in use. To perform this, a battery management system (BMS) is typically used. Each cell's voltage, current, and temperature are kept within its ideal safety parameters by the BMS. There are states of a battery, such as the state of charge (SoC) that shares the energy storage information one battery has [96] and the state of health (Soh) that describes the capability of a battery to hold on to a charge as compared to a new battery [97]. These parameters can never be examined directly, but can be analyzed with the help of voltage, current, and temperature measurements. Currently, they are accurately measured and optimized using intelligent algorithms [97]. These all-measurement techniques involve sensors to track all such parameters that share all information related to battery life. Implantable sensors that are integrated into battery cells are thus receiving more and more attention.

This approach will enable us to measure previously unmeasured quantities, learn more about the physical parameters, and comprehend the parasitic chemical activities that happen inside of the cells. This results in the improvement of dependability and advancing the security of batteries. The factors such as pressure, strain, expansion, temperature, and composition of the proposed electrolyte enable the numerous possibilities when computed using next-generation state estimation techniques [98].

Another recent area of study is batteries that can heal themselves. Unwanted chemical alterations within the cell are the cause of battery breakdown. Reversing these modifications to return the battery to its initial configuration and operation is the idea behind self-healing in batteries. Batteries with self-healing capabilities will focus on repairing damaged electrodes on their own to restore their conductivity, controlling ion movement inside the cell, and reducing the impact of parasitic side effects. Due to the difficult chemical environment that self-healing mechanisms must operate in, the field of battery technology has been sluggish to adopt them, but the subject is presently gaining ground quickly. There are some polymer-based substrates that heal themselves and are commonly known as self-healing polymer substrates (SHPS). Their primary goal is to repair all damages on electrodes and try to restore the conductivity [99]. Self-healing polymer binders, such as those used in silicon anodes, are designed to keep fractured active material particles from losing electrical contact with one another [100]. Functionalized membranes, which can trap undesirable molecules and stop them from reacting with other components in the cell, are another potential idea.

On the other hand, self-healing electrolytes have healing agents that can dissolve undesirable depositions [90]. The enclosed self-healing molecules are a promising concept for the future. These are made up of therapeutic substances housed in microcapsules. By supplying the proper stimulus, the healing chemicals can be released when necessary. The functions of sensing and self-healing should be emphasized as being closely related. Smart batteries combine both of these features so that in the first place a BMS will receive signals from the integrated sensors and analyze them. In the event that a fault is found, the BMS will signal the actuator, stimulating the proper self-healing procedure. The reliability, lifetime, user confidence, and safety of future batteries will all be maximized by this ground-breaking strategy.

One may be thinking about integrating the sensors with the battery, so for this purpose as described in [91], where an integrated conductivity and temperature (CT) micro-sensor for the conductivity high-precision measurement of electric car battery coolant was inserted, one may embed the sensor with the battery in the same manner. An inter-digital micro-electrode is made for conductivity sensing, and the temperature sensor cell is a thin-film platinum resistor. With a resolution of 0.1 S/cm, the integrated CT sensor has a respectable limit of detection. Moreover, sensors have a high-precision signal collection and processing circuit constructed for them, and a desirable full-scale measuring error is seen. Moreover, once this sensor is deployed the data can later be easily transported to a static IP using the Internet of Things and can perform several AI algorithms for further monitoring and predictive maintenance.

4. Intelligent Bidirectional V2G Systems

For the widespread use of EVs, the battery needs to be charged quickly and effectively. The average modern electric car can travel 300–400 km without recharging. There are numerous difficulties to think about: the first is having charging stations available everywhere; the second is quick charging; and the third is improving power density and specific power [101]. There are four primary forms of charging in use today. The following charger types are described in Table 1 [102].

Table 1. Different charging level systems [102].

System Level	Charging Duration	Output Nature	Location
Next-generation-based ultra-fast charging system (NXG-UFCS)	It takes approximately 8 min to charge EV for 320 km	Three-phase Vac: 210–600 AC circuit dually converted to DC circuit to EVs. Output normally ranges around 800 V and 400 kW.	Off-board 3 phase
DC fast chargers (DC-FCs)	It takes 30 min to 1 h to charge for 100–130 km of range per hour	Three-phase Vac: 210–600 AC circuit dually converted to DC circuit but with an output range of 500 A and in between 50 and 350 KW.	Off-board 3 phase
Level 2 chargers (L2C)	These are domestic chargers and are available at home to charge for 16 to 32 km/h in 4–8 h	Vac: 240 (as per US Standard, whereas 400 as per EU standards). The output ranges from 15–80 A and power in between 3.1 and 19.2 kW.	On-board single/3 phase
Level 1 chargers (L1C)	This system takes approximate 7–10 h to charge for 3–8 km and depends strongly on the type of EV model	Vac: 240 (as per US Standard, whereas 400 as per EU standards). The output ranges from 12–16 A and power in between 1.44 and 1.92 kW.	On-board single phase

In level 1 and level 2 chargers, batteries are always plugged in on-board, whereas in level 3 converters they are usually off-board systems and have enough ability for a high-power charge. Moreover, it is also seen that level 1 and 2 are sluggish when it comes to their charging time, and therefore they are available commonly in public spaces, homes, and private setups. In most of the shopping centers one may find level 3 charging systems, which are DC power in nature and very fast in charging the system [101,102]. Regarding the level 2 charging systems, they produce approximately 20 kW of AC charging and take almost 2 h; using this system the EV may travel up to 200 km. In addition to this, one may cover 200 km by utilizing the 150 kW DC charging system that may lower down the time by 15 min compared to the conventional one. Similarly, the charging system of 350 kW takes 7 min [103,104].

Regarding the three-phase topologies compared to the front-end inverters, they may include rectifiers based on diodes, matrix rectifiers, as well as Vienna rectifiers [105]. The simplest and most effective tool for power conversion is a diode rectifier. However, the output fixed voltage is affected by the three-phase supply voltage. In terms of total harmonic distortion, it is unfavorable (THD). A three-phase active front-end (AFE) rectifier tackles the THD problem by generating three-phase sine shaped input current waveforms with an enhanced power factor and efficiency and offering variable DC output voltages. Even though it might not be as well known, the Vienna rectifier is becoming more and more widespread. Out of all the three-phase conversion techniques described so far [103,106], the AFE boost rectifier can be used for off-board fast-charging systems. As the number of battery electric vehicles has expanded, so too has the prevalence of grid-connected power electronic converters (PEC). If these PECs are bidirectional, the power kept in a car can be used to either supply peak power or temporarily store electricity (V2G, or vehicle-to-grid; grid-to-vehicle, G2V). Active switches now replace diodes in the existing PEC topologies to handle the bidirectional power flow. The system design for the multiphase-bidirectional on-board charger is depicted in Figure 5.

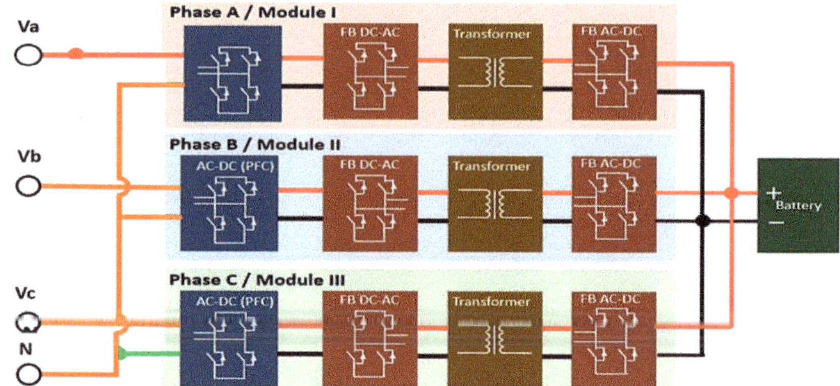

Figure 5. Architecture of the multiphase-bidirectional on-board charger system [107].

Bidirectional (V2G/G2V) On-Board Charging Systems

For opting for the off-board charger in the power electronic converter there are a few main factors that must be studied, such as high reliability, distortion-free operation, less grid interference, lower system size, weight, high efficiency, and, last but certainly not least, high efficiency. In order to be portable, light, and effective, wide band gap devices have contributed a lot, and they made switching frequency optimal as required in this domain by low gate charge and output capacitance. This was possible all because of the GaN-based power transistors. The devices such as capacitors, inductors, and transformers, which are passive in nature, were also changed to be lighter weight and smaller in size [107,108] because of the advent of WBG technology. It is noted that GaN-based transistors are high electron mobility transistors (HEMT), thus we call them GaN-HEMT in short, and they have a voltage rating up to 660 V, whereas the current ranges from 20–50 A [104,107]. These components are mostly deployed in off-board chargers (OBCs) with the output power ranging in between 3.0 kW and 20 kW. Figure 6 is illustrated in order to show two single-phase bidirectional off-board chargers with a special DC to DC stage structure as well as the same identical AC to DC structure, which is a totem pole PFC. One may also see the dual active bridge that is functional because of the galvanic-nature-based isolation and bidirectional power transformation, including zero-voltage detector and switching at both primary as well as secondary sides. This has compact-size-based components and a fixed-frequency operation, as referred to in [109].

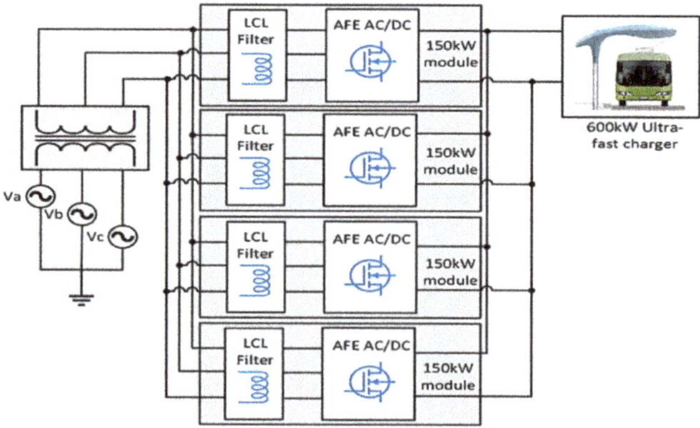

Figure 6. GaN-switch-based bidirectional OBC system topologies [107].

It is tough to achieve the full range of ZVS due to the wide range in load power. The resonant bidirectional CLLC architecture (where C is capacitance and L is inductance) shown in Figure 7b is incredibly efficient because of the zero-current switching (ZCS) on the secondary side and the zero-voltage switching (ZVS) in the main bridge. The CLLC architecture has the drawback of not being able to adjust output voltage using the series resonant frequency when it is being used for charging.

In order to solve this issue, reference [70] advises switching from frequency modulation in the DC-DC stage to DC bus voltage modulation in the PFC stage. The resonant CLLC stage will be able to function at its most effective level as a result [109,110]. A modular converter method is a suitable replacement for the development of ultra-fast charging systems.

Four AFE converters are combined and connected in parallel to generate the current design, which is suggested for the 600 kW DC ultra-fast charger [111]. This is seen in Figure 7. A comparison of silicon-based and silicon-carbide-based semiconductors has been performed for each module with a 150 kW power rating. In order to examine the effectiveness of Si (SKM400GB12T4) and SiC (CAS300M12BM2) devices at various power levels, a non-linear electro-thermal simulation model was adopted. The simulations for both scenarios contain the pertinent datasheet information. Figure 8 illustrates how SiC devices are substantially more effective as chargers than silicon-based ones. Wide band gap devices can save energy in this way because Si has a larger loss than SiC.

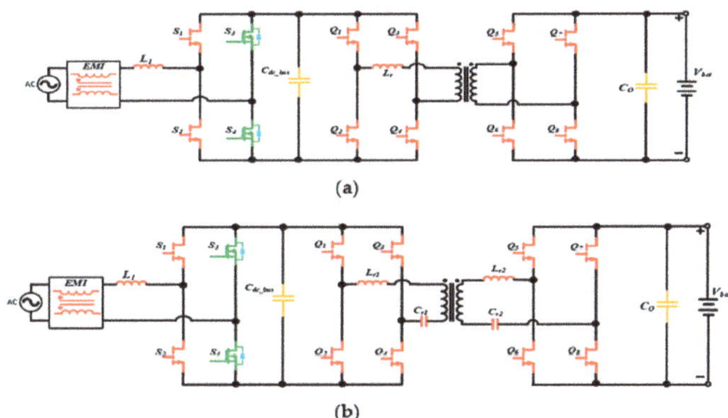

Figure 7. (**a**,**b**) Modular 600 kW DC ultra-fast charger [111].

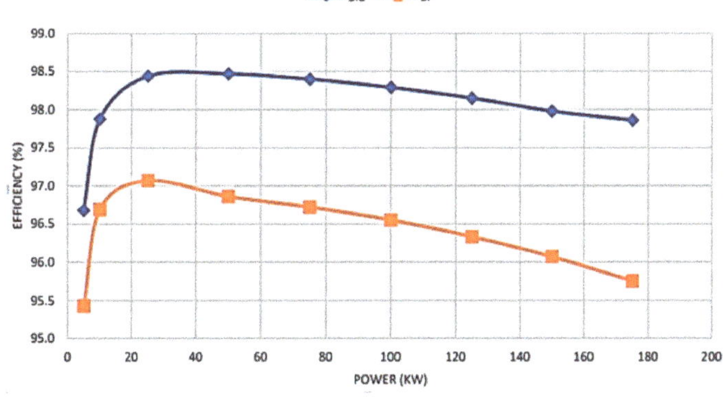

Figure 8. Si- and SiC-based high-power off-board charging system efficiency map.

5. The Transition to Climate-Neutral Transportation and Energy

Integrating renewable energy sources such as wind, solar, and hydro is essential to creating communities with sustainable energy. Even when accounting for the hourly effects of renewables' intermittency in a fully dynamic energy system, renewable energy sources have a far lower impact on climate change from a life cycle viewpoint than conventional energy sources such as oil, natural gas, or coal [112]. The increasing amount of distributed power generation equipment connected to the utility network has caused problems with power quality, safe operation, and islanding protection. In order to adhere to grid inter-connection standards, distributed generating system control must be improved [113]. For instance, transitioning to all-electric cars would only increase electricity use by 20% in Belgium [114]. Renewable energy is being used more frequently. What happens if there is neither wind nor sun, though? In these situations, we must either rely more heavily on alternate energy sources or invest more money on energy storage. Battery size plays a big role in electric car performance. Batteries in cars can be used to store extra solar- or wind-generated electricity. The phrase "smart charge management" is used to describe it. When there is a high demand for power, the stored energy can be released back into the grid. The technical term for this is vehicle-to-grid, or V2G.

An extensive cycle test revealed that using the V2H to power a house had little influence on the battery's effects of aging. The main cause of the V2G features' minimal effects on battery aging is that the discharge current needed to power a house is substantially lower than the current needed to accelerate an automobile [115]. A battery can be integrated into a Local Energy Community (LEC) in a number of advantageous ways, enabling energy to be stored while it is affordable on the wholesale market and released when it is more expensive. Capacity credit, a service offered, can help delay or reduce the need for infrastructure upgrades in the production, transmission, or distribution sectors. Batteries installed behind the meter can also help with backup power and energy cost reduction by increasing PV self-consumption in microgrids.

As energy communities and more decentralized production become more common, the power grid is projected to alter. The energy management of such systems must include electric fleet bidirectional charging systems, which can offer services that are flexible in nature, enhance self-consumption, and continue to prevent grid congestion. A vehicle-to-grid case study's techno-economic analysis can be found in reference [116]. However, for vehicles and chargers to function in a bidirectional manner, electricity must be able to be transmitted in both directions. This raises a need to initiate communication with the local grid operator, which is still an unresolved issue. The first realization shows how taking use of value streams connected to grid balancing can be facilitated by intelligently integrating electric vehicles into a grid [117]. Therefore, it is crucial to build a real laboratory where this research can be carried out. Several guidelines and specific requirements for integrating the V2G in a local energy system are provided in reference [118]. Electric vehicles are seen to emit two times less carbon dioxide (CO_2) over the course of their whole lives than gasoline or diesel engines do if we use the European electricity mix. This may be four times less if we use the electrical mix in Belgium as an example. If automobiles were fueled by renewable energy, carbon dioxide emissions might be reduced by a factor of more than ten [8,102,119]. Figure 9 displays the findings for each vehicle's ability to contribute to climate change or global warming. The BEV using Belgium's power mix receives the lowest overall grade for climate change.

The BEV also outperforms conventional gasoline and diesel vehicles in many other mid-range areas, with the exception of human toxicity. The large impact on human toxicity is brought on by the creation of auxiliary components such as batteries, motors, electronics, etc. However, when comparing the well-to-wheel (WTW) phase, which is appropriate for the Belgian limits (and urban region), it is evident that the BEV has higher ratings than all other vehicles in the investigated impact categories.

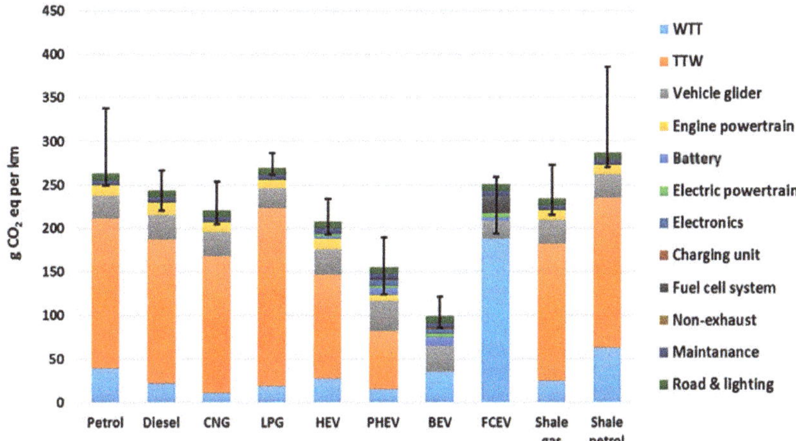

Figure 9. Results of the life cycle assessment (LCA) for climate change [8].

In light of this, reference [120] suggests a range-based LCA method that takes into account the market variability of each technology. The results reveal that, as shown in Figure 10, the BEV performs best when evaluated on an all-encompassing single score level.

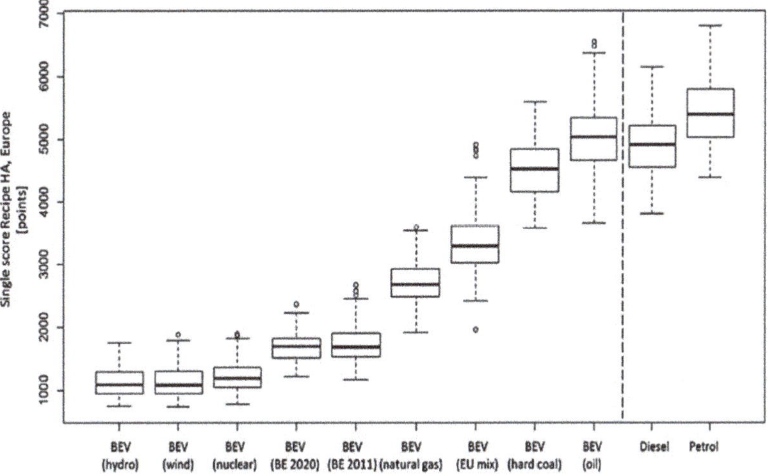

Figure 10. Results of single-score LCA [8].

6. Autonomous Electric Vehicles (AEVs)

These industries are transitioning towards greater automation together with the electrification of the energy and transportation sectors. The development of electric vehicles with high levels of automation is receiving more research attention and funding from the automotive industry as well as other technical industries. This is the main reason that the electric vehicle has to be autonomous, to bring more perks in terms of cost reduction, safety, service level, and above all environment benefits [121,122].

Synergies between AVs and EVs can be used as a result of the transition from EV to AEV. New innovations in data-driven algorithms, artificial intelligence, robust sensor technology, and smart communication are all necessary for this transformation. The mobility system and its integration into the power grid may be further optimized and its environmental impact may be decreased by addressing the fleet management and energy demand challenges [123]. Strong and rapid communication protocols are important to offer a seamless integration.

6.1. Wireless-Enabled Technology for AEVs

While embedding the highly autonomous features in a car to turn it into an autonomous vehicle, there must be a method of communication to interact with surrounding vehicles, and this communication is commonly known as vehicle-to-vehicle communication (V2V), whereas the communication in between vehicles and any infrastructure is known as vehicle-to-infrastructure (V2I). Moreover, there is vehicle-to-home (V2H) as well as V2P, which stands for vehicle-to-people communication. V2N is an acronym for vehicle-to-network. All of these protocols are illustrated in Figure 11.

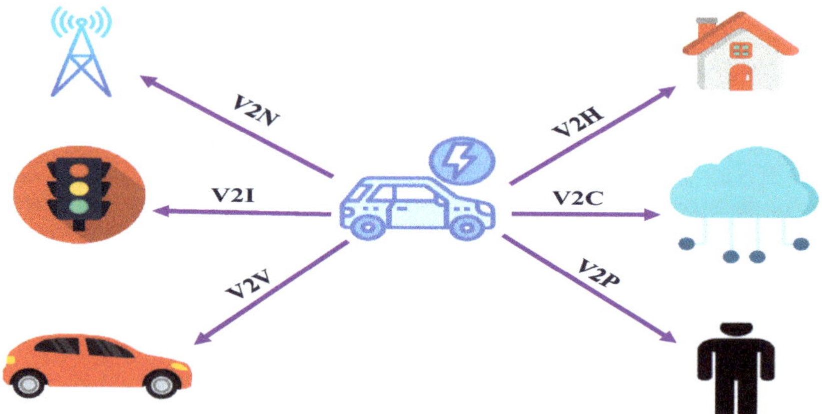

Figure 11. Vehicle-to-everything (V2E) protocol illustration.

There are numerous ways to establish this contact, each having benefits and downsides of their own. Bluetooth, 5G, and Wi-Fi are well-known wireless communication technologies. Although these radio wave technologies may occasionally offer enough bandwidth for V2V and V2I communication, it is important to take into account circumstances where this is not possible. Examples include rural areas, urban areas with poor coverage, areas with high electromagnetic interference, indoor and underground spaces such as parking lots and tunnels, etc. Light Fidelity (Li-Fi), which uses visible and infrared light for data flow, is an alternative to radio wave communication. Professor Harald Haas coined the term "Li-Fi" for the first time in 2011 [123]. Using the light from a straightforward LED desk lamp, he demonstrated how data may be delivered in the direction of a photoreceiver. By adjusting the light radiation from currently installed lighting infrastructure, such as streetlights, automobile headlights, etc., this can be accomplished. A unidirectional or bidirectional communication link with a bandwidth that can produce a data rate up to 100 times greater than Wi-Fi can be established with the use of suitable photoreceivers [123].

Figure 12 shows the technical implementation of Li-Fi. By turning the current on and off, an electrical driver controls the brightness of the light generated by the transmitter of solid-state light sources such as an LED or a laser diode. Implementing Li-Fi is rather simple because solid-state lighting is becoming more and more common in infrastructure (road lighting, traffic signals, and vehicle head- and taillights). A completely new infrastructure must be constructed in comparison to similar systems employing traditional RF-based communication (such as dedicated short-range communications, or DSRC). Since a Li-Fi transmitter can be as basic as an LED light, current lighting systems could be converted to Li-Fi transmitters. Following that, it can serve as a hub for accessing information for both automobiles and other road users (pedestrians, bikes, etc.). This results in a low implementation cost and a large number of accessible access points. With minimal effort, the current "dumb" lighting infrastructure for roads may transform into a "smart" lighting infrastructure. However, as was previously mentioned, its implementation is still difficult. However, compared to alternatives, the expenses of implementation are cheaper.

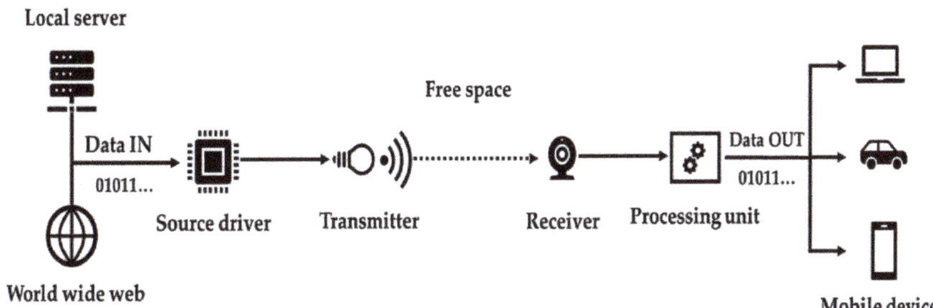

Figure 12. Implementation of a Li-Fi downlink channel on a technological level.

6.2. Shared Electric Autonomous Vehicles (SEAVs)

Given that they may be less expensive, safer, and more effective alternatives to the ridesharing and car-sharing choices available now, shared autonomous vehicles (SAV) are generating a lot of attention [124]. In addition, SEAVs, the electric version, could economically compete with current modes of transportation and have less of an impact on the environment than traditional combustion engine cars. As a result, they are thought of as a promising element of smart mobility [124]. The use of SEAVs involves a number of difficulties. Estimating passenger demand and determining the desire to use and pay for this service will be essential from an economic perspective in order to develop workable business models [124]. The vehicle supply must correspond to travel demand from the perspective of mobility. SEAVSs may increase mobility, particularly for elderly and less mobile individuals [125–127]. In this aspect, the digital divide between people—where those who are less tech-savvy and resistant to embracing new technologies are people who are socially excluded—is concerning. Due to the electric nature of AEVs, fleet management must maintain passenger service while also considering the driving range and charging requirements. For SEAV fleet charging, it is important to consider the anticipated quantity, location, and power levels of the charging stations [123]. This is still an active area of research, because studies on SEAVs that take charging aspects into account [128,129] have only included a spatial distribution or rule-based introduction up to this point; they have not looked at other factors to determine whether a location is suitable or examined grid constraints or impacts.

The widespread adoption of electric vehicles also prompts energy-related worries about the availability of electricity and the electrical infrastructure. The widespread use of EVs, however, has been demonstrated to only slightly increase the demand for electricity and presents significant potential to balance the electricity grid through a variety of ancillary services with improved bidirectional charging (vehicle-to-grid) [130]. Additionally, as discussed in the previous chapter, EVs can help accelerate the deployment of renewable energy sources (RES) by balancing their intermittent nature. The SEAV fleets offer potential in this regard because of their high degree of controllability and coordination [131]. Due to their autonomous and electric natures, which enable optimized fleet behavior, studies at this time highlight the potential of SEAV fleets (environmental, economic, and service-related). However, it does provide a challenging fleet management issue that necessitates further research as well as the creation of crucial enabling technologies, including mobility and energy demand.

7. Autonomous Electric Vehicles and Autonomous Driving Concept

This section gives a thorough overview of the enabling technologies used to make autonomous vehicles a reality, including advanced driver assistance systems and the idea of operating electric vehicles (EVs) on their own. It then offers suggestions for resolving identified problems and identifies any gaps in the existing literature.

7.1. Advanced Driver Assistance Systems (ADASs)

Prior to fully autonomous driving, the technology of advanced driver assistance systems (ADASs) is briefly discussed. To improve road safety, monitoring, braking, and alerting functions can be helped by ADASs. ADASs are capable of monitoring or assisting with parking. Streetlights, traffic data, and other connected technology can increase the safety of roads for both drivers and pedestrians, in addition to ADAS. Governments may soon mandate the installation of necessary ADASs and their components in automobiles over the next few years as ADASs continue to strive for more advantages. It is important to emphasize that the ADASs covered here are not autonomous vehicles but rather technology that aids the driver while driving. The technology in today's driver assistance systems is gradually becoming more sophisticated. The majority of systems attempt to provide adaptive cruise control, driver fatigue detection, forward collision warnings, lane-departure warnings, and parking assistance [132]. There is plethora of commercially available advanced driver assistance systems (ADAS) that have the potential to improve driving comfort and safety. Owing to age-specific performance limits, older drivers could profit a lot from such in-vehicle technology, assuming that they are purchased and used. However, at the same time, as per the findings of various market research surveys, there is much greater knowledge of ADASs than there is usage of them. In a semi-structured interview study, 32 senior citizens were polled to examine the gap between awareness and desire to utilize ADAS. There are numerous research studies, such as [132], that look at senior people's knowledge, experiences, and obstacles to using ADASs.

Backward parking is intended to be both secure and comfortable via parking assistance systems. A collision while reversing is avoided thanks to a reference that tells the driver where the car is going. When a motorist gets too close to a vehicle in front of them, forward collision avoidance systems are intended to provide them an audio and visual warning [133]. To assess whether there is a risk of collision, these systems often evaluate the distance between the two vehicles and keep an eye on their own speed as well as the speed of the vehicle in front of them [134]. LiDAR, GPS, radar, and vision-based sensors can all be used for monitoring [135,136].

The usual causes of abnormal driving are intoxication, carelessness, and/or exhaustion [136–144]. Any of these factors usually cause a motorist to act differently or move their body in a certain way. The typical behaviors of a fatigued driver are rapid and continuous blinking, head nodding or swinging, and frequent yawning [145]. On the other hand, a drunk driver frequently gets into the habit of accelerating or decelerating suddenly and reacting slowly. In some ways, reckless driving is similar to drunk driving. The motorist may be conscious but under the influence of emotional elements, which would cause them to accelerate or decelerate suddenly and go over the speed limit [145]. Consequently, a driver monitoring system can be implemented by either directly or indirectly watching the driver. Direct driver monitoring systems use a variety of sensors to track the driver's bodily movements and heart rate. Analyzing pedal and steering movements as well as responses to specific events is a part of indirect driver monitoring [146,147]. A warning mechanism will be activated when such anomalous activity is discovered.

7.2. Constraints of Autonomous Driving

Even though there has been a lot of research conducted in the field of autonomous vehicles, several topics have not been covered. First off, a sudden obstruction during the autonomous vehicle's parking trajectory has not yet been mentioned in the literature. A toddler might sprint into the parking lot to grab something, or an adult might inadvertently enter that space. For the benefit of the driver, a rear camera is placed, and a sensor that detects impediments in the rear is also installed. However, it is possible that the driver will not glance in that direction or that the sensors will not raise an alarm. When completing the parking trajectory, an autonomous vehicle should stop appropriately if such a sudden impediment appears. For instance, if the autonomous vehicle detects a balloon as a barrier, it should continue with parking rather than halting.

Systems geared toward obstacle avoidance are discussed in the most recent research [148,149]. A further factor was offered by Funke et al. [149] that takes into account unforeseen barriers such as a deer crossing the road. None of the studies, however, have addressed how autonomous vehicles ought to respond when something falls off a vehicle. In the case of a large truck delivering a mass of iron rods for construction, if the rods were to fall off and penetrate the window of the truck, it may result in deadly injuries. Although giving way to an emergency vehicle has not been mentioned in the literature, emergency vehicles do have priority in junctions [150]. Future self-driving electric vehicles must rely on both their own internal sensors and the sensors of other cars. V2V will be used to exchange measurement data in order to improve environmental awareness. Utilizing low-cost GNSS receivers [151,152], radar-camera-based traffic monitoring devices [151], microscale traffic information, and other networks [153–156], it is possible to integrate ADASs and smart lighting infrastructures.

It is an admitted fact that a driverless car has so many perks, but at the same time it can cause several issues for our society. One important issue is being responsible for an accident; either the manufacturer should be accountable for this or an insurance company [157,158]. In [158], it has been claimed that treating autonomous vehicles and human drivers equally would guarantee that autonomous vehicles would only be held responsible for negligence-related conduct. Giving a car the same rights as a human may be easier to speak than to actually do. Tort laws should apply to cars in the same manner they do to dogs [159]. The dog law and this would be comparable. Because the authors did not address how the rule might be applied to autonomous vehicles, this indicates that there is still much work to be conducted before they can be put into practice. In [160], it was argued that manufacturers should be in charge of their design and emphasized that products should go through rigorous testing before actually being distributed because the installation of such a system should not compromise road safety. It is clear that improvements must be made to the law [160–163] controlling autonomous vehicles before they can be accepted by both the general public and manufacturers.

8. Challenges and Opportunities

The most recent research obstacles for intelligent, autonomous, and connected electric vehicle technologies are examined in this section. The following details are provided [164–173]. Better decision-making capabilities for driving are provided by autonomous vehicles, which eliminate intoxication, distraction, exhaustion, and the inability to make quick decisions. Many of these elements contribute to the technologies' capacity to outperform human decision-making abilities when it comes to driving [169]. Hence, real-time responses and error avoidance represent key hurdles for AI-integrated autonomous cars. The significance of autonomous vehicle safety and performance measures has been covered in numerous research studies. These measurements ought to take into account sensor error, programming errors, unforeseen events and entities, likelihoods of cyberattacks and threats, and hardware failures. In the future, it will be crucial to develop these indicators and analyze them in a real-time setting. The comparative evaluation of autonomous driving systems is highlighted in Table 2.

Table 2. Challenges and future directions of modern intelligent vehicle technologies.

Key Findings	Challenges and Future Directions	Year
The significance of deep learning in autonomous driving was covered in this paper. Here, a number of issues with autonomous driving systems are examined and solutions using artificial intelligence and deep learning are presented.	The discussion of deep learning's function and its integration with other autonomous driving assistance systems can be expanded in this study. It includes elements of modern infrastructure, such as blockchain, cloud, and IoT technology [174].	2020
In this study, the current state of automated driving was investigated and classified, and a taxonomy for self-driving cars was developed. In addition, a hybrid architecture concept combining human and computer intelligence was developed in this work. The layout of the car itself served as a summary of autonomous driving. This effort created a taxonomy of autonomous driving technologies, similar to the self-driving car technology. We placed a higher importance on machine–human interaction and information integrity than simple driver substitution.	It is possible to add discourse and safety requirements to this effort. The proposed hybrid architecture includes a system for safety monitoring that may be expanded with other cutting-edge tools such as drones and cloud computing setups. Blockchain technology can also be used to address data security and privacy. Further performance issues can be investigated using cutting-edge networks such as 5G networks [175].	2021
In this piece, the usage of drones in autonomous systems is mostly covered. In addition, the anti-collision techniques for traffic monitoring and drone movement are covered. By adjusting the number of drones and on-road vehicles, the results are analyzed.	This work has application to the deployment and monitoring of autonomous systems in real-time. However, the relationship between drones and driverless vehicles needs to be thoroughly investigated [176].	2020
This study examined the system configurations, elements, functions, and practical situations for drones, smart UAVs, and autonomous vehicles. The key issues for research and security for AI-based threats are addressed.	This article provides a brief survey of the key technological elements and how they relate to autonomous vehicles, driving, and systems. It is possible to expand on this work by having in-depth discussions about technological issues [177].	2020
This paper presents blockchain-based architecture that supports the networks and autonomous vehicles' safety and security.	For greater credibility, the work can be expanded to incorporate smart contracts for various systems and subsystems [178].	2020

Cyberattacks fall under a number of areas, including those that target control systems, driving system components, communications across vehicle-to-everything networks, and risk assessment and survey systems. Sensor attacks, mobile-application-based vehicle information system assaults, IoT-infrastructure-based attacks, physical attacks, and side-channel attacks are the main threat types that need to be investigated and examined. Moreover, cybersecurity uses artificial intelligence for attack identification. Another intriguing feature is autonomy architecture. Autonomous systems that integrate sensors, actuators, control mechanisms, a vehicle's environment for monitoring, external control variables, speed, visibility, and object identification are crucial subsystems to pay attention to and investigate in architecture.

The cost of communication will rise as the number of autonomous vehicles rises. This results in packet delay or loss, which indirectly reduces performance or increases communication error. Human life depends on autonomous vehicles and their implementation. The drawbacks of previous efforts include the lack of in-depth research of current trends such as the use of deep learning and IoT. Furthermore, it is crucial to discuss intelligent tools and software, which are not covered in the works that have already been published. Moreover, improvements in effective simulation are needed. To create autonomous vehicles, object identification, path planning, sensors, and cloud computing should all be enhanced.

Path planning and motion control for autonomous vehicles can be determined using a predictive model. A more advanced AI-based model for AVs is required. Each element of real-time architecture must be taken care of. For instance, object detection and object tracking are necessary for scene recognition [179]. Current AV architectures do not provide a start-to-end representation [180]. System errors and scalability management should be able to be handled by the AVs' architecture. As AVs must communicate with other cars in real-time while also perceiving their environment, real-time architecture is necessary. AI-based methods can accomplish this. Infrastructure and devices act as the primary agents in AVs, and they must cooperate for accuracy [181]. The SAE categorizes automation levels on a scale of 0 to 5, where 0 denotes no automation and 5 denotes complete performance. To reach level 5, businesses and researchers are working very hard [81]. According to SAEJ 3016, the following component classes are necessary for architecture:

- In the operational class, vehicle control is the main concern.
- Second is the tactical class, where object identification and tracking as well as path planning are planned.
- Third, but certainly no less important, is the strategic class, where one may consider destination planning.

Design, development, validation, and real-time monitoring of AVs have all considerably benefited from AI. AI is a useful tool for perception, path planning, and decision making. AVs employ AI in the following ways:

- Based on a predictive algorithm, autonomous vehicles choose their own routes.
- AVs use real-time information from various sensors in an intelligent way.
- Autonomous cars make decisions about their speed and course by studying the past.
- In addition to this there are several future directions provided, such as the following:
- An intentional attack on the AI system that interferes with its operation may put autonomous vehicles in danger of being destroyed. Attacks against stop signs, such as placing stickers on them to make them more challenging to identify, is an example of such attacks. As a result of these modifications, artificial intelligence may erroneously detect objects, resulting in the autonomous vehicle behaving in a way that puts humans in danger. Thus, there is a need to explore the RFID or IoT-based solutions that use artificial intelligence to solve these challenges.
- It is observed that self-driving cars will revolutionize our lives. There is a need for legislators to create legislation that benefits the country's economy and social structure. Studies have examined an AV's potential to become a "killer app" with dramatic consequences. AVs will have substantial impacts over time, even if they are still in development. Thus, there is a need to study the safety precautions before accepting them in real environments.
- Deep neural networks (DNNs) enable self-driving cars to learn how to move around their surroundings independently. Human brains are similar to DNNs because of they both learn via trial and error. There is no hard and fast rule regarding autonomous driving and how many DNNs are required. Thus, there is a need to conduct an in-depth study in the future.
- A real autonomous driving on-road environment requires millions of interactions between vehicles, people, and devices. To handle such an extensive infrastructure, there is a need for high-end infrastructure, which may be costly. Thus, there is a need to study how artificial intelligence can efficiently utilize the infrastructure for smooth autonomous experiences.
- In the future, more intelligent tools and software should be developed to implement better path planning and object detection in autonomous vehicles. Data communication should be of higher velocity, as real-time decisions are to be made [182–184].
- In autonomous systems, the machine learning system monitors machine activity to predict problems. The solution reduces unplanned downtime costs, extends asset life, and increases operational efficiency. There is a need to identify the best machine learning algorithms and approaches that monitor a machine or its activities. This task can be explored in the future.
- Early diagnosis of vasculature via fundus imaging may be able to prevent retinopathies such as glaucoma, hypertension, and diabetes, among others, from developing [185–187]. The overall purpose of this study is to create a new way for combining the benefits of

old template-matching techniques with those of more current deep learning methods in order to achieve greater efficiency. A U-shaped fully connected convolutional neural network is used to train the segmentation of vessels and backgrounds in pixels of images (Unet). Likewise, other advanced technologies such as blockchain and quantum can be explored for AV mobile information systems [182–185]. The wireless sensor network is used in autonomous vehicles for information communication [186–192].

9. Conclusions

This manuscript proposes all of the cutting-edge discoveries related to electric vehicular technology and innovation. In addition to this is the brief study associated with the charging systems and their respective levels. The synergy between the shared autonomous electric vehicles and the current market is also described, and safety issues are addressed as well. This leads us to the conclusion that there is an immediate need to improvise the present advanced driving assistance systems, and this aspect is also covered in this review paper. Current developments in the battery technology and their system interfaces and cutting-edge solid-state battery evolution theory have been presented. Batteries will become more reliable and secure with the aid of this cutting-edge technology, self-healing batteries, and the integration of embedded sensors within the cell. The usage of a digital twin (DT) will allow for higher-reliability powertrain design, while also improving the economics and dependability of EVs. New trends and directions for innovative, practical, and reasonably priced powertrains are thus given.

Regarding autonomous driving systems, drivers will not have to manage such a complex chore any longer, preventing any potential harm when parking a car and reducing traffic congestion and fuel consumption. As was previously indicated, notions in the literature have been formed based on the full adoption of autonomous vehicles, which may not occur very soon. It has not been discussed how an autonomous vehicle should react to a careless driver. Tailgating, driving against traffic, speeding, neglecting to use turn signals, proceeding through red lights without stopping, and failing to surrender the right of way are all examples of reckless driving. Additionally, motorbike and autonomous vehicle interactions have not been taken into account by researchers, who have instead concentrated on four-wheel-drive interactions. It is difficult to determine how an autonomous vehicle should deal with a mode of transportation where there is a substantial danger of fatalities for motorcyclists. Modern technologies now have the ability to analyze driving behavior, which can help to prevent anomalous driving habits. The devices are able to control the lateral motion of an EV during unusual behavior. NVIDIA has established a new paradigm for autonomous driving software with the successful demonstration of neural-network-based autonomous driving. Autonomous lateral control is self-driving automobiles' major problem. In terms of offering a full software stack for autonomous driving, an end-to-end model appears to be quite promising. This technology is one of several steps toward the realization of self-driving automobiles, even though it is not yet ready to be offered as a feature on the market. The work discussed in the paper [190] focuses on the application of an end-to-end paradigm. With the goal of illuminating deep learning and the software needed for neural network training, the subtleties of building an effective end-to-end model are highlighted. The model showed a 96.62% autonomy for a multilane track, like the one that was employed for training in this research contribution [190]. The model successfully maneuvered the car on single-lane, uncharted tracks 89.02% of the time. The findings show that end-to-end learning and behavioral cloning can be used with artificial intelligence to enable autonomous driving in novel and uncharted environments.

There are several reasons why EVs are attractive, among such reasons being the coordination between carbon footprints and the power grids employing various renewable sources. It is being investigated whether coordinated charging of electric vehicles has the potential to reduce the CO_2 emissions associated with their charging by charging only when the grid's carbon intensity (gCO_2/kWh) is low and absorbing excess wind generation during periods when it would otherwise be curtailed. A time-coupled linearized optimal power flow formulation, based on plugging-in periods generated from a sizable travel dataset, is described as a way of scheduling charge events that seeks the lowest carbon intensity of charging while respecting EV and network restrictions [191,192]. Another reason is of course the efficiency of autonomous vehicles that has been increased significantly with the advent of artificial intelligence. Thus, an outline of autonomous vehicles is also included in this manuscript. The key components of an autonomous vehicle that enable data gathering and transmission are sensors. An improved system for lane keeping, lane change, and obstacle recognition is made possible by this information. However, various sensors have a number of limitations. Techniques for image processing could reduce costs; however, they are susceptible to climatic and environmental factors. Therefore, more work is required to either increase the reliability of inexpensive sensors or lower the cost of high-reliability sensors for mass production. In addition, the areas of research needed for autonomous intelligent vehicles were also noted. With the advent of technology day by day, road safety measures have been increased and still are top concerns while designing any advanced driver assistance product. The common issues on roads associated with society directly affect the driverless vehicles and their concept. This is the main reason that there is still no legislation on it. The future concern must be in this regard to set a main focus over the major penetration of autonomous shared electric vehicle and their co-existence with normal vehicles on the same roads.

Author Contributions: G.E.M.A. performed writing, studied the literature, and compiled all facts. N.E.O., M.A.M., S.A.B.M.Z. and V.S.A. reviewed the entire paper. K.K., V.S.A. and N.E.O. contributed the information about wide bandgap (WBG) power electronics and charging technology and drafted some of the sections on the battery technology. G.E.M.A., K.K., S.A.B.M.Z. and V.S.A. also wrote information on autonomous electric vehicles and shared autonomous electric vehicle and Li-Fi technology. All authors have equally contributed to this research review article. All authors have read and agreed to the published version of the manuscript.

Funding: This research received no external funding.

Data Availability Statement: Data sharing not applicable.

Acknowledgments: We thank the Centre for Automotive Research and Electric Mobility (CAREM) Universiti Teknologi PETRONAS, Malaysia, for supplying all the necessary data and resources for this project. Moreover, we also acknowledge the support of the co-authors from the Laboratory of Mechanical, Computer, Electronics and Telecommunications, Faculty of Sciences and Technology, Hassan First University, Morocco, and the Electrical Engineering Department, Faculty of Engineering, Minia University, Minia, Egypt.

Conflicts of Interest: The author declares no conflict of interest.

References

1. Chan, C.C.; Wong, Y.S.; Bouscayrol, A.; Chen, K. Powering sustainable mobility: Roadmaps of electric, hybrid, and fuel cell vehicles [point of view]. *Proc. IEEE* **2009**, *97*, 603–607. [CrossRef]
2. Lebeau, K.; Van Mierlo, J.; Lebeau, P.; Mairesse, O.; Macharis, C. Consumer attitudes towards battery electric vehicles: A large-scale survey. *Int. J. Electr. Hybrid Veh.* **2013**, *5*, 28. [CrossRef]
3. Bloomberg NEF. BloombergNEF's 2019 Battery Price Survey BNEF. Available online: https://about.bnef.com/blog/battery-packprices-fall-as-market-ramps-up-with-market-average-at-156-kwh-in-2019/ (accessed on 3 February 2023).
4. Berckmans, G.; Messagie, M.; Smekens, J.; Omar, N.; Vanhaverbeke, L.; Van Mierlo, J. Cost Projection of State of the Art Lithium-Ion Batteries for Electric Vehicles Up to 2030. *Energies* **2017**, *10*, 1314. [CrossRef]
5. Vijayagopal, R.; Rousseau, A. Benefits of Electrified Powertrains in Medium- and Heavy-Duty Vehicles. *World Electr. Veh. J.* **2020**, *11*, 12. [CrossRef]
6. Simeu, S.K.; Brokate, J.; Stephens, T.; Rousseau, A. Factors Influencing Energy Consumption and Cost-Competitiveness of Plug-in Electric Vehicles. *World Electr. Veh. J.* **2018**, *9*, 23. [CrossRef]
7. Islam, E.S.; Moawad, A.; Kim, N.; Rousseau, A. Vehicle Electrification Impacts on Energy Consumption for Different Connected-Autonomous Vehicle Scenario Runs. *World Electr. Veh. J.* **2020**, *11*, 9. [CrossRef]
8. Messagie, M.; Boureima, F.-S.; Coosemans, T.; Macharis, C.; Mierlo, J.V. A Range-Based Vehicle Life Cycle Assessment Incorporating Variability in the Environmental Assessment of Different Vehicle Technologies and Fuels. *Energies* **2014**, *7*, 1467–1482. [CrossRef]

9. Marmiroli, B.; Messagie, M.; Dotelli, G.; Van Mierlo, J. Electricity Generation in LCA of Electric Vehicles: A Review. *Appl. Sci.* **2018**, *8*, 1384. [CrossRef]

10. Rangaraju, S.; De Vroey, L.; Messagie, M.; Mertens, J.; Van Mierlo, J. Impacts of electricity mix, charging profile, and driving behavior on the emissions performance of battery electric vehicles: A Belgian case study. *Appl. Energy* **2015**, *148*, 496–505. [CrossRef]

11. Islam, S.; Iqbal, A.; Marzband, M.; Khan, I.; Al-Wahedi, A.M. State-of-the-art vehicle-to-everything mode of operation of electric vehicles and its future perspectives. *Renew. Sustain. Energy Rev.* **2022**, *166*, 112574. [CrossRef]

12. Yong, J.Y.; Ramachandaramurthy, V.K.; Tan, K.M.; Mithulananthan, N. A review on the state-of-the-art technologies of electric vehicle, its impacts and prospects. *Renew. Sustain. Energy Rev.* **2015**, *49*, 365–385.

13. Shariff, S.M.; Iqbal, D.; Alam, M.S.; Ahmad, F. A State of the Art Review of Electric Vehicle to Grid (V2G) technology. *IOP Conf. Ser. Mater. Sci. Eng.* **2019**, *561*, 012103. [CrossRef]

14. Alam, F.; Mehmood, R.; Katib, I.; Albogami, N.N.; Albeshri, A. Data Fusion and IoT for Smart Ubiquitous Environments: A Survey. *IEEE Access* **2017**, *5*, 9533–9554. [CrossRef]

15. Munoz, R.; Vilalta, R.; Yoshikane, N.; Casellas, R.; Martinez, R.; Tsuritani, T.; Morita, I. Integration of IoT, Transport SDN, and Edge/Cloud Computing for Dynamic Distribution of IoT Analytics and Efficient Use of Network Resources. *J. Light. Technol.* **2018**, *36*, 1420–1428. [CrossRef]

16. Frustaci, M.; Pace, P.; Aloi, G.; Fortino, G. Evaluating Critical Security Issues of the IoT World: Present and Future Challenges. *IEEE Internet Things J.* **2018**, *5*, 2483–2495. [CrossRef]

17. Ngu, A.H.; Gutierrez, M.; Metsis, V.; Nepal, S.; Sheng, Q.Z. IoT middleware: A survey on issues and enabling technologies. *IEEE Internet Things J.* **2017**, *4*, 1–20. [CrossRef]

18. Kannan, M.; Mary, L.W.; Priya, C.; Manikandan, R. Towards smart city through virtualized and computerized car parking system using arduino in the internet of things. In Proceedings of the 2020 International Conference on Computer Science, Engineering and Applications (ICCSEA), Gunupur, India, 13–14 March 2020; pp. 1–6.

19. Kuutti, S.; Fallah, S.; Katsaros, K.; Dianati, M.; Mccullough, F.; Mouzakitis, A. A Survey of the State-of-the-Art Localization Techniques and Their Potentials for Autonomous Vehicle Applications. *IEEE Internet Things J.* **2018**, *5*, 829–846. [CrossRef]

20. Kong, L.; Khan, M.K.; Wu, F.; Chen, G.; Zeng, P. Millimeter-wave wireless communications for IoT-cloud supported autonomous vehicles: Overview, design, and challenges. *IEEE Commun. Mag.* **2017**, *55*, 62–68.

21. Honnaiah, P.J.; Maturo, N.; Chatzinotas, S. Foreseeing semi-persistent scheduling in mode-4 for 5G enhanced V2X communication. In Proceedings of the 2020 IEEE 17th Annual Consumer Communications & Networking Conference (CCNC), Las Vegas, NV, USA, 10–13 January 2020; pp. 1–2.

22. Li, L.; Liu, Y.; Wang, J.; Deng, W.; Oh, H. Human dynamics based driver model for autonomous car. *IET Intell. Transp. Syst.* **2016**, *10*, 545–554. [CrossRef]

23. Andresen, L.; Brandemuehl, A.; Honger, A.; Kuan, B.; Vödisch, N.; Blum, H.; Reijgwart, V.; Bernreiter, L.; Schaupp, L.; Chung, J.J.; et al. Accurate mapping and planning for autonomous racing. In Proceedings of the 2020 IEEE/RSJ International Conference on Intelligent Robots and Systems (IROS), Las Vegas, NV, USA, 24 October–24 January 2020; pp. 4743–4749.

24. Bensekrane, I.; Kumar, P.; Melingui, A.; Coelen, V.; Amara, Y.; Chettibi, T.; Merzouki, R. Energy Planning for Autonomous Driving of an Over-Actuated Road Vehicle. *IEEE Trans. Intell. Transp. Syst.* **2020**, *22*, 1114–1124. [CrossRef]

25. Choi, Y.-J.; Hur, J.; Jeong, H.-Y.; Joo, C. Special issue on V2X communications and networks. *J. Commun. Netw.* **2017**, *19*, 205–208. [CrossRef]

26. Chen, S.; Hu, J.; Shi, Y.; Peng, Y.; Fang, J.; Zhao, R.; Zhao, L. Vehicle-to-Everything (v2x) Services Supported by LTE-Based Systems and 5G. *IEEE Commun. Stand. Mag.* **2017**, *1*, 70–76. [CrossRef]

27. Bai, B.; Chen, W.; Ben Letaief, K.; Cao, Z. Low Complexity Outage Optimal Distributed Channel Allocation for Vehicle-to-Vehicle Communications. *IEEE J. Sel. Areas Commun.* **2010**, *29*, 161–172. [CrossRef]

28. Zhang, R.; Cheng, X.; Yao, Q.; Wang, C.-X.; Yang, Y.; Jiao, B. Interference Graph-Based Resource-Sharing Schemes for Vehicular Networks. *IEEE Trans. Veh. Technol.* **2013**, *62*, 4028–4039. [CrossRef]

29. Du, L.; Dao, H. Information Dissemination Delay in Vehicle-to-Vehicle Communication Networks in a Traffic Stream. *IEEE Trans. Intell. Transp. Syst.* **2014**, *16*, 66–80. [CrossRef]

30. Mei, J.; Zheng, K.; Zhao, L.; Teng, Y.; Wang, X. A Latency and Reliability Guaranteed Resource Allocation Scheme for LTE V2V Communication Systems. *IEEE Trans. Wirel. Commun.* **2018**, *17*, 3850–3860. [CrossRef]

31. Belanovic, P.; Valerio, D.; Paier, A.; Zemen, T.; Ricciato, F.; Mecklenbrauker, C.F. On Wireless Links for Vehicle-to-Infrastructure Communications. *IEEE Trans. Veh. Technol.* **2009**, *59*, 269–282. [CrossRef]

32. Liu, N.; Liu, M.; Cao, J.; Chen, G.; Lou, W. When transportation meets communication: V2P over VANETs. In Proceedings of the 2010 IEEE 30th International Conference on Distributed Computing Systems, Genoa, Italy, 21–25 June 2010; pp. 567–576.

33. Lee, S.; Kim, D. An Energy Efficient Vehicle to Pedestrian Communication Method for Safety Applications. *Wirel. Pers. Commun.* **2015**, *86*, 1845–1856. [CrossRef]

34. Merdrignac, P.; Shagdar, O.; Nashashibi, F. Fusion of Perception and V2P Communication Systems for the Safety of Vulnerable Road Users. *IEEE Trans. Intell. Transp. Syst.* **2016**, *18*, 1740–1751. [CrossRef]

35. Campolo, C.; Molinaro, A.; Iera, A.; Menichella, F. 5G Network Slicing for Vehicle-to-Everything Services. *IEEE Wirel. Commun.* **2017**, *24*, 38–45. [CrossRef]
36. Abboud, K.; Omar, H.A.; Zhuang, W. Interworking of DSRC and Cellular Network Technologies for V2X Communications: A Survey. *IEEE Trans. Veh. Technol.* **2016**, *65*, 9457–9470. [CrossRef]
37. Wei, Q.; Wang, L.; Feng, Z.; Ding, Z. Wireless Resource Management in LTE-U Driven Heterogeneous V2X Communication Networks. *IEEE Trans. Veh. Technol.* **2018**, *67*, 7508–7522. [CrossRef]
38. Naik, G.; Choudhury, B.; Park, J.-M. IEEE 802.11bd & 5G NR V2X: Evolution of Radio Access Technologies for V2X Communications. *IEEE Access* **2019**, *7*, 70169–70184. [CrossRef]
39. Saiteja, P.; Ashok, B. Critical review on structural architecture, energy control strategies and development process towards optimal energy management in hybrid vehicles. *Renew. Sustain. Energy Rev.* **2022**, *157*, 112038. [CrossRef]
40. Chidambaram, K.; Ashok, B.; Vignesh, R.; Deepak, C.; Ramesh, R.; Narendhra, T.M.; Usman, K.M.; Kavitha, C. Critical analysis on the implementation barriers and consumer perception toward future electric mobility. *Proc. Inst. Mech. Eng. Part D J. Automob. Eng.* **2022**, 09544070221080349. [CrossRef]
41. Dueholm, J.V.; Kristoffersen, M.S.; Satzoda, R.K.; Moeslund, T.B.; Trivedi, M.M. Trajectories and Maneuvers of Surrounding Vehicles with Panoramic Camera Arrays. *IEEE Trans. Intell. Veh.* **2016**, *1*, 203–214. [CrossRef]
42. Han, L.; Zheng, K.; Zhao, L.; Wang, X.; Shen, X. Short-Term Traffic Prediction Based on DeepCluster in Large-Scale Road Networks. *IEEE Trans. Veh. Technol.* **2019**, *68*, 12301–12313. [CrossRef]
43. Shabir, B.; Khan, M.A.; Rahman, A.U.; Malik, A.W.; Wahid, A. Congestion Avoidance in Vehicular Networks: A Contemporary Survey. *IEEE Access* **2019**, *7*, 173196–173215. [CrossRef]
44. MacHardy, Z.; Khan, A.; Obana, K.; Iwashina, S. V2X access technologies: Regulation, research, and remaining challenges. *IEEE Commun. Surv. Tutor.* **2018**, *20*, 1858–1877.
45. Hu, Q.; Luo, F. Review of secure communication approaches for in-vehicle network. *Int. J. Automot. Technol.* **2018**, *19*, 879–894.
46. Masini, B.M.; Bazzi, A.; Zanella, A. A Survey on the Roadmap to Mandate on Board Connectivity and Enable V2V-Based Vehicular Sensor Networks. *Sensors* **2018**, *18*, 2207. [CrossRef] [PubMed]
47. Wang, X.; Mao, S.; Gong, M.X. An Overview of 3GPP Cellular Vehicle-to-Everything Standards. *GetMobile Mob. Comput. Commun.* **2017**, *21*, 19–25. [CrossRef]
48. Chen, L.; Englund, C. Cooperative intersection management: A survey. *IEEE Trans. Intell. Transp. Syst.* **2015**, *17*, 570–586. [CrossRef]
49. Dixit, S.; Fallah, S.; Montanaro, U.; Dianati, M.; Stevens, A.; Mccullough, F.; Mouzakitis, A. Trajectory planning and tracking for autonomous overtaking: State-of-the-art and future prospects. *Annu. Rev. Control* **2018**, *45*, 76–86. [CrossRef]
50. Bresson, G.; Alsayed, Z.; Yu, L.; Glaser, S. Simultaneous localization and mapping: A survey of current trends in autonomous driving. *IEEE Trans. Intell. Veh.* **2017**, *2*, 194–220. [CrossRef]
51. Bousselham, M.; Benamar, N.; Addaim, A. A new security mechanism for vehicular cloud computing using fog computing system. In Proceedings of the 2019 International Conference on Wireless Technologies, Embedded and Intelligent Systems (WITS), Fez, Morocco, 3–4 April 2019; pp. 1–4.
52. Mekki, T.; Jabri, I.; Rachedi, A.; ben Jemaa, M. Vehicular cloud networks: Challenges, architectures, and future directions. *Veh. Commun.* **2017**, *9*, 268–280. [CrossRef]
53. Boukerche, A.; De Grande, R.E. Vehicular cloud computing: Architectures, applications, and mobility. *Comput. Netw.* **2018**, *135*, 171–189. [CrossRef]
54. Yang, Q.; Zhu, B.; Wu, S. An Architecture of Cloud-Assisted Information Dissemination in Vehicular Networks. *IEEE Access* **2016**, *4*, 2764–2770. [CrossRef]
55. Meneguette, R.I.; Boukerche, A.; de Grande, R. SMART: An Efficient Resource Search and Management Scheme for Vehicular Cloud-Connected System. In Proceedings of the IEEE Global Communications Conference (GLOBECOM), Washington, DC, USA, 4–8 December 2016; pp. 1–6. [CrossRef]
56. De Souza, A.B.; Rego, P.A.L.; de Souza, J.N. Exploring computation offloading in vehicular clouds. In Proceedings of the 2019 IEEE 8th International conference on cloud networking (CloudNet), Coimbra, Portugal, 4–6 November 2019; pp. 1–4.
57. Sharma, V.; You, I.; Yim, K.; Chen, R.; Cho, J.H. BRIoT: Behavior rule specification-based misbehavior detection for IoT-embedded cyber-physical systems. *IEEE Access* **2019**, *7*, 118556–118580. [CrossRef]
58. Salahuddin, M.A.; Al-Fuqaha, A.; Guizani, M. Software-Defined Networking for RSU Clouds in Support of the Internet of Vehicles. *IEEE Internet Things J.* **2014**, *2*, 133–144. [CrossRef]
59. Tran, D.-D.; Vafaeipour, M.; El Baghdadi, M.; Barrero, R.; Van Mierlo, J.; Hegazy, O. Thorough state-of-the-art analysis of electric and hybrid vehicle powertrains: Topologies and integrated energy management strategies. *Renew. Sustain. Energy Rev.* **2020**, *119*, 109596. [CrossRef]
60. Hannan, M.A.; Hoque, M.D.M.; Hussain, A.; Yusof, Y.; Ker, A.P.J. State-of-the-Art and Energy Management System of Lithium-Ion Batteries in Electric Vehicle Applications: Issues and Recommendations. *IEEE Access Spec. Sect. Adv. Energy Storage Technol. Appl.* **2018**, *6*, 19362–19378. [CrossRef]

61. Chen, K.; Bouscayrol, A.; Lhomme, W. Energetic Macroscopic Representation and Inversion-based Control: Application to an Electric Vehicle with an Electrical Differential. *J. Asian Electr. Veh.* **2008**, *6*, 1097–1102. [CrossRef]
62. Chan, C.C.; Bouscayrol, A.; Chen, K. Electric, Hybrid, and Fuel-Cell Vehicles: Architectures and Modeling. *IEEE Trans. Veh. Technol.* **2009**, *59*, 589–598. [CrossRef]
63. Koot, M.; Kessels, J.T.; De Jager, B.; Heemels, W.; Van den Bosch, P.; Steinbuch, M. Energy management strategies for vehicular electric power systems. *IEEE Trans. Veh. Technol.* **2005**, *54*, 771–782. [CrossRef]
64. Hofman, T.; Steinbuch, M.; Van Druten, R.; Serrarens, A. Rule-based energy management strategies for hybrid vehicles. *Int. J. Electr. Hybrid Veh.* **2007**, *1*, 71. [CrossRef]
65. Madni, A.M.; Madni, C.C.; Lucero, S.D. Leveraging Digital Twin Technology in Model-Based Systems Engineering. *Systems* **2019**, *7*, 7. [CrossRef]
66. Wu, B.; Widanage, W.D.; Yang, S.; Liu, X. Battery digital twins: Perspectives on the fusion of models, data and artificial intelligence for smart battery management systems. *Energy AI* **2020**, *1*, 100016. [CrossRef]
67. Microsemi, P.P.G. Gallium Nitride (GaN) Versus Silicon Carbide (SiC) in the High Frequency (RF) and Power Switching Applications. Digi-Key. 2014. Available online: https://www.digikey.cz/Site/Global/Layouts/DownloadPdf.ashx?pdfUrl=40 39FA9B108F42229FEF9E84EE758607 (accessed on 20 March 2023).
68. Rasool, H.; El Baghdadi, M.; Rauf, A.M.; Zhaksylyk, A.; Hegazy, O. A Rapid Non-Linear Computation Model of Power Loss and Electro Thermal Behaviour of Three-Phase Inverters in EV Drivetrains. In Proceedings of the 2020 International Symposium on Power Electronics, Electrical Drives, Automation and Motion (SPEEDAM), Sorrento, Italy, 24–26 June 2020; pp. 317–323.
69. Keshmiri, N.; Wang, D.; Agrawal, B.; Hou, R.; Emadi, A. Current Status and Future Trends of GaN HEMTs in Electrified Transportation. *IEEE Access* **2020**, *8*, 70553–70571. [CrossRef]
70. Sewergin, A.; Wienhausen, A.H.; Oberdieck, K.; De Doncker, R.W. Modular bidirectional full-SiC DC-DC converter for automotive applications. In Proceedings of the 2017 IEEE 12th International Conference on Power Electronics and Drive Systems (PEDS), Honolulu, HI, USA, 12–15 December 2017; pp. 277–281.
71. Rui, R. Power Stage of 48V BSG Inverter. Infineon Appl. Note. 2018. Available online: https://www.infineon.com/dgdl/Infineon-20180802_AN-Power_stage_of_48V_BSG_inverter_V2.2-AN-v01_00-EN.pdf?fileId=5546d46265487f7b0165a3863b8e5bcf (accessed on 20 March 2023).
72. Liu, Z.; Li, B.; Lee, F.C.; Li, Q. High-Efficiency High-Density Critical Mode Rectifier/Inverter for WBG-Device-Based On-Board Charger. *IEEE Trans. Ind. Electron.* **2017**, *64*, 9114–9123. [CrossRef]
73. Rasool, H.; Zhaksylyk, A.; Chakraborty, S.; El Baghdadi, M.; Hegazy, O. Optimal design strategy and electro-thermal modelling of a high-power off-board charger for electric vehicle applications. In Proceedings of the 2020 Fifteenth International Conference on Ecological Vehicles and Renewable Energies (EVER), Monte-Carlo, Monaco, 10–12 September 2020; pp. 1–8.
74. Chakraborty, S.; Vu, H.-N.; Hasan, M.M.; Tran, D.-D.; El Baghdadi, M.; Hegazy, O. DC-DC Converter Topologies for Electric Vehicles, Plug-in Hybrid Electric Vehicles and Fast Charging Stations: State of the Art and Future Trends. *Energies* **2019**, *12*, 1569. [CrossRef]
75. Lu, L.; Han, X.; Li, J.; Hua, J.; Ouyang, M. A review on the key issues for lithium-ion battery management in electric vehicles. *J. Power Sources* **2013**, *226*, 272–288. [CrossRef]
76. Fuchs, G.; Lunz, B.; Leuthold, M.; Sauer, D.U. Technology Overview on Electricity Storage. *ISEA Aachen Juni.* **2012**, 26. Available online: https://sei.info.yorku.ca/files/2013/03/Sauer2.pdf (accessed on 20 March 2023).
77. Li, M.; Lu, J.; Chen, Z.; Amine, K. 30 years of lithium-ion batteries. *Adv. Mater.* **2018**, *30*, 1800561.
78. Sun, Y.-K.; Myung, S.-T.; Park, B.-C.; Prakash, J.; Belharouak, I.; Amine, K. High-energy cathode material for long-life and safe lithium batteries. *Nat. Mater.* **2009**, *8*, 320–324. [CrossRef]
79. Philippot, M.; Alvarez, G.; Ayerbe, E.; Van Mierlo, J.; Messagie, M. Eco-Efficiency of a Lithium-Ion Battery for Electric Vehicles: Influence of Manufacturing Country and Commodity Prices on GHG Emissions and Costs. *Batteries* **2019**, *5*, 23. [CrossRef]
80. Schmuch, R.; Wagner, R.; Hörpel, G.; Placke, T.; Winter, M. Performance and cost of materials for lithium-based rechargeable automotive batteries. *Nat. Energy* **2018**, *3*, 267–278. [CrossRef]
81. Xie, J.; Lu, Y.-C. A retrospective on lithium-ion batteries. *Nat. Commun.* **2020**, *11*, 2499. [CrossRef] [PubMed]
82. Gopalakrishnan, R.; Goutam, S.; Oliveira, L.M.; Timmermans, J.-M.; Omar, N.; Messagie, M.; Bossche, P.V.D.; van Mierlo, J. A Comprehensive Study on Rechargeable Energy Storage Technologies. *J. Electrochem. Energy Convers. Storage* **2016**, *13*. [CrossRef]
83. Berckmans, G.; De Sutter, L.; Marinaro, M.; Smekens, J.; Jaguemont, J.; Wohlfahrt-Mehrens, M.; van Mierlo, J.; Omar, N. Analysis of the effect of applying external mechanical pressure on next generation silicon alloy lithium-ion cells. *Electrochim. Acta* **2019**, *306*, 387–395. [CrossRef]
84. Edström, K. BATTERY 2030+. Inventing the Sustainable Batteries of the Future. Research Needs and Future Actions. Available online: https://battery2030.eu/digitalAssets/860/c_860904-l_1-k_roadmap-27-march.pdf (accessed on 3 February 2021).
85. Ev, I.G. *Outlook to Electric Mobility*; International Energy Agency (IEA): Paris, France, 2019.
86. Pasta, M.; Armstrong, D.; Brown, Z.L.; Bu, J.; Castell, M.R.; Chen, P.; Cocks, A.; Corr, S.A.; Cussen, E.J.; Darnbrough, E.; et al. 2020 roadmap on solid-state batteries. *J. Phys. Energy* **2020**, *2*, 032008. [CrossRef]
87. Randau, S.; Weber, D.A.; Kötz, O.; Koerver, R.; Braun, P.; Weber, A.; Ivers-Tiffée, E.; Adermann, T.; Kulisch, J.; Zeier, W.G.; et al. Benchmarking the performance of all-solid-state lithium batteries. *Nat. Energy* **2020**, *5*, 259–270. [CrossRef]

88. Albertus, P.; Babinec, S.; Litzelman, S.; Newman, A. Status and challenges in enabling the lithium metal electrode for high-energy and low-cost rechargeable batteries. *Nat. Energy* **2017**, *3*, 16–21. [CrossRef]
89. Gao, Y.; Rojas, T.; Wang, K.; Liu, S.; Wang, D.; Chen, T.; Wang, H.; Ngo, A.T.; Wang, D. Low-temperature and high-rate-charging lithium metal batteries enabled by an electrochemically active monolayer-regulated interface. *Nat. Energy* **2020**, *5*, 534–542. [CrossRef]
90. Forsyth, M.; Porcarelli, L.; Wang, X.; Goujon, N.; Mecerreyes, D. Innovative Electrolytes Based on Ionic Liquids and Polymers for Next-Generation Solid-State Batteries. *Acc. Chem. Res.* **2019**, *52*, 686–694. [CrossRef]
91. Chen, X.; Wang, X.; Sun, W.; Jiang, C.; Xie, J.; Wu, Y.; Jin, Q. Integrated interdigital electrode and thermal resistance micro-sensors for electric vehicle battery coolant conductivity high-precision measurement. *J. Energy Storage* **2023**, *58*, 106402. [CrossRef]
92. Kerman, K.; Luntz, A.; Viswanathan, V.; Chiang, Y.-M.; Chen, Z. Review—Practical Challenges Hindering the Development of Solid State Li Ion Batteries. *J. Electrochem. Soc.* **2017**, *164*, A1731–A1744. [CrossRef]
93. Garbayo, I.; Struzik, M.; Bowman, W.J.; Pfenninger, R.; Stilp, E.; Rupp, J.L. Glass-Type Polyamorphism in Li-Garnet Thin Film Solid State Battery Conductors. *Adv. Energy Mater.* **2018**, *8*, 1702265.
94. Smekens, J.; Gopalakrishnan, R.; Steen, N.V.D.; Omar, N.; Hegazy, O.; Hubin, A.; Van Mierlo, J. Influence of Electrode Density on the Performance of Li-Ion Batteries: Experimental and Simulation Results. *Energies* **2016**, *9*, 104. [CrossRef]
95. Krauskopf, T.; Mogwitz, B.; Rosenbach, C.; Zeier, W.G.; Janek, J. Diffusion Limitation of Lithium Metal and Li–Mg Alloy Anodes on LLZO Type Solid Electrolytes as a Function of Temperature and Pressure. *Adv. Energy Mater.* **2019**, *9*, 1902568. [CrossRef]
96. Truchot, C.; Dubarry, M.; Liaw, B.Y. State-of-charge estimation and uncertainty for lithium-ion battery strings. *Appl. Energy* **2014**, *119*, 218–227. [CrossRef]
97. Li, Y.; Liu, K.; Foley, A.M.; Zülke, A.; Berecibar, M.; Nanini-Maury, E.; Van Mierlo, J.; Hoster, H.E. Data-driven health estimation and lifetime prediction of lithium-ion batteries: A review. *Renew. Sustain. Energy Rev.* **2019**, *113*, 109254. [CrossRef]
98. De Sutter, L.; Berckmans, G.; Marinaro, M.; Wohlfahrt-Mehrens, M.; Berecibar, M.; Van Mierlo, J. Mechanical behavior of Silicon-Graphite pouch cells under external compressive load: Implications and opportunities for battery pack design. *J. Power Sources* **2020**, *451*, 227774. [CrossRef]
99. Campanella, A.; Döhler, D.; Binder, W.H. Self-healing in supramolecular polymers. *Macromol. Rapid Commun.* **2018**, *39*, 1700739. [CrossRef]
100. Wang, C.; Wu, H.; Chen, Z.; McDowell, M.T.; Cui, Y.; Bao, Z. Self-healing chemistry enables the stable operation of silicon microparticle anodes for high-energy lithium-ion batteries. *Nat. Chem.* **2013**, *5*, 1042–1048. [CrossRef] [PubMed]
101. Langer, E. *Liquid Cooling in Electric Vehicles—What to Know to Keep EVs on the Go By*; CPC: Preston, UK, 2019.
102. Habib, S.; Khan, M.M.; Abbas, F.; Tang, H. Assessment of electric vehicles concerning impacts, charging infrastructure with unidirectional and bidirectional chargers, and power flow comparisons. *Int. J. Energy Res.* **2018**, *42*, 3416–3441. [CrossRef]
103. Van Mierlo, J.; Berecibar, M.; El Baghdadi, M.; De Cauwer, C.; Messagie, M.; Coosemans, T.; Jacobs, V.A.; Hegazy, O. Beyond the State of the Art of Electric Vehicles: A Fact-Based Paper of the Current and Prospective Electric Vehicle Technologies. *World Electr. Veh. J.* **2021**, *12*, 20. [CrossRef]
104. Dusmez, S.; Cook, A.; Khaligh, A. Comprehensive analysis of high quality power converters for level 3 off-board chargers. In Proceedings of the 2011 IEEE Vehicle Power and Propulsion Conference, Chicago, IL, USA, 6–9 September 2011; pp. 1–10. [CrossRef]
105. Salgado-Herrera, N.; Anaya-Lara, O.; Campos-Gaona, D.; Medina-Rios, A.; Tapia-Sanchez, R.; Rodriguez-Rodriguez, J. Active Front-End converter applied for the THD reduction in power systems. In Proceedings of the 2018 IEEE Power & Energy Society General Meeting (PESGM), Portland, OR, USA, 5–10 August 2018; pp. 1–5.
106. Kesler, M.; Kisacikoglu, M.C.; Tolbert, L.M. Vehicle-to-Grid Reactive Power Operation Using Plug-In Electric Vehicle Bidirectional Offboard Charger. *IEEE Trans. Ind. Electron.* **2014**, *61*, 6778–6784. [CrossRef]
107. Vu, H.-N.; Abdel-Monem, M.; El Baghdadi, M.; Van Mierlo, J.; Hegazy, O. Multi-Objective Optimization of On-Board Chargers Based on State-of-the-Art 650V GaN Power Transistors for the Application of Electric Vehicles. In Proceedings of the 2019 IEEE Vehicle Power and Propulsion Conference (VPPC), Hanoi, Vietnam, 14–17 October 2019; pp. 1–6.
108. Yilmaz, M.; Krein, P.T. Review of Battery Charger Topologies, Charging Power Levels, and Infrastructure for Plug-In Electric and Hybrid Vehicles. *IEEE Trans. Power Electron.* **2013**, *28*, 2151–2169. [CrossRef]
109. Xue, L.; Shen, Z.; Boroyevich, D.; Mattavelli, P.; Diaz, D. Dual Active Bridge-Based Battery Charger for Plug-in Hybrid Electric Vehicle with Charging Current Containing Low Frequency Ripple. *IEEE Trans. Power Electron.* **2015**, *30*, 7299–7307. [CrossRef]
110. Li, B.; Lee, F.C.; Li, Q.; Liu, Z. Bi-directional on-board charger architecture and control for achieving ultra-high efficiency with wide battery voltage range. In Proceedings of the 2017 IEEE Applied Power Electronics Conference and Exposition (APEC), Tampa, FL, USA, 26–30 March 2017; pp. 3688–3694.
111. Zhaksylyk, A.; Rasool, H.; Geury, T.; El Baghdadi, M.; Hegazy, O. Masterless Control of Parallel Modular Active front-end (AFE) Systems for Vehicles and Stationary Applications. In Proceedings of the 2020 Fifteenth International Conference on Ecological Vehicles and Renewable Energies (EVER), Monte-Carlo, Monaco, 10–12 September 2020; pp. 1–6.
112. Blaabjerg, F.; Teodorescu, R.; Liserre, M.; Timbus, A.V. Overview of Control and Grid Synchronization for Distributed Power Generation Systems. *IEEE Trans. Ind. Electron.* **2006**, *53*, 1398–1409. [CrossRef]

113. Messagie, M.; Mertens, J.; Oliveira, L.; Rangaraju, S.; Sanfelix, J.; Coosemans, T.; Van Mierlo, J.; Macharis, C. The hourly life cycle carbon footprint of electricity generation in Belgium, bringing a temporal resolution in life cycle assessment. *Appl. Energy* **2014**, *134*, 469–476.

114. Van Mierlo, J. The world electric vehicle journal, the open access journal for the e-mobility scene. *World Electr. Veh. J.* **2018**, *9*, 1.

115. Li, Y.; Messagie, M.; Berecibar, M.; Hegazy, O.; Omar, N.; Van Mierlo, J. The impact of the vehicle-to-grid strategy on lithium-ion battery ageing process. In Proceedings of the 31st International Electric Vehicle Symposium & Exhibition (EVS 31), Kobe, Japan, 1–3 October 2018.

116. Van Kriekinge, G.; De Cauwer, C.; Van Mierlo, J.; Coosemans, T.; Messagie, M. Techno-economical assessment of vehicle-to-grid in a microgrid: Case study. In Proceedings of the 33th International Electric Vehicle Symposium and Exhibition (EVS 2020), Portland, OR, USA, 14–17 June 2020; pp. 14–17.

117. Syed, A.; Crispeels, T.; Jahir Roncancio Marin, J.; Cardellini, G.; De Cauwer, C.; Coosemans, T.; Van Mierlo, J.; Messagie, M. A Novel Method to Value the EV-Fleet's Grid Balancing Capacity. In Proceedings of the 33th International Electric Vehicle Symposium and Exhibition (EVS 2020), Portland, OR, USA, 14–17 June 2020; pp. 14–17.

118. De Cauwer, C.; Van Kriekinge, G.; Van Mierlo, J.; Coosemans, T.; Messagie, M. Integration of Vehicle-to-Grid in Local Energy Systems: Concepts and Specific Requirements. In Proceedings of the 33th International Electric Vehicle Symposium and Exhibition (EVS 2020), Portland, OR, USA, 14–17 June 2020; pp. 14–17.

119. Hooftman, N.; Messagie, M.; Van Mierlo, J.; Coosemans, T. The Paris Agreement and Zero-Emission Vehicles in Europe: Scenarios for the Road towards a Decarbonised Passenger Car Fleet. In *Towards User-Centric Transport in Europe 2: Enablers of Inclusive, Seamless and Sustainable Mobility*; Springer, 2020; pp. 151–168. Available online: https://link.springer.com/chapter/10.1007/978-3-030-38028-1_11 (accessed on 20 March 2023). [CrossRef]

120. Messagie, M.; Coosemans, T.; Van Mierlo, J. The Need for Uncertainty Propagation in Life Cycle Assessment of Vehicle Technologies. In *Towards User-Centric Transport in Europe 2: Enablers of Inclusive, Seamless and Sustainable Mobility*; IEEE Xplorer: 2019; pp. 1–7. Available online: https://ieeexplore.ieee.org/abstract/document/8952350/ (accessed on 20 March 2023). [CrossRef]

121. Narayanan, S.; Chaniotakis, E.; Antoniou, C. Shared autonomous vehicle services: A comprehensive review. *Transp. Res. Part C Emerg. Technol.* **2020**, *111*, 255–293.

122. Loeb, B.; Kockelman, K.M. Fleet performance and cost evaluation of a shared autonomous electric vehicle (SEAVS) fleet: A case study for Austin, Texas. *Transp. Res. Part A Policy Pract.* **2019**, *121*, 374–385.

123. Haas, H. Wireless Data from Every Light Bulb. Available online: https://www.ted.com/talks/harald_haas_wireless_data_from_every_light_bulb/transcript (accessed on 13 February 2023).

124. Golbabaei, F.; Yigitcanlar, T.; Bunker, J. The role of shared autonomous vehicle systems in delivering smart urban mobility: A systematic review of the literature. *Int. J. Sustain. Transp.* **2020**, *15*, 731–748. [CrossRef]

125. Maurer, M.; Gerdes, J.C.; Lenz, B.; Winner, H. *Autonomous Driving: Technical, Legal and Social Aspects*; Springer Nature: Berlin/Heidelberg, Germany, 2016.

126. Fagnant, D.J.; Kockelman, K. Preparing a nation for autonomous vehicles: Opportunities, barriers and policy recommendations. *Transp. Res. Part A Policy Pract.* **2015**, *77*, 167–181.

127. Cohen, T.; Cavoli, C. Automated vehicles: Exploring possible consequences of government (non)intervention for congestion and accessibility. *Transp. Rev.* **2019**, *39*, 129–151. [CrossRef]

128. Chen, T.D.; Kockelman, K.M.; Hanna, J.P. Operations of a shared, autonomous, electric vehicle fleet: Implications of vehicle & charging infrastructure decisions. *Transp. Res. Part A Policy Pract.* **2016**, *94*, 243–254. [CrossRef]

129. Iacobucci, R.; McLellan, B.; Tezuka, T. Modeling shared autonomous electric vehicles: Potential for transport and power grid integration. *Energy* **2018**, *158*, 148–163. [CrossRef]

130. Tan, K.M.; Ramachandaramurthy, V.K.; Yong, J.Y. Integration of electric vehicles in smart grid: A review on vehicle to grid technologies and optimization techniques. *Renew. Sustain. Energy Rev.* **2016**, *53*, 720–732. [CrossRef]

131. Rangaraju, S. Environmental Performance of Battery Electric Vehicles: Implications for Future Integrated Electricity and Transport System. Ph.D. Thesis, 2018. Available online: https://researchportal.vub.be/en/publications/environmental-performance-of-battery-electric-vehicles-implicatio (accessed on 20 March 2023).

132. Trübswetter, N.; Bengler, K. Why should I use ADAS? Advanced driver assistance systems and the elderly: Knowledge, experience and usage barriers. In *Driving Assesment Conference*; University of Iowa: Iowa, IA, USA, 2013; Volume 7.

133. Eichelberger, A.H.; McCartt, A.T. Toyota drivers' experiences with dynamic radar cruise control, pre-collision system, and lane-keeping assist. *J. Saf. Res.* **2016**, *56*, 67–73. [CrossRef] [PubMed]

134. Hubele, N.; Kennedy, K. Forward collision warning system impact. *Traffic Inj. Prev.* **2018**, *19*, S78–S83. [CrossRef] [PubMed]

135. Patra, S.; Veelaert, P.; Calafate, C.T.; Cano, J.-C.; Zamora, W.; Manzoni, P.; González, F. A Forward Collision Warning System for Smartphones Using Image Processing and V2V Communication. *Sensors* **2018**, *18*, 2672. [CrossRef]

136. Motamedidehkordi, N.; Amini, S.; Hoffmann, S.; Busch, F.; Fitriyanti, M.R. Modeling tactical lane-change behavior for automated vehicles: A supervised machine learning approach. In Proceedings of the 2017 5th IEEE International Conference on Models and Technologies for Intelligent Transportation Systems (MT-ITS), Naples, Italy, 26–28 June 2017; pp. 268–273.

137. Yan, Z.; Yang, K.; Wang, Z.; Yang, B.; Kaizuka, T.; Nakano, K. Intention-Based Lane Changing and Lane Keeping Haptic Guidance Steering System. *IEEE Trans. Intell. Veh.* **2020**, *6*, 622–633. [CrossRef]

138. Katzourakis, D.I.; Lazic, N.; Olsson, C.; Lidberg, M.R. Driver Steering Override for Lane-Keeping Aid Using Computer-Aided Engineering. *IEEE/ASME Trans. Mechatron.* **2015**, *20*, 1543–1552. [CrossRef]
139. Shen, D.; Yi, Q.; Li, L.; Tian, R.; Chien, S.; Chen, Y.; Sherony, R. Test Scenarios Development and Data Collection Methods for the Evaluation of Vehicle Road Departure Prevention Systems. *IEEE Trans. Intell. Veh.* **2019**, *4*, 337–352. [CrossRef]
140. Sternlund, S.; Strandroth, J.; Rizzi, M.; Lie, A.; Tingvall, C. The effectiveness of lane departure warning systems—A reduction in real-world passenger car injury crashes. *Traffic Inj. Prev.* **2017**, *18*, 225–229. [PubMed]
141. Abdullahi, A.; Akkaya, S. Adaptive cruise control: A model reference adaptive control approach. In Proceedings of the 2020 24th International Conference on System Theory, Control and Computing (ICSTCC), Sinaia, Romania, 8–10 October 2020; pp. 904–908.
142. Li, Y.; Li, Z.; Wang, H.; Wang, W.; Xing, L. Evaluating the safety impact of adaptive cruise control in traffic oscillations on freeways. *Accid. Anal. Prev.* **2017**, *104*, 137–145. [CrossRef] [PubMed]
143. Plessen, M.G.; Bernardini, D.; Esen, H.; Bemporad, A. Spatial-Based Predictive Control and Geometric Corridor Planning for Adaptive Cruise Control Coupled with Obstacle Avoidance. *IEEE Trans. Control Syst. Technol.* **2017**, *26*, 38–50. [CrossRef]
144. Hu, J.; Xu, L.; He, X.; Meng, W. Abnormal Driving Detection Based on Normalized Driving Behavior. *IEEE Trans. Veh. Technol.* **2017**, *66*, 6645–6652. [CrossRef]
145. Adochiei, I.-R.; Stirbu, O.-I.; Adochiei, N.-I.; Pericle-Gabriel, M.; Larco, C.-M.; Mustata, S.-M.; Costin, D. Drivers' drowsiness detection and warning systems for critical infrastructures. In Proceedings of the 2020 International Conference on e-Health and Bioengineering (EHB), Iasi, Romania, 29–30 October 2020; pp. 1–4.
146. Saito, Y.; Itoh, M.; Inagaki, T. Driver Assistance System with a Dual Control Scheme: Effectiveness of Identifying Driver Drowsiness and Preventing Lane Departure Accidents. *IEEE Trans. Human Mach. Syst.* **2016**, *46*, 660–671. [CrossRef]
147. Yin, J.-L.; Chen, B.-H.; Lai, K.-H.R.; Li, Y. Automatic Dangerous Driving Intensity Analysis for Advanced Driver Assistance Systems from Multimodal Driving Signals. *IEEE Sens. J.* **2017**, *18*, 4785–4794. [CrossRef]
148. Chen, Y.; Peng, H.; Grizzle, J. Obstacle Avoidance for Low-Speed Autonomous Vehicles with Barrier Function. *IEEE Trans. Control Syst. Technol.* **2017**, *26*, 194–206. [CrossRef]
149. Funke, J.; Brown, M.; Erlien, S.M.; Gerdes, J.C. Collision Avoidance and Stabilization for Autonomous Vehicles in Emergency Scenarios. *IEEE Trans. Control Syst. Technol.* **2016**, *25*, 1204–1216. [CrossRef]
150. Viriyasitavat, W.; Tonguz, O.K. Priority management of emergency vehicles at intersections using self-organized traffic control. In Proceedings of the 2012 IEEE Vehicular Technology Conference (VTC Fall), Quebec City, QC, Canada, 3–6 September 2012; pp. 1–4.
151. Masini, B.M.; Zanella, A.; Pasolini, G.; Bazzi, A.; Zabini, F.; Andrisano, O.; Mirabella, M.; Toppan, P. Toward the integration of ADAS capabilities in V2X communications for cooperative driving. In Proceedings of the 2020 AEIT International Conference of Electrical and Electronic Technologies for Automotive (AEIT AUTOMOTIVE), Associazione Italiana di Elettrotecnica, Elettronica, Milano, Italy, 18–20 November 2020; pp. 1–6.
152. Narula, L.; Wooten, M.J.; Murrian, M.J.; LaChapelle, D.M.; Humphreys, T.E. *ADAS Enhanced by 5G Connectivity: Volumes 1 and 2*; No. D-STOP/2018/139; University of Texas at Austin, Center for Transportation Research: Texas, TX, USA, 2018.
153. Bazzi, A.; Masini, B.M.; Zanella, A.; De Castro, C.; Raffaelli, C.; Andrisano, O. Cellular aided vehicular named data networking. In Proceedings of the 2014 International Conference on Connected Vehicles and Expo (ICCVE), Vienna, Austria, 3–7 November 2014; pp. 747–752.
154. Masini, B.M.; Silva, C.M.; Balador, A. The Use of Meta-Surfaces in Vehicular Networks. *J. Sens. Actuator Netw.* **2020**, *9*, 15. [CrossRef]
155. Guanetti, J.; Kim, Y.; Borrelli, F. Control of connected and automated vehicles: State of the art and future challenges. *Annu. Rev. Control* **2018**, *45*, 18–40. [CrossRef]
156. Rios-Torres, J.; Malikopoulos, A.A. A Survey on the Coordination of Connected and Automated Vehicles at Intersections and Merging at Highway On-Ramps. *IEEE Trans. Intell. Transp. Syst.* **2016**, *18*, 1066–1077. [CrossRef]
157. Birnbacher, D.; Birnbacher, W. Fully Autonomous Driving: Where Technology and Ethics Meet. *IEEE Intell. Syst.* **2017**, *32*, 3–4. [CrossRef]
158. Greenblatt, N.A. Self-driving cars and the law. *IEEE Spectr.* **2016**, *53*, 46–51. [CrossRef]
159. Urooj, S.; Feroz, I.; Ahmad, N. Systematic literature review on user interfaces of autonomous cars: Liabilities and responsibilities. In Proceedings of the 2018 International Conference on Advancements in Computational Sciences (ICACS), Lahore, Pakistan, 19–21 February 2018; pp. 1–10.
160. Borenstein, J.; Herkert, J.; Miller, K. Self-Driving Cars: Ethical Responsibilities of Design Engineers. *IEEE Technol. Soc. Mag.* **2017**, *36*, 67–75. [CrossRef]
161. Fournier, T. Will my next car be a libertarian or a utilitarian? Who will decide? *IEEE Technol. Soc. Mag.* **2016**, *35*, 40–45.
162. Lin, P. The Ethics of Autonomous Cars. The Atlantic. Available online: http://www.theatlantic.com/technology/archive/2013/10/theethics-of-autonomous-cars/280360/ (accessed on 14 February 2023).
163. Ma, A. China Has Started Ranking Citizens with a Creepy 'Social Credit' System—Here's What You Can Do Wrong, and the Embarrassing, Demeaning Ways They Can Punish You. Business Insider US. Available online: https://www.businessinsider.sg/china-social-credit-system-punishments-and-rewards-explained-2018-4/?r=US&IR=T (accessed on 12 February 2023).
164. Chan, T.K.; Chin, C.S.; Chen, H.; Zhong, X. A Comprehensive Review of Driver Behavior Analysis Utilizing Smartphones. *IEEE Trans. Intell. Transp. Syst.* **2019**, *21*, 4444–4475. [CrossRef]

165. Solanke, T.U.; Ramachandaramurthy, V.K.; Yong, J.Y.; Pasupuleti, J.; Kasinathan, P.; Rajagopalan, A. A review of strategic charging–discharging control of grid-connected electric vehicles. *J. Energy Storage* **2020**, *28*, 101193. [CrossRef]

166. Zou, Y.; Zhao, J.; Gao, X.; Chen, Y.; Tohidi, A. Experimental results of electric vehicles effects on low voltage grids. *J. Clean. Prod.* **2020**, *255*, 120270. [CrossRef]

167. Arena, F.; Pau, G.; Severino, A. An Overview on the Current Status and Future Perspectives of Smart Cars. *Infrastructures* **2020**, *5*, 53. [CrossRef]

168. Das, H.S.; Rahman, M.M.; Li, S.; Tan, C.W. Electric vehicles standards, charging infrastructure, and impact on grid integration: A technological review. *Renew. Sustain. Energy Rev.* **2020**, *120*, 109618. [CrossRef]

169. Pappalardo, G.; Cafiso, S.; Di Graziano, A.; Severino, A. Decision Tree Method to Analyze the Performance of Lane Support Systems. *Sustainability* **2021**, *13*, 846. [CrossRef]

170. Ghahari, S.; Assi, L.; Carter, K.; Ghotbi, S. *The Future of Hydrogen Fueling Systems for Fully Automated Vehicles*; American Society of Civil Engineers (ASCE): Reston, VA, USA, 2019; pp. 66–76.

171. Lin, T.Y.; Maire, M.; Belongie, S.; Hays, J.; Perona, P.; Ramanan, D.; Dollár, P.; Zitnick, C.L. Microsoft coco: Common objects in context. In Proceedings of the Computer Vision–ECCV 2014: 13th European Conference, Zurich, Switzerland, 6–12 September 2014; Springer International Publishing: Berlin/Heidelberg, Germany, 2014; pp. 740–755.

172. Zhang, S.; Benenson, R.; Omran, M.; Hosang, J.; Schiele, B. Towards reaching human performance in pedestrian detection. *IEEE Trans. Pattern Anal. Mach. Intell.* **2017**, *40*, 973–986.

173. Li, B. 3D fully convolutional network for vehicle detection in point cloud. In Proceedings of the 2017 IEEE/RSJ International Conference on Intelligent Robots and Systems (IROS), Vancouver, BC, Canada, 24–28 September 2017; pp. 1513–1518. [CrossRef]

174. Grigorescu, S.; Trasnea, B.; Cocias, T.; Macesanu, G. A survey of deep learning techniques for autonomous driving. *J. Field Robot.* **2019**, *37*, 362–386. [CrossRef]

175. Ning, H.; Yin, R.; Ullah, A.; Shi, F. A Survey on Hybrid Human-Artificial Intelligence for Autonomous Driving. *IEEE Trans. Intell. Transp. Syst.* **2021**, *23*, 6011–6026. [CrossRef]

176. Kumar, A.; Krishnamurthi, R.; Nayyar, A.; Luhach, A.K.; Khan, M.S.; Singh, A. A novel Software-Defined Drone Network (SDDN)-based collision avoidance strategies for on-road traffic monitoring and management. *Veh. Commun.* **2020**, *28*, 100313. [CrossRef]

177. Kim, H.; Ben-Othman, J.; Mokdad, L.; Son, J.; Li, C. Research challenges and security threats to AI-driven 5G virtual emotion applications using autonomous vehicles, drones, and smart devices. *IEEE Netw.* **2020**, *34*, 288–294.

178. Wang, Y.; Su, Z.; Zhang, K.; Benslimane, A. Challenges and Solutions in Autonomous Driving: A Blockchain Approach. *IEEE Netw.* **2020**, *34*, 218–226. [CrossRef]

179. Kato, S.; Takeuchi, E.; Ishiguro, Y.; Ninomiya, Y.; Takeda, K.; Hamada, T. An Open Approach to Autonomous Vehicles. *IEEE Micro* **2015**, *35*, 60–68. [CrossRef]

180. Novickis, R.; Levinskis, A.; Kadikis, R.; Fescenko, V.; Ozols, K. Functional architecture for autonomous driving and its implementation. In Proceedings of the 2020 17th Biennial Baltic Electronics Conference (BEC), Tallinn, Estonia, 6–8 October 2020; pp. 1–6.

181. Martínez-Díaz, M.; Soriguera, F. Autonomous vehicles: Theoretical and practical challenges. *Transp. Res. Procedia* **2018**, *33*, 275–282. [CrossRef]

182. Sharma, S.; Gupta, S.; Gupta, D.; Juneja, S.; Gupta, P.; Dhiman, G.; Kautish, S. Deep Learning Model for the Automatic Classification of White Blood Cells. *Comput. Intell. Neurosci.* **2022**, *2022*, 7384131. [CrossRef] [PubMed]

183. Samad, A.; Alam, S.; Mohammed, S.; Bhukhari, M.U. Internet of vehicles (IoV) requirements, attacks and countermeasures. In Proceedings of the 12th INDIACom; INDIACom-2018; 5th International Conference on "Computing for Sustainable Global Development" IEEE Conference, New Delhi, India, 13–15 March 2018.

184. Atakishiyev, S.; Salameh, M.; Yao, H.; Goebel, R. Explainable Artificial Intelligence for Autonomous Driving: A Comprehensive Overview and Field Guide for Future Research Directions. *arXiv* **2021**, arXiv:2112.11561.

185. Shuaib, M.; Hassan, N.H.; Usman, S.; Alam, S.; Bhatia, S.; Mashat, A.; Kumar, A.; Kumar, M. Self-Sovereign Identity Solution for Blockchain-Based Land Registry System: A Comparison. *Mob. Inf. Syst.* **2022**, *2022*, 8930472. [CrossRef]

186. Singh, S.; Malik, A.; Kumar, R.; Singh, P.K. A proficient data gathering technique for unmanned aerial vehicle-enabled heterogeneous wireless sensor networks. *Int. J. Commun. Syst.* **2021**, *34*, e4956. [CrossRef]

187. Sharma, A.; Singh, P.K. UAV-based framework for effective data analysis of forest fire detection using 5G networks: An effective approach towards smart cities solutions. *Int. J. Commun. Syst.* **2021**, e4826. [CrossRef]

188. Jain, S.; Ahuja, N.J.; Srikanth, P.; Bhadane, K.V.; Nagaiah, B.; Kumar, A.; Konstantinou, C. Blockchain and autonomous vehicles: Recent advances and future directions. *IEEE Access* **2021**, *9*, 130264–130328.

189. Kumar, A.; de Jesus Pacheco, D.A.; Kaushik, K.; Rodrigues, J.J. Futuristic view of the internet of quantum drones: Review, challenges and research agenda. *Veh. Commun.* **2022**, *36*, 100487.

190. Sharma, S.; Tewolde, G.; Kwon, J. Behavioral cloning for lateral motion control of autonomous vehicles using deep learning. In Proceedings of the 2018 IEEE International Conference on Electro/Information Technology (EIT), Rochester, MI, USA, 3–5 May 2018; pp. 0228–0233.

191. Dixon, J.; Bukhsh, W.; Edmunds, C.; Bell, K. Scheduling electric vehicle charging to minimise carbon emissions and wind curtailment. *Renew. Energy* **2020**, *161*, 1072–1091. [CrossRef]
192. Haddadian, G.; Khalili, N.; Khodayar, M.; Shahiedehpour, M. Security-constrained power generation scheduling with thermal generating units, variable energy resources, and electric vehicle storage for V2G deployment. *Int. J. Electr. Power Energy Syst.* **2015**, *73*, 498–507.

MDPI

St. Alban-Anlage 66

4052 Basel

Switzerland

www.mdpi.com

Energies Editorial Office

E-mail: energies@mdpi.com

www.mdpi.com/journal/energies